Advances in Anatomy, Embryology and Cell Biology publishes critical reviews and state-of-the- art surveys on all aspects of anatomy and of developmental, cellular and molecular biology, with a special emphasis on biomedical and translational topics.

The series publishes volumes in two different formats:

• Contributed volumes, each collecting 5 to 15 focused reviews written by leading experts
• Single-authored or multi-authored monographs, providing a comprehensive overview of their topic of research

219
Advances in Anatomy, Embryology and Cell Biology

More information about this series at
http://www.springer.com/series/102

Winnok H. De Vos •
Sebastian Munck •
Jean-Pierre Timmermans

Focus on Bio-Image Informatics

Editors
Winnok H. De Vos
Dept. of Veterinary Sciences
University of Antwerp
Antwerpen
Antwerpen, Belgium

Sebastian Munck
Dept. of Human Genetics
KU Leuven
Leuven, Belgium

Jean-Pierre Timmermans
University of Antwerp
Antwerp, Belgium

ISSN 0301-5556 ISSN 2192-7065 (electronic)
Advances in Anatomy, Embryology and Cell Biology
ISBN 978-3-319-28547-4 ISBN 978-3-319-28549-8 (eBook)
DOI 10.1007/978-3-319-28549-8

Library of Congress Control Number: 2016936721

Printed on acid-free paper

This Springer imprint is published by Springer Nature
The registered company is Springer International Publishing AG Switzerland

EDITORIAL

From Pictures to Knowledge, Bio-Image Informatics Coming of Age

W. H. De Vos[1], S. Munck[2,3], J. P. Timmermans[1]

1 Laboratory of Cell Biology and Histology, Department of Veterinary Sciences, University of Antwerp, Groenenborgerlaan 151, Antwerp 2020, Belgium

2 VIB Bio Imaging Core and Center for the Biology of Disease, Herestraat 49, Box 602, Leuven 3000, Belgium

3 KU Leuven, Center for Human Genetics, Herestraat 49, Box 602, Leuven 3000, Belgium

Microscopy bridges a gap in systems biology. Meticulous observation and faithful documentation of biological phenomena have laid the foundation for modern cell and developmental biology. However, research in life science has long suffered from a gap between data acquisition and analysis possibilities. Fuelled by the rapid rise of large-scale *omics* approaches, exhaustive molecular inventories of various biological model systems have been generated. Despite significant descriptive and correlative value, these rich molecular data sets do not allow inferring causal relationships at the scale of most (cell) biological processes since they lack adequate spatiotemporal resolution. Furthermore, by averaging across large populations of cells, underlying patterns originating from dynamic oscillations, the presence of distinct subpopulations or stochastic events may be obscured (Spiller et al. 2010). To fully understand how different molecules interact in space and time with cellular specificity, they must be studied in individual living cells or organisms. Microscopy – in particular fluorescence microscopy – has been instrumental in filling in this need. Indeed, microscopic visualisation enables interrogating the spatiotemporal whereabouts and interactions of (macro)molecular complexes, cells and even small organisms at the relevant scales. Unfortunately, the technique has long suffered from an anecdotic and descriptive image, mostly due to its low throughput. This has now changed. Far-reaching automation and standardisation of image acquisition has transformed microscopy into a robust and quantitative data acquisition paradigm, paving the way for so-called systems microscopy (Lock and Strömblad 2010). With the explosive growth in both the number and complexity of the acquired images, the challenge now lies in the conversion of pictures into relevant quantitative descriptors that can be scrutinised with statistical power.

Enter bio-image informatics. To meet the growing demand for unbiased extraction of information content from biological images, a new field is rapidly gaining shape. This field, which is referred to as bio-image informatics (BII), unites computational techniques from computer vision, image processing, modelling and data mining and seamlessly connects with other disciplines like cell biology, biochemistry and biophysics (Peng 2008). Its chief task is to develop robust strategies for automated image reconstruction, visualisation and interpretation, starting from raw image data sets and biological knowledge. The variety and multidimensionality of image data sets that modern systems microscopy screens produce call for novel, integrated approaches. A remarkable and unique feature in this context is the proliferation and active maintenance of various powerful open-source image analysis platforms (Eliceiri et al. 2012). Growing communities of both developers and users are acting synergistically to bring new features to the software platforms and to increase interoperability between the different platforms. Moreover, public sharing of large image data sets and benchmarking tools, in combination with 'grand challenges' to tackle the burning issues, is rapidly pushing the field forward.

The future is bigger. BII is becoming a mature science, with indispensable value for hypothesis testing and hypothesis generation based on imaging experimentation. The microscopy community is progressively redefining the limits in two ostensibly juxtaposed directions: one towards increased sensitivity and resolution and the other towards ever-expanding dimensionality. Clearly, image data sets will only grow in size and complexity. Thus, the challenge for future bio-image informaticians will be to extract information from high-dimensional data sets, interrelate multimodal image information and adaptively represent enormously large images across scales.

This special edition of *Advances in Anatomy, Embryology, and Cell Biology* contains a series of manuscripts highlighting the developments and applications in the field of BII. The issue starts with an overview of the major trends in the field of image analysis followed by two chapters on essential image-processing tasks, namely, image restoration and image segmentation (i.e. the detection of image features). Subsequently, the utility of BII is exemplified in three different biological contexts, namely, to gauge the spatial distribution of proteins at the plasma membrane, to assess neuronal connectivity and to provide insights into cellular redox biology. Next, the operation of two image-processing platforms (KNIME and Ilastik) is explained for typical image analysis tasks, providing a convenient guide for the non-expert cell biologist. Finally, an overview is given of benchmarking efforts within the community, and the issue concludes with a perspective on the analysis of large-scale data sets using distributed image computing. Together, this work provides an excellent reference for both junior and experienced researchers, and it highlights the growth potential for BII.

References

Eliceiri KW et al (2012) Biological imaging software tools. Nat Meth 9:697–710

Lock JG, Strömblad S (2010) Systems microscopy: an emerging strategy for the life sciences. Exper Cell Res 316:1438–1444

Peng H (2008) Bioimage informatics: a new area of engineering biology. Bioinformatics 24:1827–1836

Spiller DG, Wood CD, Rand DA, White MRH (2010) Measurement of single-cell dynamics. Nature 465:736–745

Proposal Advances Anatomy, Embryology and Cell Biology: Focus on Bio-Image Informatics

Meticulous observation and faithful documentation of small-scale biological phenomena have laid the foundation for modern cell and developmental biology. Invaluable information has been garnered on the morphological rearrangements that accompany crucial decision points such as cell division, differentiation and embryonic development. Even today, interpretation and annotation of microscopy images offer an elegant and convincing way of proving scientific observations.

However, in an era of *omics*, life science is becoming evermore quantitative. This also holds true for microscopy. On the quest towards quantitative biology and systems microscopy, manual microscopic documentation makes way for the standardised, high-throughput workflows that typify molecular platforms. And whilst the complexity of the biological processes under investigation inflates image data set dimensions, an unbiased assessment of the image content becomes an equally important challenge. A consequent need for novel strategies of image warehousing, reconstruction and automated analysis has sparked the development of a new discipline, bio-image informatics, in which systems biology, modelling and computational analyses unite to provide robust, spatiotemporally defined information on the building blocks of life.

This volume of *Advances Anatomy Embryology and Cell Biology* focuses on the emerging field of bio-image informatics, presenting novel and exciting ways of handling and interpreting large image data sets. A collection of focused reviews written by key players in the field highlights the major directions and provides an excellent reference work for both young and experienced researchers.

Contents

Chapter 1
Seeing Is Believing: Quantifying Is Convincing: Computational Image Analysis in Biology

Ivo F. Sbalzarini

Abstract Imaging is center stage in biology. Advances in microscopy and labeling techniques have enabled unprecedented observations and continue to inspire new developments. Efficient and accurate quantification and computational analysis of the acquired images, however, are becoming the bottleneck. We review different paradigms of computational image analysis for intracellular, single-cell, and tissue-level imaging, providing pointers to the specialized literature and listing available software tools. We place particular emphasis on clear categorization of image-analysis frameworks and on identifying current trends and challenges in the field. We further outline some of the methodological advances that are required in order to use images as quantitative scientific measurements.

1.1 Introduction

"Seeing is believing" is an old saying in microscopy. With the classical biochemical methods being increasingly complemented by imaging techniques, however, the subjective interpretation of what one sees in a microscopy image gets in the way of scientific reproducibility and logical deduction. Arguments such as "I saw it that way" or "it looked like" become less tolerable as conclusions are based on image data. Positing that the images are correctly acquired and free of artifacts (North 2006), one hence wishes to reduce, or at least quantify, viewer bias and subjective expectations and beliefs by extracting reproducible numbers from the images.

Together with the need to process ever-larger sets of images at high throughput, reproducible quantification motivates the use of computational image analysis

---

I.F. Sbalzarini (✉)
Chair of Scientific Computing for Systems Biology, Faculty of Computer Science, TU Dresden, Dresden, Germany

MOSAIC Group, Center for Systems Biology Dresden, Dresden, Germany

Max Planck Institute of Molecular Cell Biology and Genetics, Pfotenhauerstr. 108, D-01307 Dresden, Germany
e-mail: ivos@mpi-cbg.de

W.H. De Vos et al. (eds.), *Focus on Bio-Image Informatics*, Advances in Anatomy, Embryology and Cell Biology 219, DOI 10.1007/978-3-319-28549-8_1

(Eils and Athale 2003; Myers 2012). Having a computer software do the image analysis renders the results reproducible. If the same software is run twice on the same image, the same result is produced. Manual analysis, however, is often not reproducible, as different people would quantify the image differently and even the same person might attribute slightly different numbers to the same object upon different repetitions of the analysis. Computational analysis also increases the throughput, as thousands of images can be processed, potentially even in parallel on a computer cluster. A third reason for using computational image analysis it that algorithms can detect minute pixel variations that the human eye cannot see (Danuser 2011). Finally, results from computational image analysis, such as cell shapes and fluorescence distributions, can be directly used to build systems models and computer simulations of biological processes (Sbalzarini 2013). Such simulations can then test whether the hypothesized, simulated mechanism is *sufficient* to produce the experimentally observed behavior. Perturbation experiments can then show whether it is also *necessary*. Image analysis is hence the first step toward a systems understanding of spatiotemporal biological processes.

Image analysis is a large and complex field, intersecting with *image processing* and *computer vision*. Image processing is a branch of signal processing, interpreting images as multi-dimensional continuous or discrete signals. Computer vision is the branch of artificial intelligence that tries to teach computers to "see," i.e., to interpret images. Computer vision has a 40-year history and is a well-researched field. Importing techniques from computer vision can help solve problems in biological image analysis (Danuser 2011). Nevertheless, computer vision is not a panacea for bio-image analysis, because computer vision has evolved with different images and goals in mind. The focus in computer vision is on interpreting complex scenes from images with good resolution and signal, i.e., conditions under which also the human eye operates. Such images are typically acquired with digital photo cameras and show objects that are much larger than the wavelength of the light used to image them. This has the important consequences that diffraction effects can be neglected and that imaging noise can be modeled as Gaussian, as the photon count per pixel is high. These assumptions pose challenges to computational analysis that are different from those for images acquired in microscopy. Microscopy images are typically characterized by low signal-to-noise ratios and significant diffraction artifacts. Moreover, the noise is frequently not Gaussian. Low-signal detectors such as (EM-)CCD and CMOS cameras, as well as photo-multiplier tubes and photodiodes, produce dominant Poisson noise, which is overlaid with the Gaussian noise from the electronics. Finally, microscopy frequently acquires 3D or 4D images, whereas digital photography is mostly limited to 2D. The specifics of biological images and their analysis have given rise to the new discipline of *bio-image informatics* (Peng 2008; Myers 2012).

Despite significant advances in the past years, bio-image informatics is still in its early days, and many challenges remain to be addressed. This includes the development of algorithms and software that better utilize the available computer resources in order to allow high-throughput studies and high resolution with large multi-dimensional image data. Second, the topic of uncertainty quantification needs to be addressed, which has so far been poorly dealt with in bio-imaging. If the goal is to be quantitative, i.e., to use imaging and image analysis as *measurements* in the scientific sense (Dietrich 1991), one has to know and quantify the measurement errors and their propagation and amplification along the analysis pipeline. Simply having an algorithm that tells us "here is a nucleus" is not of much use in big-data studies, and it prohibits statistical tests on the results. We need to know the probability that there is a nucleus, and the probability that the algorithm failed or produced a wrong detection. Third, we need to develop versatile frameworks and algorithms that can be adapted to different applications without having to re-write the software on a case-by-case basis. This requires theoretical and algorithmic frameworks that are adapted to the specifics of biological images and provide us with systematic and principled ways of including prior knowledge about the imaging process and the imaged objects into the analysis. Fourth, the new algorithms need to be made available to the community as user-friendly, open-source software. These four current challenges equally apply to all image-analysis paradigms and all imaging modalities.

Light microscopy is probably the prevalent imaging modality in biology today, as it allows live-cell imaging and real-time observation (Royer et al. 2015) of dynamic processes in cells and tissues. Focusing on fluorescence microscopy, we discuss the above four challenges and show for each of them where the field currently stands and what remains to be addressed. First, however, we outline the different paradigms of image analysis, providing a scaffold to structure the discussion. We close this article by highlighting different design approaches and current trends in bio-image analysis software tools, and by summarizing and generalizing our observations.

1.2 Computational Bio-image Analysis

Bio-image informatics enables us to address biological questions that could not be addressed otherwise, or only at a much higher cost (Peng 2008; Myers 2012). These questions are naturally posed in terms of biological entities and concepts, such as "do cells in the vicinity of a dividing cell have a higher propensity to divide next during tissue growth?" Bio-image analysis has to bridge the gap between the biological question and the image data. Questions like what it means for a cluster of pixels in an image to be considered a "cell," how "vicinity" is measured over the pixel grid of an image, and what cell division "looks" like in the table of pixel-intensity numbers need to be addressed and formulated as algorithmic recipes that can be programmed into a computer. This entails addressing the "what is where" inference problem over images, just as computer vision does. Quantitative bio-image analysis additionally needs to address the "how much of what is where" problem. This is harder, as more ambiguities exist. Due to the diffraction limit, for example, it is not always possible to distinguish the diffraction-limited bright spot created by a 50 nm object with high fluorophore concentration from that created by a 200 nm object with lower fluorophore concentration. This becomes an issue when one is interested in quantifying the concentration of the labeled protein in small subcellular structures (e.g., endosomes) as a biologically meaningful readout (Helmuth et al. 2009). Unique answers can only be found when including problem-specific prior knowledge and calibration into the analysis.

1.2.1 From Specimen to Pixels to Objects to Meaning

Image analysis is a *data representation problem*, as illustrated in Fig. 1.1. The information for the final conclusion is already contained in the original specimen, albeit in a very different data representation.

Multiple steps are required working one's way through the data representation hierarchy from the specimen to meaning. The first step is image acquisition. The specimen can be imaged in many different ways, for example, using different imaging modalities, different microscopes, different magnifications, different view-

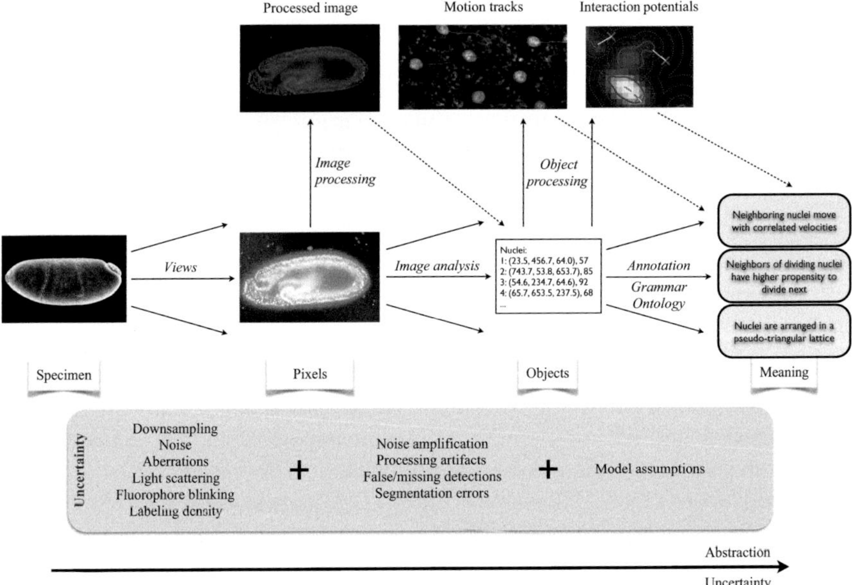

Fig. 1.1 Data representation and uncertainties in image analysis. The same specimen can be imaged in different views, giving rise to different images. Image analysis extracts quantitative information from the image, which can then be interpreted to form conclusions. Every step increases the level of abstraction and adds uncertainties and errors that often remain uncharacterized. *Dashed arrows* indicate routes of additional processing. (Image sources: "Specimen" image from FlyBase.org; "Pixels" image courtesy of Pavel Tomancak, MPI-CBG; "processed image," "motion tracks," and "interaction potentials" by Yuanhao Gong, Pietro Incardona, and Jo Helmuth, respectively, all MOSAIC Group)

ing angles, and different fluorescent markers. We call a specific imaging setup a *view*, leading to a digital image represented as a table of numerical pixel-intensity values. This clearly amounts to information loss, as many different images could be produced from one and the same specimen. Moreover, the intensity values in the pixel matrix are not easily related to actual fluorophore concentrations, since the microscope optics have an imperfect impulse-response function (Hecht 2001), called the *point spread function* (PSF), the excitation light intensity may be unknown, and light is absorbed and scattered as it propagates through the sample. In order to bridge to biological meaning, the matrix of pixels hence needs to be interpreted in terms of the objects represented in the image, which can be done using different approaches, as discussed in the following.

All approaches are commonly referred to as *image analysis*, which aims to extract semantic meaning from images. Image analysis hence takes an image as input and produces object representations as output, for example, a list of nuclei positions and sizes, or a 3D representation of cell shapes in a tissue. After image analysis, the information is hence no longer encoded in pixel matrices, but explicitly available as biologically tangible objects. This is in contrast to *image processing*, which transforms one image into another image that has, for example, less noise or blur. Examples include image deconvolution and contrast enhancement. Image processing is a sub-field of signal processing and operates within the image domain. *Image restoration* is a sub-field of image processing that aims to transform an image into one where the uncertainties and errors introduced by the image-acquisition process are reduced.

Image analysis often implies delineating objects represented in the image, a process called *image segmentation*. *Object detection* is a sub-task of segmentation that finds occurrences of the objects of interest in the image, and maybe counts them, but does not quantify their shapes. *Spatial pattern analysis* uses the detected and/or segmented objects to ask the question whether their spatial distribution is random or follows a certain pattern (Lagache et al. 2013). *Interaction analysis* is a special case of spatial pattern analysis, asking whether the distribution of one type of objects (e.g., viruses) is independent of the distribution of another type of objects (e.g., endosomes) and, if not, what hypothetical interaction between the two best explains their observed relative distribution (Helmuth et al. 2010; Lagache et al. 2015). An important analysis in time-lapse video is *motion tracking*, aiming at following moving objects over time and extracting their trajectories. This requires determining object correspondences across time points and is usually done after object detection or segmentation. Given the detected or segmented objects in each frame of a movie, tracking answers the question which detection in one frame corresponds to which detection in the next frame in the sense that the two are images of the same real-world object at different time points (Bar-Shalom and Blair 2000). A wealth of tracking methods exist in biological imaging, both for particle tracking (compared and reviewed in Chenouard et al. 2014) and for cell tracking (compared and reviewed in Maška et al. 2014). Most of them have by now been integrated in standard software packages. The extracted trajectories of moving objects are rich sources of information about dynamics, types of motion (Sbalzarini and Koumoutsakos 2005; Wieser et al. 2008; Ruprecht et al. 2011), and motion patterns (Helmuth et al. 2007). This has, for example, been used to analyze virus motion on and inside infected host cells (Ewers et al. 2005; Helmuth et al. 2007; Yamauchi et al. 2011) and to analyze the mobility of single molecules in plasma membranes (Wieser and Schütz 2008). For spatially extended objects, one can also track the deformations of their outlines. This involves determining which point on an outline corresponds to which point on the later outline. Solutions based on mechanical ball-and-spring models (Machacek and Danuser 2006) and level-set methods (Shi and Karl 2005; Machacek and Danuser 2006) have successfully been applied. Level-set methods (Sethian 1999) have also been used to track high-resolution outlines of polarizing and migrating keratocytes in phase-contrast movies (Ambühl et al. 2012), and to segment and track fluorescent HeLa and CHO cells (Dzyubachyk et al. 2010). This allows quantifying cells, their shapes, and temporal dynamics.

From this quantitative information about the shapes, positions, spatial distributions, and motion of the imaged objects, the researcher needs to derive biological meaning and new knowledge. Such meaning may come in the form of *annotations* of the objects found in the image (e.g., "this bright blob of pixels is a nucleus"), *grammar* (e.g., "nuclei are inside cells"), or *semantics* (e.g., "if a nucleus looks condensed and bright, the cell is entering mitosis"). This high-level interpretation of the image is application-specific and necessarily includes prior knowledge about what has been imaged. Otherwise, an image of fluorescently labeled virus particles on a cell membrane would be hard to distinguish from a photograph of the starry sky at night.

Including prior knowledge and interpreting the data inevitably introduce uncertainty. Indeed, errors and uncertainties are introduced at *every* stage of the image-analysis process and are propagated downstream. This includes uncertainties in the specimen itself, such as unknown labeling densities and blinking fluorophores (Annibale et al. 2011). Additional uncertainties are introduced by the view, i.e., the image-acquisition process. This includes light scattering in the sample, aberrations from the optics, and noise from the photon-detection process and the electronics in the camera. Image processing and analysis are also not perfect and may amplify noise, introduce false detections, quantification errors, and missed detections. This is not limited to computational image processing; also humans make mistakes when interpreting and quantifying images. Finally, the interpretation of the resulting information is subject to uncertainties, as we are often implicitly assuming a model that may not be correct. Ideally, all of these uncertainties and errors should be known and quantified, and their influence on downstream results should be bounded. Else it is impossible to decide whether an observed difference between two analyzed images is due to biological differences in the specimen, or just due to analysis artifacts.

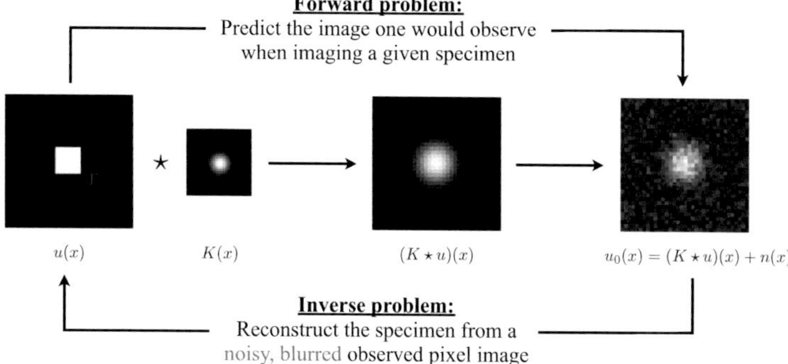

Fig. 1.2 The forward and the inverse problem in fluorescence microscopy: the specimen $u(x)$ is imaged using optics with a certain point spread function $K(x)$, yielding a diffraction-limited, blurred image of the specimen. This image is then digitized onto the finite pixel grid of the camera, and the photon-counting process in the detector, as well as the camera electronics, introduce noise $n(x)$. Modeling this image-formation process, i.e., predicting the image for a known or hypothesized specimen is the *forward problem*. The *inverse problem* consists in reconstructing an unknown specimen or its delineating boundary Γ (*red line*) from a particular observed image $u_0(x)$. (Image source: Grégory Paul, MOSAIC Group)

1.2.2 The Forward and the Inverse Problem

It is common to distinguish between the *forward problem* and the *inverse problem* in imaging, as illustrated in Fig. 1.2 for the case of fluorescence microscopy. The forward problem consists of finding a predictive model, the forward model, of the image-formation process. For a known or hypothesized specimen, this model predicts the image. The inverse problem entails reconstructing the specimen from a given observed image. Due to blur, noise, and other uncertainties introduced during image acquisition, there is usually no unique solution to the inverse problem, or its solution is unstable, which is why the inverse problem is called "ill-posed." A solution can only be found by including application-specific prior knowledge to *regularize* the problem. This prior knowledge can, for example, be the limit curvature of lipid membranes, or the PSF of the microscope. The solution space of the inverse problem is then restricted to those solutions that are compatible with the prior knowledge, eventually leading to a unique answer when sufficient prior knowledge is included.

The question arises, however, how much prior knowledge is required. In the absence of a closed theory, the pragmatic approach in bio-image analysis is to match the analysis aims and tools to the level of detail required by the biological question. Questions such as whether a co-localization study should account for fluorophore blinking and chromatic aberration, or not, are largely decided opportunistically with the final aim of the analysis in mind. This can be seen as a heuristic way of deciding which prior knowledge to include into the analysis.

1.2.3 Bayesian, or Not?

Image analysis considers the inverse problem and is therefore an inference task. The goal is to infer shapes, locations, and distributions of the imaged objects from the acquired images. As in any inference task, there are two views of the problem: frequentist and Bayesian. Frequentist inference draws conclusions from the data by looking at frequencies of occurrence. For example, thresholding considers a pixel to be part of an object if its intensity is in the upper 10 % of all pixels in the image. Bayesian inference draws conclusions that have high probability of explaining the data, given the prior expectations.

While both approaches include prior knowledge, e.g., about how the image has been acquired or the experimental design, the Bayesian approach formalizes the prior knowledge in the mathematical form of a Bayesian *prior*. *Prior knowledge* and a Bayesian *prior* are hence not the same, and in frequentist inference the former is present without the latter. In Bayesian inference, the inclusion of prior knowledge is not necessarily limited to the prior either, but may also enter other terms, such as the likelihood.

Bayesian inference is based on Bayes' theorem, as illustrated in Fig. 1.3. Applied to image segmentation, the theorem states that the segmentation that is most likely to produce the observed image when run through the given forward model is obtained by maximizing a quantity called the *posterior*, which is proportional to ($\propto$) the product of two known and computable terms: The first term is called the *likelihood* and it quantifies how likely it would be to observe the given image if the hypothetical segmentation were true. This is often done by measuring the difference between the observed image and the one predicted by the forward model for the hypothetical segmentation. The smaller this difference, the more likely the segmentation is. The second term is called the *prior* and it measures the *a priori* probability that the hypothetical segmentation is correct, irrespective of the image observed. This could, e.g., attribute lower probability to membrane segmentations that are highly curved, formalizing our prior knowledge that lipid membranes tend to form smooth shapes.

All terms in Bayesian inference have the meaning of probabilities. However, given that in image analysis and experimentation it is often more natural to talk about *evidence* rather than *probability*, a theory like the Dempster–Shafer evidence theory might provide a more appropriate interpretation (Shafer 1976). The key

Fig. 1.3 Illustration of Bayes' theorem: we seek the segmentation that has the highest posterior probability of being correct given the observed image (*blue*). This can be achieved by maximizing the product of the likelihood of observing the image given the forward model output (*red*), and the prior knowledge about the imaged objects (*green*)

difference is that evidence does not have to sum up to 1, as probability does. Probability is the "chance" of there being two touching cells in the image, as opposed to a single cell. Evidence is the "degree of belief" that there is one or two cells. If the image is so blurry that one cannot decide whether it is one or two cells, one could give a low value to the probability of there being two cells. Since probabilities have to sum to 1, however, this implies a high probability that there is one cell. The statement hence inevitably becomes: "I am very certain that there is only one cell." This is not the same as: "I cannot decide whether there is one or two." If one cannot decide, this means there is neither compelling evidence for one nor for two cells. Since evidence does not have to sum to 1, one could hence simultaneously give low evidence to *both* possibilities.

In summary, there are three inference frameworks for image analysis: frequentist inference, Bayesian inference, and evidence theory. The first two currently dominate the field.

1.3 Image-Analysis Paradigms

Irrespective of the inference framework used, there are different philosophies and approaches to image analysis. Each of them has its own way of interpreting images and of including prior knowledge, and comes with its own set of advantages and caveats. The approach implemented in a given software largely defines what the software is in principle able to do, and what not. We discuss the three most prominent paradigms below, focusing on how to convert an image, stored as a matrix of pixel values, into quantitative information about the features and objects represented in it.

1.3.1 The Filter-Based Paradigm

The filter-based approach to image analysis consists in applying a series of arithmetic operations to the pixel-intensity values in order to isolate or reveal objects of interest, or compute object segmentation masks. Prior knowledge is included in the filter design. In order to detect bright spots in an image, one could, for example, run a band-pass filter over the image to reduce noise and background, and then use a relative threshold filter to select all local maxima that are brighter than a given threshold (Crocker and Grier 1996). More advanced approaches to spot detection use multi-scale wavelet filters (Olivo-Marin 2002).

Filters are classified as linear and non-linear, shift-invariant and shift-variant, and discrete (digital) and continuous (analog). Linear filters only compute linear combinations of the pixels in the input image, for example, weighted sums or differences. Non-linear filters apply arbitrary non-linear operations. Shift-invariant filters always perform the same operations regardless of *where* in the image they are applied. For example, they always compute the average of all neighbors of a pixel, regardless of which pixel is at the center. Shift-variant filters perform different operations depending on where they are shifted to, e.g., using a different PSF in different parts of the image. Discrete filters operate on discrete pixel lattices with discrete intensity levels, whereas continuous filters can also be evaluated at sub-pixel locations and may produce non-integer intensity values. Since digital images are discrete by nature, continuous filters are often based on assuming certain continuous basis functions, like splines or Bézier curves. Any linear shift-invariant filter amounts to convolving the input image with a filter kernel and can hence be efficiently computed as a convolution.

Most filter-based approaches are found in image processing, including prominent examples such as the fast Fourier transform (Cooley and Tukey 1965), wavelet transforms (Chan and Shen 2005), thresholding (Otsu 1975), edge-detection (Canny 1986), anisotropic diffusion filters (Perona and Malik 1990), and image naturalization (Gong and Sbalzarini 2014) (see Fig. 1.4). A special case are the discrete filters used in mathematical morphology (Najman and Talbot 2010). In image analysis, a famous filter-based method is watershed segmentation (Najman and Schmitt 1996), which is a combination of linear shift-invariant filters to determine the seeds for a subsequent watershed transform (Meyer et al. 1997) from mathematical

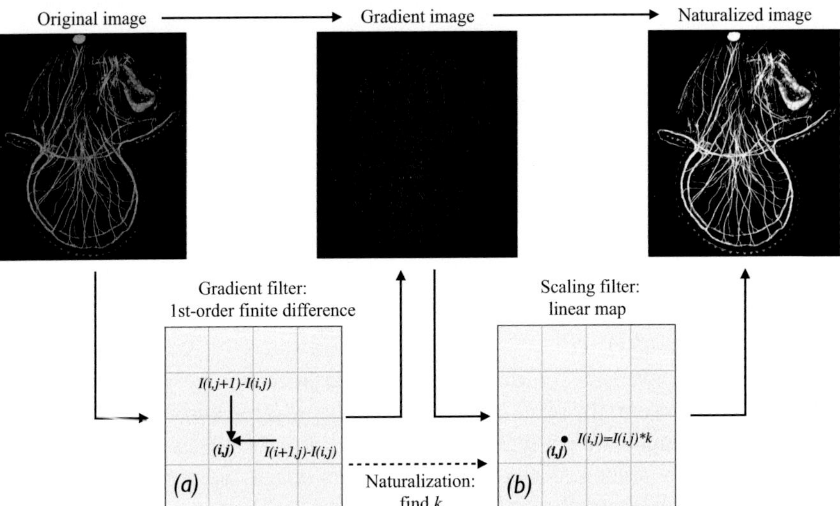

Fig. 1.4 Example of a typical filter-based workflow: image naturalization for denoising and contrast enhancement (Gong and Sbalzarini 2014). (**a**) The first filter computes the gradient of the input image by applying forward finite differences, subtracting each pixel intensity $I(i, j)$ from its top and right neighbors. (**b**) The second filter scales each pixel intensity by a value k that is automatically determined so that the naturalized image has a gradient histogram that fits the one expected from natural-scene images (Gong and Sbalzarini 2014). (Original image from: American Microscopical Society, Winner of the 2004 Buchsbaum Prize (amicros.org); Lee&Matus, U Hawaii, confocal image of *Pilidium* larva of the nemertean *Cerebratulus sp.*)

morphology. Filter-based approaches to motion tracking notably include those based on pixel cross-correlations (Willert and Gharib 1991) and split/merge data-association filters, which are however often augmented with Bayesian model-based approaches for multi-hypothesis tracking (Genovesio and Olivo-Marin 2004). Approaches combining Gaussian filtering with mathematical morphology and thresholding are routinely used for single-particle detection (Sbalzarini and Koumoutsakos 2005) and filament segmentation (Ruhnow et al. 2011).

Due to their explicit nature, filter-based approaches are computationally fast. They are, however, typically specialized. Filter-based approaches are designed specifically to the task. They provide less flexibility to adapt to different tasks than machine-learning and model-based methods do. Moreover, filter-based approaches typically have a large number of parameters that need to be adjusted and tuned. Depending, for example, on how one sets the threshold in a thresholding filter, one can get any result one wants. Often, there is no good *a priori* criterion to tune the parameters, leaving us with arguments like "it gave me what I wanted" or "it looked best" that fundamentally go against the idea of image quantification. Finally, filter-based image analysis yields a label image (e.g., a binary segmentation mask), from which objects and object information yet need to be extracted. While this can be as straightforward as finding connected components, it can also be more sophisticated, like in *Largest Contour Segmentation* (LCS) (Manders et al. 1996) where multiple segmentations/objects are found for different thresholds and combined afterwards.

1.3.2 The Machine-Learning Paradigm

The machine-learning approach is based on detecting patterns in numerical features computed from the image (reviewed by Shamir et al. 2010). This can mean classifying each pixel to be either part of an object or not. However, the approach is not limited to pixels, and also patches and whole images can be classified, e.g., whether they contain cells or not. This classification is done based on features that are computed for each pixel or over the whole patch/image. The simplest feature is the (average) intensity. More advanced features include texture (Li et al. 2003; Orlov et al. 2008), gradients (Orlov et al. 2008), and shape (Etyngier et al. 2007). The machine-learning algorithm could, e.g., classify all bright regions with rough texture as belonging to a nucleus, or all images that contain curve-like shapes as images of filaments. Machine learning can also be used to classify spatial patterns without prior segmentation (Huang and Murphy 2004). The machine-learning approach is frequently combined with the filter-based approach by either computing features of a filtered image where, for example, edges have been enhanced or by computing features using filters.

Machine learning can follow either an unsupervised or a supervised approach (Cherkassky and Mulier 1998; Duda et al. 2000; Bishop 2007). In *unsupervised learning* the pixels/images are grouped according to their features using, e.g., clustering techniques. This yields "sub-populations" that have similar features within, but different ones across. Frequently, the assumption is that pixels or images of the same sub-population show the same type of objects, e.g., nuclei. In the *supervised* approach, the classification is learned from examples. The algorithm first has to be "trained" using pre-classified examples from each class. This typically means segmenting or classifying a number of images by hand in order to train the algorithm. The number of examples required for training depends on the number of features and the learning algorithm used. Prior knowledge is included both in the design of the learning algorithm and in the choice of features used for classification (Hong et al. 2008). Image information that is not captured by the selected set of features is lost. In the supervised approach, prior knowledge is additionally included in the training examples selected.

Machine learning is particularly popular for complex images, like electron microscopy images, MRI, and X-ray images, and histological sections. In these images, texture and context often play an important role in detecting objects, hampering the design of generic filters or models. There, machine learning has, for example, been used to segment brain MRI images using supervised artificial neural networks (Reddick et al. 1997), to segment tumors in MRI images (Zhou et al. 2005), and to detect microcalcifications in mammograms (El-Naqa et al. 2002). Classification of image patches or whole images has, e.g., been used to classify sub-cellular patterns without previous cell segmentation (Huang and Murphy 2004). This approach is particularly prevalent in histology (McCann et al. 2012) and pathology (Fuchs et al. 2008, 2009; Orlov et al. 2010), where entire images are often scored, e.g., for lymphoma detection (Orlov et al. 2010). This approach is illustrated in Fig. 1.5, where supervised classification of image patches is used to detect different tissue types, followed by classifying the overall histological score for the whole image. Frequently used image feature sets include *weighted neighbor distances* (WND) (Orlov et al. 2008), *scale-invariant features* (SIFT) (Lowe 1999), *binarized statistical image features* (BSIF) (Kannala and Rahtu 2012), and *basic image features* (BIF) (Crosier and Griffin 2010). State-of-the-art supervised learning algorithms for image analysis include random forests (Breiman 2001), regression tree fields (Jancsary et al. 2012), and deep neural networks ("deep learning") (Ciresan et al. 2012). Machine-learning approaches to motion tracking started with using Support Vector Machines (Schölkopf and Smola 2002), a popular supervised classification method, to track optical flows (Avidan 2004). Later, this was generalized to *Relevance Vector Machines* that predict displacements rather than detecting flow, hence leading to a model-based Bayesian learning paradigm (Williams et al. 2005).

Machine-learning methods provide more flexibility than filter-based methods, albeit at the expense of higher computational time. Supervised approaches are particularly flexible, as they can be trained by example to solve a variety of image-analysis problems. Having to manually annotate and select the training samples, however, is additional effort. Moreover, the final analysis depends on the chosen training data, hence introducing additional user bias that is not present in other methods. Usually, one wishes to keep the feature set as small as possible, as the computational cost of machine-learning algorithms grows with the number of

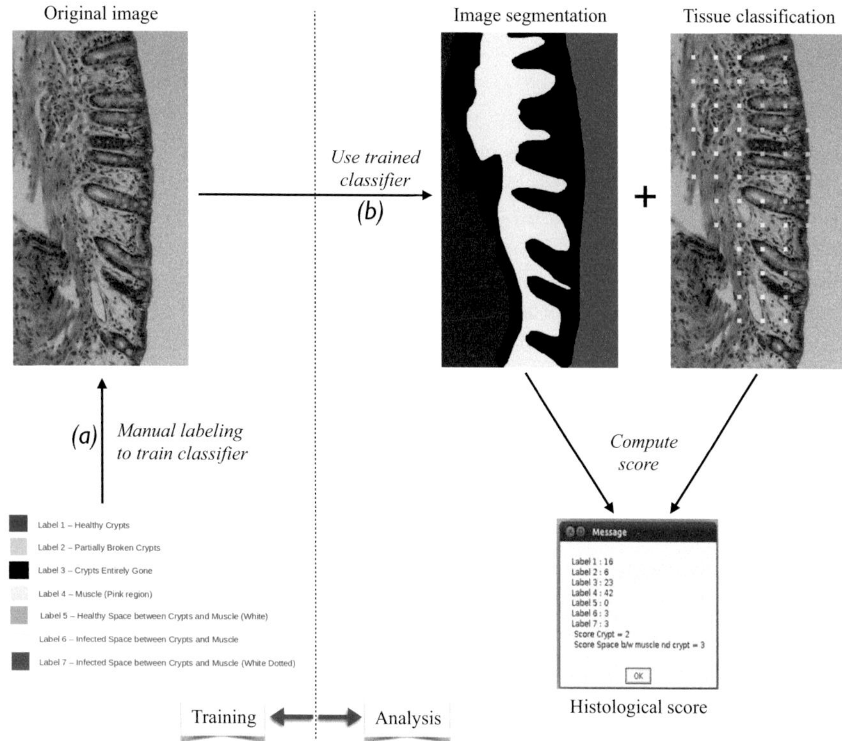

Fig. 1.5 Example of a typical machine-learning workflow: automatic scoring of histology sections for colitis detection. (**a**) The machine-learning classifier is trained by the user manually labeling the tissues in several example locations. (**b**) After training, the classifier can be used to segment the image and automatically assign tissue class labels everywhere. From these, the final histological score is computed. The example shown here uses the Random Forest classifier from *WEKA* (Hall et al. 2009) on the *WND-CHARM* (Orlov et al. 2008) image features. Prior to training, 22 out of the 1025 features were determined to be important for the problem (feature selection). (Original image from: Institute of Physiology, University Hospital Zurich, mouse colitis histology section; Segmentation and classification by Dheeraj Mundhra, MOSAIC Group)

features used. In a supervised approach, the amount of training data needed also increases with the feature count. Deciding which features to include is a hard problem known as *feature selection*. Finally, like in filter-based approaches, pixel-level classification yields a labeled output image from which objects and their properties still need to be extracted. This image-to-object transformation implies additional prior knowledge and can also be done using machine learning.

1.3.3 The Model-Based Paradigm

The model-based approach does not operate on the pixels of the observed image, but rather estimates a model of the imaged scenery that is most likely to explain the observed image. The image is hence only used as a gold standard to compare with. A key ingredient is *how* a hypothetical segmentation or scenery is compared with the image, and how the result of this comparison is used to iteratively refine and improve the former. Approaches range from comparing image intensities (Kass et al. 1988) or gradients (Lin et al. 2003) to including a predictive model of the image-formation process (Helmuth et al. 2009). The latter is illustrated in Fig. 1.6, where the segmentation is iteratively updated until the output of the forward problem (see section "The Forward and the Inverse Problem") fits the observed image as closely as possible.

Model-based image analysis requires up to three models to be specified: the object model, the imaging model, and the noise model. In many cases, not all three are present, or some are implicitly assumed to be, e.g., the identity map or a Gaussian. The *imaging model* describes the noise-free image-formation process, predicting the image one would *expect* (over the statistical distribution of the noise) to see when imaging a particular scenery. A simple imaging model for a fluorescence microscope is the convolution with the PSF (Linfoot and Wolf 1956; Zhang et al. 2007), neglecting any noise. For phase-contrast microscopy, the imaging model is more complex (Yin et al. 2010). If the image-formation process is not to be accounted for, the imaging model can also be the identity map. If the comparison is not done on pixel intensities, but on other features, such as intensity gradients, the imaging model computes these features (Lin et al. 2003). The hypothetical sceneries that are scored by the imaging model correspond to different realizations of the object model. The *object model* defines the admissible sceneries and parameterizes

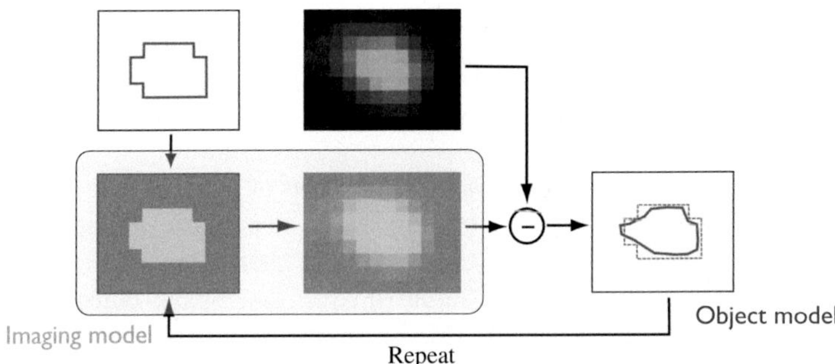

Fig. 1.6 The model-based approach: an imaging model is used to predict the expected image for a given, hypothetical segmentation. This predicted image is then compared with the actually observed image and the segmentation is adjusted to minimize the difference between the two. The specific metric used to quantify the difference implicitly defines the noise model one assumes. The object model defines what shapes, deformation, and intensity distributions are admissible for the segmentation (credit: Jo Helmuth, MOSAIC Group)

them. A simple object model for a nucleus could be a sphere, parameterized by its center location and radius. For each realization of this model, i.e., concrete values for center and radius, the imaging model predicts how the image features of that nucleus would look like (e.g., bright spot around the projection of the sphere with some diffraction blur at the boundary). The imaging and object models can be arbitrarily complex and may even include physics-based numerical simulations. For example, segmenting cardiac deformation from ultrasound images has been done using a finite-element simulation of the mechanics of the myocardium as an object model (Papademetris et al. 1999). Finally, the *noise model* specifies the statistical distribution of the imaging noise, hence providing statistical significance to the comparison between imaging model and data. In the simplest case, the noise model defines how to compare the imaging-model output with the image data (Paul et al. 2013). Assuming Gaussian noise on the data, one would, for example, compare images by the sum of squared pixel-intensity differences. Other noise models lead to different comparison metrics (Chesnaud et al. 1999; Martin et al. 2004; Paul et al. 2013). In many cases, the noise model is not explicit, but implicitly assumed, e.g., in the way the imaging model is evaluated, or in the features of the image used for the comparison.

Model-based approaches are mostly classified with respect to the model assumptions. Examples include piecewise constant object models that assume the intensity within each object to be uniform (Fig. 1.7). Piecewise smooth object models allow for intensity gradients within an object (Fig. 1.8). Deconvolving imaging models account for the PSF of the microscope (Fig. 1.7). Data-driven models try to fit fea-

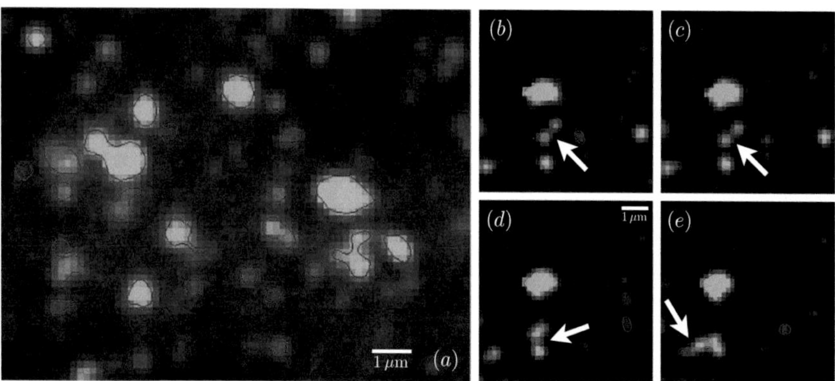

Fig. 1.7 Example of model-based image segmentation, accounting for the Point Spread Function (PSF) of the microscope to get high-resolution outlines of small intracellular objects (Helmuth et al. 2009; Helmuth and Sbalzarini 2009). The images show maximum-intensity projections of confocal z-stacks of fluorescently labeled Rab5, a protein localizing to endosomes. (**a**) The *red* outlines show the resulting object boundaries using deconvolving active contour segmentation (Helmuth et al. 2009; Helmuth and Sbalzarini 2009). (**b**)–(**e**) Time-lapse sequence of reconstructed Rab5-GFP outlines illustrating a fusion event (*arrow*). (Raw images: Greber lab, University of Zurich; segmentations: Jo Helmuth, MOSAIC Group)

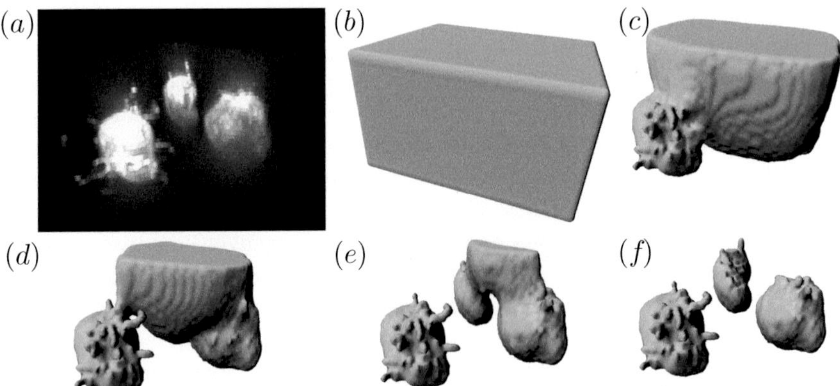

Fig. 1.8 3D model-based segmentation of germ cells in a zebrafish embryo. (**a**) The raw 3D confocal image stack, showing three cells with a fluorescent membrane staining. The intensity is inhomogeneous, with the background over the *top* of the cells brighter than the interior at the *bottom* of the cells. (**b**) Initial segmentation provided by the user to the algorithm from Cardinale et al. (2012). (**c**)–(**f**) Evolution of the outline as computed by the algorithm, converging to the final segmentation using a piecewise smooth object model and a fluorescence imaging model. (Raw image: Mohammad Goudarzi, University of Münster; segmentations: Janick Cardinale, MOSAIC Group)

tures (e.g., gradients) learned from data. A second classification is with respect to the algorithm used to determine the best scenery. Examples include statistical estimators (Zhu and Yuille 1996), variational solvers (Chan and Shen 2005), sampling schemes and random fields (Geman and Geman 1984), combinatorial optimizers (Blake et al. 2011), graph-based optimizers (Boykov et al. 2001), dynamic programming (Nilufar and Perkins 2014), and greedy gradient descent (Kass et al. 1988).

A dynamic-programming approach was, for example, used for whisker tracing (filament segmentation) in behavioral videos of mice, by Bayesian inference over a whisker object model (called "detector" therein) (Clack et al. 2012). Additionally including an imaging model, model-based image analysis has been used to determine deconvolved segmentations without computing a deconvolution (Helmuth and Sbalzarini 2009; Helmuth et al. 2009). Because the imaging model accounts for PSF blur and imaging noise, the segmentations in Fig. 1.7 jointly solve the deconvolution and segmentation tasks (Paul et al. 2011, 2013). This is beneficial when segmenting small objects near the diffraction limit, like the Rab5-GFP endosome domains in Fig. 1.7. The model-based approach also allows including physical properties of the imaged objects, such as bending stiffnesses of lipid membranes, hence ensuring that the image-analysis result corresponds to a physically feasible membrane configuration. This can be useful, e.g., when studying cell-edge dynamics in polarizing and migrating keratocytes (Ambühl et al. 2012). Efficient 3D methods for model-based image segmentation are also available (Boykov et al. 2001; El-Zehiry and Elmaghraby 2009; Cardinale et al. 2012) and topological constraints on the objects can be accounted for (Cardinale et al. 2012) (Fig. 1.8). Data-driven models have been used to robustly segment dense and touching nuclei in fluorescence microscopy images using a gradient model (Lin et al. 2003).

Model-based approaches are also popular for motion tracking (Kalaidzidis 2007, 2009), for example, based on a model of how the objects move (Crocker and Grier 1996), using approximate graph matching to fit a model to the data (Vallotton et al. 2003), using approximate combinatorial optimization methods for model fitting (Sbalzarini and Koumoutsakos 2005; Ruhnow et al. 2011; Jaqaman et al. 2008), using Kalman filters based on linear state-space models (Li et al. 2006, 2007), using particle filters based on non-linear state-space models (Hue et al. 2002; Smal et al. 2008; Cardinale et al. 2009), using Bayesian probabilistic models for multi-target tracking (Genovesio and Olivo-Marin 2004) and multiple hypothesis tracking (Cox and Hingorani 1996), and using integer-programming optimization over graph-based motion and appearance models (factor graphs) (Schiegg et al. 2013).

Model-based image analysis includes prior knowledge via the imaging, noise, and object models. While only the latter constitutes a prior in the Bayesian sense, all encode prior knowledge. Postulating the wrong models leads to wrong results. Therefore, the model-based approach is particularly suited to clear-cut cases, like fluorescence microscopy, where the object and imaging models are suggested by physics. However, even when using appropriate models, the resulting optimization problem can be difficult to solve and is often restricted to local optimization starting from a user-specified initial segmentation (Kass et al. 1988; Helmuth et al. 2009; Helmuth and Sbalzarini 2009; Cardinale et al. 2012) (Fig. 1.8). This is relaxed in globally optimal methods, which are independent of initialization and guarantee that there is no other result that would explain the image better than the one found (Pock et al. 2009; Brown et al. 2011; Paul et al. 2011, 2013). While this result may still be wrong, it uses all the information available in the image and represents the best-possible result under the assumed models (Rizk et al. 2014). This is a strong statement that is much harder to make in the filter and machine-learning paradigms. Another important advantage of model-based methods is that they are physics-based and the same algorithm can be used for a variety of cases by swapping the model. Switching from fluorescence to phase-contrast images can be as easy as replacing the imaging model accordingly, leaving the algorithm unchanged. Finally, model-based approaches directly yield objects and object properties. They hence unite image labeling and image-to-object transformation into a single step.

1.4 Challenges in Computational Bio-image Analysis

There are currently four major challenges in computational bio-image analysis: large and multi-dimensional data, uncertainty quantification, more generic algorithms, and collaborative open-source software. In addition, there are several challenges in related areas, such as image databases (Swedlow et al. 2003), annotation systems (Peng et al. 2010), and gold standards for testing and benchmarking of algorithms (Rajaram et al. 2012; Vebjorn et al. 2012). Together with the advancements in optics, microscopy, and labeling techniques, these developments will enable unprecedented image-based studies.

1.4.1 Large and Multi-dimensional Data

In bio-image analysis, big data comes in two flavors: many images or large images. The former is typically the case in high-throughput screens (Collinet et al. 2010) and can be dealt with by distributing the images over multiple computers for analysis. The latter is a feature of multi-dimensional and high-resolution imaging techniques, such as imaging mass spectrometry (Stoeckli et al. 2001) and light-sheet microscopy (Huisken et al. 2004; Engelbrecht and Stelzer 2006), and requires solutions within a single image. This can, for example, be done by multi-scale image representations, such as scale-space approaches (Witkin 1984) and super-pixels (Xu and Corso 2012), akin to the "Google Maps" zooming function.

However, the question arises as to what should be done if the imaging equipment delivers data at a faster rate than what can be written to hard disks. Recent microscopes with CMOS cameras, for example, deliver 3D images at a rate of 1 GB/s, per camera. A setup that uses two cameras hence produces almost 173 TB of data a day (Reynaud et al. 2014). This is faster than any hard disk or other permanent storage system could archive the images and raises data-handling and storage issues that have so far been confined to the high-performance computing and particle physics communities (Tomer et al. 2012; Weber and Huisken 2012). Storing all raw images that come from such microscopes is infeasible. Using lossless data compression techniques, however, fast networks are able to stream the data directly into a computer cluster, where it can be distributed across multiple computers for analysis. Only the analysis results are then stored, e.g., the positions of all nuclei or the shapes, sizes, and locations of all cells in the tissue. If the analysis is to be repeated, it is quicker to image another sample than to archive the raw data, read it back, and re-run the analysis. This trend is also observed in large computer simulations running on supercomputing systems and is known there as the "data gap" (Sbalzarini 2010). Analysis results and visualizations are hence determined at runtime, and if later a new variable is to be measured or a new feature to be computed, the whole simulation is re-run.

Not storing the raw data comes with two requirements: (1) The analysis software needs to run in real time, possibly distributed across multiple computers. (2) The confidence intervals and uncertainties of the analysis results need to be known and stored along with the results.

The first requirement is of technical nature. Individual computer processor cores are not getting faster any more at the rate they used to. Instead, chip manufacturers pack multiple cores into each processor and have them operate in parallel. Leveraging this speedup, however, requires that algorithms are designed and implemented with parallelism in mind. This is often not straightforward and

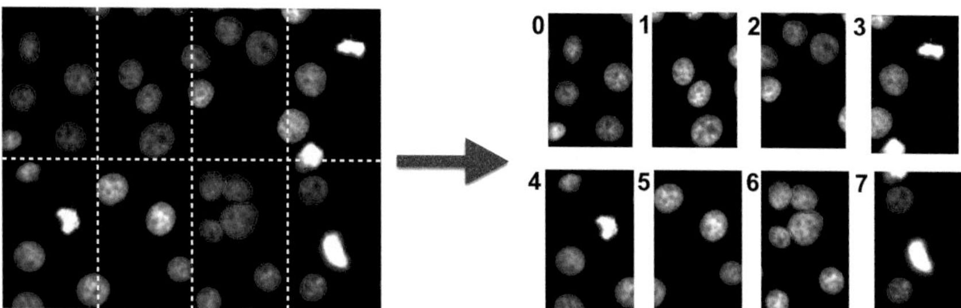

Fig. 1.9 Domain-decomposition approach to deal with big image data. A large image is subdivided into several smaller sub-images that are each given to a different computer or cluster node for processing. The different computers (0–7 in this example) communicate with each other over the network whenever segmentations cross sub-image boundaries. This ensures that the overall segmentation is the same as that which would have been obtained on a single computer. The distributed result, however, is computed faster (here: about eight times faster) and using less memory on each computer. (Raw image: Dr. Liberali, University of Zurich; distributed segmentation by Yaser Afshar, MOSAIC Group)

requires re-thinking many image-analysis algorithms. One way that is currently uncommon is to distribute each image. Instead of having one processor core analyze image after image, the images are divided into smaller sub-images that are scattered across multiple cores (see Fig. 1.9). Each core then works on its part of the image and they exchange information with the other cores over a computer network in order to collectively solve the global analysis task in a fraction of the time it would take a single core to do so (Nicolescu and Jonker 2000). This requires image-analysis algorithms that can be divided into concurrent work packages with as little interdependences as possible (Seinstra et al. 2002). Every interdependence between work packages requires communication between the processor cores, incurring additional overhead. Modern computer hardware, such as graphics processing units (GPUs) and heterogeneous many-core processors, have more than a thousand parallel cores that need to be kept busy and orchestrated. It is hence essential that we develop algorithms that map well onto such computer architectures (Galizia et al. 2015), like the example of motion tracking using parallel distributed particle filters (Demirel et al. 2014a) implemented using the PPF software library (Demirel et al. 2014c).

The second requirement when not storing the raw data is uncertainty quantification. If one only stores the analysis results, but not the raw images, it is impossible to later go back and check whether there really was a nucleus in that strange outlier image, or not. We would never know how much of the data is noise, and what is signal. Storing proper confidence intervals would, however, tell us that the probability that there actually was a nucleus is, e.g., 10 %. So we know that the data point is not to be trusted, because the algorithm was not sure about what it "sees" in that image. Storing the analysis results with their associated uncertainties enables statistical significance tests in order to decide whether an observed difference between samples is real, or not. This links to the challenge of uncertainty quantification.

1.4.2 Uncertainty Quantification

The previous example describes a situation where uncertainty quantification is indispensable. However, it is useful in far more cases. An algorithm for segmenting cells in a tissue could, for example, automatically detect regions in the image where it cannot determine a confident segmentation and flag the user to look specifically at those regions when manually post-processing the result. This would greatly reduce the proofreading overhead. One could then also specify a confidence level and instruct the algorithm to only flag cases where the result is less than, say, 95 % likely to be correct. If a mutant or knock-down then shows less than 5 % difference in the readout, we know that this is not significant and could as well be explained by image-analysis errors. Clearly, uncertainty quantification is desirable and useful.

Unfortunately, uncertainty quantification is a hard problem and has therefore not received much attention in image analysis so far. It is often difficult to express uncertainty "scores" as true probabilities, because the normalization is unknown. Evidence theory (Shafer 1976) could hence provide a more straightforward way of expressing uncertainty (see section "Bayesian, or Not?"). Regardless of their expression and interpretation, however, uncertainties in biological image analysis mainly arise from three sources:

1. The noise in the raw image is propagated through the computational analysis pipeline, leading to (potentially amplified) noise and uncertainty in the analysis result.
2. The algorithm may terminate with a solution that is not the best possible one, leaving some uncertainty about how far from the best solution it is.
3. The prior knowledge on which the algorithm is based may not adequately describe reality, leading to uncertainty about how much of the result is due to this inadequacy.

Source (1) is particularly prevalent in fluorescence microscopy, where the imaging noise is often significant. This noise may be further amplified by the image analysis. Consider, e.g., an edge detector that computes differences between pixels, or a watershed filter that compares which of two pixels is brighter. If both pixel intensities are noisy to within $\pm 10\,\%$, the difference is noisy to within $\pm 20\,\%$ and the watershed may go down the wrong way. The noise is hence amplified, leading to results that are less reliable than the original data. A famous example of a noise-amplifying process is deconvolution. Source (2) is mostly important in machine-learning and model-based methods, where the final estimated classification or scenery may not be the *global* optimum over *all* possibilities, but only a local optimum over a subset of tested possibilities. Source (3) is again relevant to all three paradigms, since all of them include prior knowledge in the filter design, feature selection, or model specification that could be wrong or inadequate.

Ideally, we quantify the uncertainty in the final result due to the combined effects of all three sources, or at least provide an upper bound for it. Unfortunately, this problem is hard (Halpern 2005). Ground truth is not available, synthetic benchmark images frequently do not share the intricacies of real images, a theoretical framework for inference over images is lacking, and theoretical performance guarantees are not available for many algorithms. Nevertheless, several promising approaches can be identified.

The first approach includes efforts to generate hand-segmented benchmark image collections (Vebjorn et al. 2012). The accuracy and robustness of algorithms can be tested on these image collections and scored against the manual gold standard. This approach works as long as the images that the algorithms are later going to be applied to are similar to the benchmark images. Since they are not going to be exactly the same, though, this introduces uncertainty about the uncertainty quantification. Moreover, the manual gold standard is not free of human error. Both points are somewhat relaxed by using synthetic ground truth (Rajaram et al. 2012). There, ground truth is known without uncertainty, and the forward model used to generate the synthetic images can be adapted to different acquisition conditions. The key difficulty, however, is to provide sufficiently realistic (in shape, noise distribution, fluorophore blinking, etc.) ground truth and forward models. Using inappropriate models again leads to uncertainty in the uncertainty estimate, according to source (3) above. Realistic shapes can, for example, be generated by sampling from shape spaces learned from training images (Murphy 2012). This, however, introduces uncertainty with respect to the training data chosen for learning the shape space.

The above approaches to uncertainty quantification are data-centric. There are, however, also algorithm-centric approaches that relax the data dependency to some extent. Source (1) can, for example, be addressed by error-propagation analysis of the involved algorithms. The conceptual idea is to re-run the analysis for different random input perturbations and see how the results vary. This is commonplace in scientific computing, where a wealth of efficient methods have been developed, including spectral uncertainty quantification (Le Maître and Knio 2010), simplex stochastic collocation (Witteveen and Iaccarino 2012), and generalized polynomial chaos expansion (Xiu and Karniadakis 2002; Xiu 2009). However, these are still rarely used in image analysis, with exceptions like *OMEGA*, which uses error propagation in particle-tracking analysis (https://github.com/OmegaProject).

Most algorithm-centric approaches so far have focused on source (2) by quantifying the residual discrepancy between the model output and the real image. These approaches do not require ground truth, but rather measure model-fitting errors. They hence only provide a *lower* bound on the real uncertainty and usually do not yield proper probabilities, but rather evidence (Shafer 1976). The simplest approach is to use the residual value of the posterior probability as a proxy for result certainty (Paul et al. 2013) (Fig. 1.10a). More sophisticated approaches use Markov-chain Monte Carlo sampling (Geman and Geman 1984; Chang and Fisher III 2011) or approximate Bayesian computation (Marjoram et al. 2003) to sample from the probability density of the posterior and get an idea of the distribution of potential results. This has, e.g., been used to provide uncertainty estimates in microtubule tracking (Cardinale et al. 2009) (Fig. 1.10b), model-based segmentation (Cardinale 2013) (Fig. 1.10c), and multi-scale approaches (Kohli et al. 2010). In addition, theoretical performance guarantees are available for some optimizers and estimators

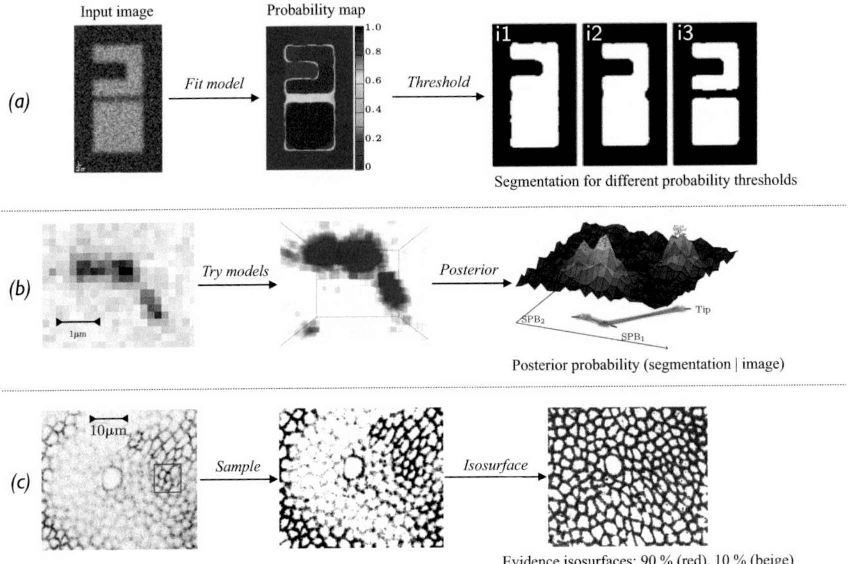

Fig. 1.10 Approaches to uncertainty quantification in image analysis. In all approaches, the input image (*left*) is first transformed to a probability or evidence map, from which the result is then computed. (**a**) In the approach of Paul et al. (2013), the residual of a globally optimal model fit is used to estimate a probability map for each pixel to be part of an object. Thresholding this map at different probability cutoffs gives different alternative segmentations for different confidence levels (i1–i3). (**b**) The approach of Cardinale et al. (2009) applied to tracking microtubule tips using particle filters (Demirel et al. 2014b). The particle filter tries many different possible model fits in order to form a cloud of possible tip localizations (*middle*). Using the resulting particle representation of the posterior (Demirel et al. 2014b), the most likely tip positions and their variances can be extracted. (**c**) The approach of Cardinale (2013) uses Markov-chain Monte Carlo sampling to sample many possible segmentations from the model posterior. This yields an unnormalized evidence map (darker means higher evidence) from which iso-surfaces can be computed. These iso-surfaces contain the correct segmentation with the indicated evidence. (Image sources: input image in (**a**) from Grégory Paul, MOSAIC Group, synthetic test image; input image in (**b**) from Barral lab, ETH Zurich, fluorescently labeled spindle-pole bodies (SPB) and spindle tip in dividing *S. cerevisiae*; input image in (**c**) from Basler lab, University of Zurich, fluorescently labeled membranes in a *D. melanogaster* wing imaginal disc; result in (**a**) by Grégory Paul, MOSAIC Group; results in (**b**) and (**c**) by Janick Cardinale, MOSAIC Group)

used in model-based methods. Examples include graph cuts (Boykov et al. 2001), Markov random fields (Geman and Geman 1984), and photometric estimators based on information theory (Paul et al. 2013). While these error bounds do not provide a probability distribution, they give an idea of the interval within which the correct result must lie. These intervals, however, are in the model-fitting energy and not in object space. If the energy is flat (i.e., has a small gradient), the result might be arbitrarily wrong and still have similar energy. This problem is addressed by the concept of *diversity solutions*, which are alternative segmentations or analysis results that are all about equally likely to be true, but may look very different (Ramakrishna and Batra 2012; Batra et al. 2012). Using diversity solutions, a segmentation algorithm could, for example, express its uncertainty about two overlapping blobs of high intensity being two individual touching objects, or one fused object. Ideally, globally optimal methods are guaranteed to find the best solution and are hence free of source (2) uncertainty (Pock et al. 2009; Brown et al. 2011). This, however, is only possible for simple object and imaging models, trading off uncertainties of source (3).

Source (3) is a classic issue in machine learning, called *model misspecification error*. To our knowledge, it has so far not been addressed in image analysis. One way to do so could be to combine machine-learning and model-based methods on the same problem. Looking at the discrepancy between the results could provide an estimate of how much uncertainty is explicable by modeling errors.

Finally, rather than asking how well a given algorithm performs on an image, one could ask how well *any* algorithm could *possibly* perform. For example, with what uncertainty is one able to quantify the center of a point source from a fluorescence microscopy image given the finite number of photons recorded? These are questions about absolute, often information-theoretic, bounds. For point localization, the problem has been solved using the concept of Cramér–Rao bounds, providing a lower bound on the estimation error any algorithm must necessarily make, given the photon count (Ober et al. 2015). For two- and three-dimensional objects, however, no such bounds are known yet. The situation is considerably more complex there, since neighboring photon sources are correlated through the (unknown) geometry of the sample.

1.4.3 Generic Algorithms

A current shortcoming in bio-image analysis is the tendency to treat every problem as a special case and develop a new algorithm or software for each project, to solve exactly the specific problem of that project. While this case-by-case approach and the associated "whatever works" mentality mostly lead to the desired results, they are wasteful and not scalable. Not only does it take a long time to come up with and implement a new analysis algorithm, it is a recipe for reinventing the wheel. Research groups hire image-processing specialists and computer programmers who often reinvent or re-implement what was already there in another group, and central image-processing facilities (if existent at all) drown in unrelated requests and do not find time to provide more general, unifying solutions. One of the most precious features of an algorithm is its generality. A strong trend in the field hence goes toward developing and implementing algorithms that are more generic and that are applicable to more than just one case or imaging modality. This also includes collections of canned algorithm building blocks and software libraries that can be used by computer programmers to more rapidly build workflow solutions from proven components (see also section "Collaborative Open-Source Software").

On the algorithmic level, there are three main axes of generality: (1) combining multiple tasks into one, (2) extending the class of problems that a given algorithm can deal with, and (3) rendering an algorithm parameter free.

The first point could, e.g., include combining image restoration with segmentation and photometry. An example are deconvolving active contours (Helmuth and Sbalzarini 2009) (see Fig. 1.7) that combine image deconvolution with segmentation by directly providing segmentation results that are compatible with the microscope's PSF. Along the same lines, image denoising, deblurring, and segmentation have been combined into a single step using the concept of Sobolev gradients (Jung et al. 2009). Segmentation has also been combined with denoising, deconvolution, and inpainting into a single level-set or split-Bregman model-based algorithm (Paul et al. 2013, 2011), as implemented in the Squassh plug-in for *Fiji* and *ImageJ* (Rizk et al. 2014). Jointly solving the image restoration (e.g., denoising, deconvolution, dehazing, and inpainting) and segmentation problems leads to better results than doing so sequentially (Paul et al. 2013). The reason is that while both individual problems are ill-posed, they naturally regularize each other when considered jointly. Computing a deconvolution, e.g., is ill-posed because there is a multitude of results that map to the same image when convolved with the microscope PSF. The result is hence not unique, and depending on how the parameters of the deconvolution algorithm are tuned, different results can be obtained. When segmenting at the same time, however, the deconvolution method does not have to produce a complete image, but only has to work in the limited solution space of segmentations. Since this space is smaller, the ambiguity is reduced.

The second point is mostly addressed by machine-learning or model-based frameworks. Both provide principled ways of adapting to new situations. This is less obvious in filter-based methods, where the problem-specific prior knowledge is implicitly included in the filter design. Changing to a new problem (e.g., from segmenting fluorescence images to segmenting phase-contrast images) would require one to re-design the filter. Machine-learning and model-based frameworks "externalize" the prior knowledge and allow one to change it without changing the core of the algorithm. In machine learning, this may be as simple as re-training the algorithm using a new set of training images (e.g., phase-contrast instead of fluorescence). In a model-based approach, the object model and/or imaging model can be replaced to adapt the algorithm to different problems. This is illustrated in Fig. 1.11.

The third point aims at rendering algorithms parameter free. Most image-analysis algorithms have a number of user-adjustable parameters. Only few algorithms work across a spectrum of problems without requiring parameter tuning. These parameter-free algorithms are particularly popular because they are easy to use and deliver robust performance across many applications. Examples include Otsu thresholding (Otsu 1975) and image naturalization (Gong and Sbalzarini 2014). There is a clear trend in the field to reduce the number of parameters of an algorithm with the ideal goal of rendering it more versatile and easier to use.

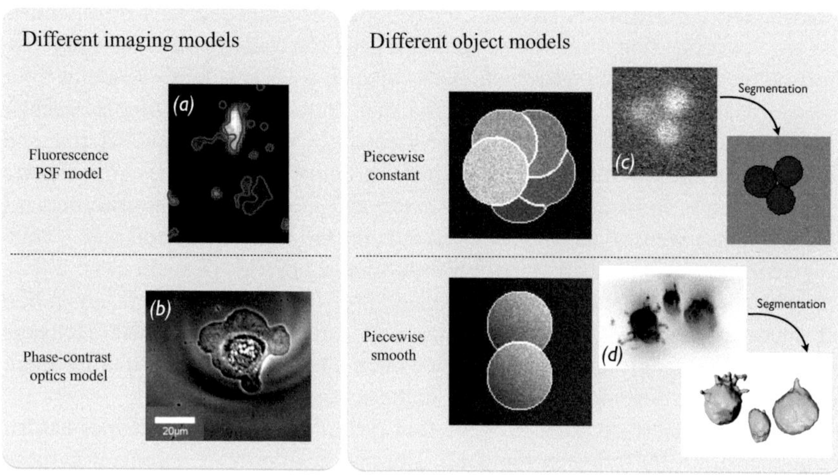

Fig. 1.11 Flexibility of model-based analysis by replacing the model in the same algorithm. *Left*: the model-based Snake method (Kass et al. 1988) can be used to segment (**a**) fluorescence images of Rab5-EGFP endosomes in HER911 cells (Helmuth and Sbalzarini 2009), as well as (**b**) phase-contrast images of polarizing fish epidermal keratocytes (Ambühl et al. 2012) by changing the imaging model. Changing the object model allows segmenting piecewise constant or piecewise smooth objects (Cardinale et al. 2012), such as (**c**) fluorescently labeled cells with cell-specific uniform staining (Cardinale 2013) and (**d**) fluorescently labeled zebrafish primordial germ cells with highly non-uniform signal (Cardinale et al. 2012). The synthetic "ice-cream" images illustrate the concept of piecewise constant and piecewise smooth object intensities. (Image sources: raw image in (**a**) from Greber lab, University of Zurich; raw image in (**b**) from Verkhovsky lab, EPFL; raw image in (**c**) from BCS Group, TU Darmstadt, and raw image in (**d**) from Mohammad Goudarzi, University of Münster; segmentation in (**a**) by Jo Helmuth, MOSAIC Group; segmentation in (**b**) by Mark Ambühl, Verkhovsky lab, EPFL, and ice-cream images and all segmentations in (**c**) and (**d**) by Janick Cardinale, MOSAIC Group)

1.4.4 Collaborative Open-Source Software

In addition to providing generic and flexible algorithms, another proven remedy against reinventing or re-implementing existing methods is to share software and create public repositories of open software modules. This motivates the trend for creating and maintaining collaborative open-source software for bio-image informatics (Swedlow and Eliceiri 2009). However, the need for open-source development reaches deeper than merely alleviating the implementation overhead for new projects. It is a fundamental prerequisite for reproducible science. Closed-source software is a black box that often does not provide enough information about the algorithms implemented. Open-source software is one way of rendering image analysis transparent, but its coordinated development and long-term maintenance come with their own set of challenges, in particular with respect to project coordination and funding (Cardona and Tomancak 2012).

Open-source software is frequently developed in academic labs by scientists who are not professional software engineers. This has traditionally had a negative effect on the usability and user-friendliness of such software. While guidelines for software usability are available, enforcing them remains challenging (Carpenter et al. 2012). Open-source projects also frequently start as individual research projects with specific biological questions in mind. Many pieces of software organically grew from there, becoming more and more generic, but the original application they were conceived for often remains the focus of the software. While this provides a rich landscape of software tools and libraries (Eliceiri et al. 2012), each with a specific application niche, it also raises the question of how integration and interoperability between the various tools can be achieved. Data exchange between different tools in order to combine them into workflows is one of the main challenges for the developer community in the coming years.

From a user-interface point of view, four different design philosophies can be distinguished, as illustrated in Fig. 1.12: The first is to provide a general-purpose command-line or scripting language with a large collection of toolboxes and subroutines that the user can combine for image analysis. This is the approach taken by tools like *R*, *Octave* (an open-source *MATLAB* look-alike), *ScyPi*, and *PIL*. These tools are particularly flexible and generic, are well suited for batch processing of large image collections, but require the user to have basic scripting skills. A second design philosophy is to provide an interactive graphical user interface, mostly combined with a plug-in architecture for third-party developers to contribute their algorithms. Examples include *ImageJ* (Abramoff et al. 2004; Schneider et al. 2012), *Fiji* (Schindelin 2008; Schindelin et al. 2012), *Icy* (de Chaumont et al. 2011, 2012), *Vaa3D* (Peng et al. 2010, 2014), *bisque* (Kvilekval et al. 2010), *OMERO* (Swedlow and Eliceiri 2009), *FARSIGHT* (Roysam et al. 2008), *CellCognition* (Held et al. 2010), *MorphoGraphX* (de Reuille et al. 2015), and *BioImageXD* (Kankaanpää et al. 2012). A third design approach puts the analysis workflow center stage, frequently specified using a graphical data-flow language. Examples of this kind are *CellProfiler* (Carpenter et al. 2006; Lamprecht et al. 2007), the workflow engine *KNIME* (Berthold et al. 2008), the workflow engine *LONI Pipeline* (Rex et al. 2003) and the image-processing environment *MiPipeline* (Nandy 2015) based thereon, and the image-processing environment *Anima* (Rantanen et al. 2014) based on the workflow engine *ANDURIL* (Ovaska et al. 2010). The fourth approach is to implement large collections of generic image-analysis algorithms in well-tested software libraries that provide an API for developing user programs. This is the most generic approach, but requires the user to have programming skills. Popular examples include the libraries *ITK* (Ibanez et al. 2005) and *VIGRA* (Köthe 1999) for image analysis and processing, *OpenCV* (Bradski and Kaehler 2008) for computer vision, and *OpenGM* (Andres et al. 2012) for machine learning.

Virtually all of these software projects implement filter-based analysis. However, many have their own specialization. *ImageJ* (Abramoff et al. 2004; Schneider et al. 2012) is, for example, particularly well suited for 2D microscopy image analysis. *CellCognition* (Held et al. 2010) caters to time-lapse cell culture imaging, *CellProfiler* (Carpenter et al. 2006; Lamprecht et al. 2007) was originally developed

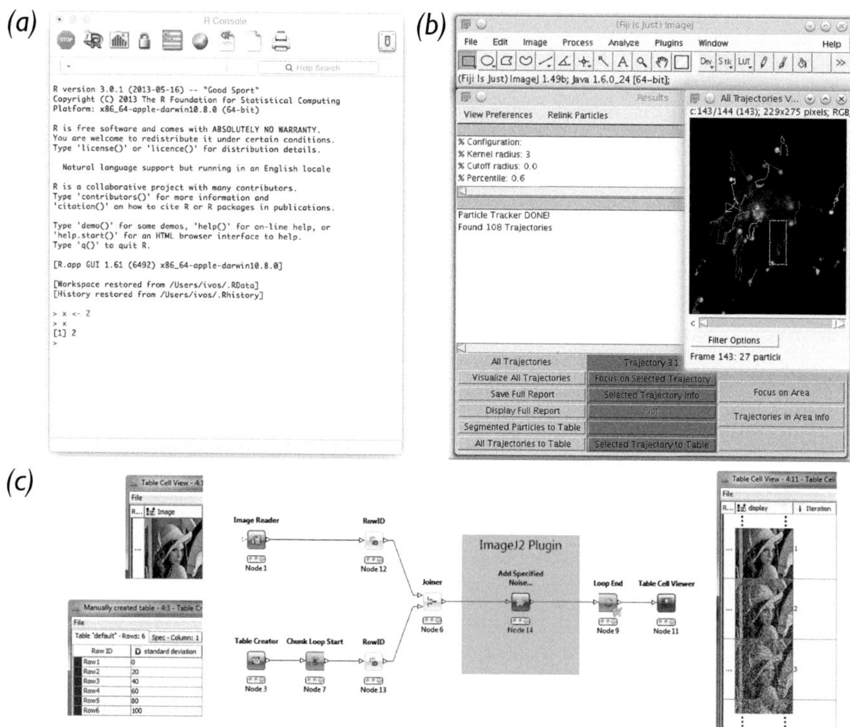

Fig. 1.12 Different user-interface philosophies. (**a**) The scripting interface of *R* offers access to thousands of functions, but requires scripting skills. (**b**) The point-and-click graphical user interface of *Fiji* (Schindelin 2008; Schindelin et al. 2012) requires no programming skills, but offers limited flexibility. (**c**) The workflow design interface of *KNIME* (Berthold et al. 2008), showing a workflow for image processing with an *ImageJ2*-plug-in integrated. This approach requires programmatic thinking and offers intermediate flexibility. (Image credits: (**a**) and (**b**) own screenshots; (**c**) from knime.org.)

for image-based high-throughput screens (Snijder et al. 2009), *MorphoGraphX* (de Reuille et al. 2015) originated as a tool for plant tissue morphogenesis, *Vaa3D* (Peng et al. 2010, 2014) and *BioImageXD* (Kankaanpää et al. 2012) started as interactive 3D visualization tools and are particularly strong at big image data visualization, and *bisque* (Kvilekval et al. 2010) and *OMERO* (Swedlow and Eliceiri 2009) have their particular strength in primarily being image databases. In addition to these generic tools, there are many specialized and often application-specific tools available, such as *PackingAnalyzer* (Farhadifar et al. 2007; Aigouy et al. 2010) to segment cell membranes in developing epithelial tissues with fluorescent membrane staining, *FIESTA* (Ruhnow et al. 2011) to segment fluorescently labeled filaments, and *OMEGA* (https://github.com/OmegaProject) for virus particle tracking with uncertainty quantification.

Software packages specifically supporting model-based image analysis include *Icy* (de Chaumont et al. 2011, 2012) for fluorescence microscopy images and *itk-SNAP* (Yushkevich et al. 2006) for medical images. The *BioImageSuite* (Duncan et al. 2004) supports model-based image segmentation using Markov random fields (Geman and Geman 1984). The *FARSIGHT* toolkit (Roysam et al. 2008) and *BioImageXD* (Kankaanpää et al. 2012) make available several model-based methods from the *ITK* library (Ibanez et al. 2005). Model-based image-analysis plug-ins are also available for *ImageJ* (Kaynig et al. 2010) and *bisque* (Bertelli et al. 2007). The *MOSAICsuite* implements the model-based segmentation methods *Squassh* (Rizk et al. 2014) and *RegionCompetition* (Cardinale et al. 2012) in a plug-in for *Fiji* and *ImageJ*, along with model-based spatial pattern and interaction analysis (Shivanandan et al. 2013).

Unsupervised machine-learning approaches are notably implemented in the *BioImageSuite* (Duncan et al. 2004). Supervised machine-learning segmentation is, for example, implemented in the software *ilastik* (Sommer et al. 2011). *WND-CHARM* (Orlov et al. 2008) uses texture features to classify images without segmenting them. The generic machine-learning library *WEKA* (Hall et al. 2009) is used to provide supervised trainable segmentation in *Fiji*. Again, many application-specific tools exist, for example, the tool *PHANTAST* (Jaccard et al. 2014) for machine-learning-based segmentation of phase-contrast images of adherent cell cultures, which is also available as a plug-in for *Fiji* and *ImageJ*.

A frequent use of machine learning is also to post-process results obtained by other image-analysis means. This is the approach taken by software tools such as *CellProfiler Analyst* (Jones et al. 2008), *CellClassifier* (Rämö et al. 2009), and *CecogAnalyzer* (Held et al. 2010). More specialized examples include a machine-learning tool to classify different mitochondrial morphologies in wide-field fluorescence microscopy images (Reis et al. 2012), to classify sub-cellular patterns (Huang and Murphy 2004), and a tool to classify cell cycle states after filter-based segmentation (Wang et al. 2008).

1.5 Conclusions and Discussion

Image analysis in biology is moving from seeing and observing to quantifying and modeling. Interpreting images as scientific measurements, rather than as mere visualizations, brings the need for uncertainty quantification, error analysis, statistical inference frameworks, etc. This raises a number of exciting theoretical and algorithmic questions.

We outlined these questions, focusing on the rapidly developing field of light microscopy, and described three conceptually different paradigms of image analysis, along with popular software tools implementing them. In practice, of course, these approaches are often mixed. It is common, e.g., to use filter-based approaches to compute image features and then use a machine-learning approach on those features in order to detect objects or classify them. Likewise, filters are often included in the forward models of model-based approaches. In fact, filter-based approaches are in some sense also model-based, albeit with an implicit model that is often not evident. Denoising an image using a moving least squares filter (Lancaster and Salkauskas 1981), for example, is equivalent to maximizing a Gaussian noise likelihood. Another example is the formal link that has been established between the model-based graph-cut framework and the filter-based watershed transform (Couprie et al. 2011). Finally, one can design filters that compute approximate solutions for model-based problems (Gong 2015). This is not surprising, since ultimately also model-based and machine-learning algorithms are discretized in the computer and hence amount to filters. There is also a blurry boundary between the model-based and machine-learning paradigms. A classical machine-learning approach classifies pixels into "object" vs. "background", based on previously computed features for each pixel. When features use neighborhood information around a pixel, however, there is a conceptual link to Markov random fields (Geman and Geman 1984) and their model-based Bayesian solution using graph cuts (Delong et al. 2011).

We identified and discussed four main challenges in today's bio-image analysis community: big data, uncertainty quantification, generic algorithms, and collaborative software. To some extent, they mutually entail each other. A big-data project that stores only the final analysis result, for example, critically depends on uncertainty quantification; and both generic algorithms and collaborative software need to be combined to render problem-solving more efficient and prevent reinventing the wheel. The four challenges are hence best addressed jointly.

Addressing these challenges would enable us to work with images like we routinely do with genome and proteome sequences. We could compare images, search image databases by content, and do statistical inference over images. This requires image distance metrics, semantic grammars and annotation, automatic inference systems, query by image content, and probabilistic frameworks over image spaces. All of these are open research areas, and much progress is needed in order to provide robust and generic solutions. The ultimate goal of image analysis is to not operate on the pixel matrix of an image, but on the information represented in the image, independent of the view and the imaging modality chosen. This links syntax and semantics of images on the level of biological meaning, in order to support queries like "find all images of yeast cells in M-phase with gene Cdc11 knocked out." Currently, this only works if the images were manually annotated before (Swedlow and Eliceiri 2009).

However, even upstream of image annotation open problems remain. One of them is that there are no good forward models for some imaging modalities, including electron microscopy and dark-field microscopy. Another problem is that object models are often *ad hoc* and not true to the biophysics of the sample. While there are occasional works that use physics-based predictive object models (Papademetris et al. 1999; Papademetris 2000), the computational cost of these models hampers their application. These models are also often black box with no gradient or structural information available that the optimization algorithm could exploit. This points to the problem that many machine-learning and model-based image-analysis frameworks make implicit assumptions about the features, training data, or models that are used with them. For example, they assume the model to be convex, linear, Gaussian, or separable, which may not be the case for a physics-based simulation. This still requires progress in black-box optimization algorithms (Müller 2010).

Besides black-box optimization, it could be promising to combine machine-learning and model-based approaches. This would, e.g., make it possible to use machine learning to *learn* the imaging and object models from user-annotated examples, and then use these learned models in a model-based analysis. This would solve the computational cost issue, since machine-learning models are quick to evaluate, and also relax the black-box optimization problem because most machine-learning models have an analytical structure with computable gradients. Conversely, model-based approaches could be used to provide ample amounts of simulated training data, to train and validate machine-learning approaches (Murphy 2012).

A natural way forward is the co-design and co-evolution of mathematical theories of images and inference over images, versatile computer algorithms for image analysis that have few parameters, software implementations thereof that parallelize well and are user friendly, and the biological application defining the level of detail and prior knowledge. These four ingredients need to be balanced and inter-connected. Building on the achievements in the community so far, a quantum leap in computational bio-image analysis and understanding could lie ahead.

Acknowledgements I thank all members of the MOSAIC Group for their creative and scientific contributions and for providing multiple test images and illustration cases for this manuscript. Particular thanks go to Dr. Grégory Paul, Dr. Janick Cardinale, and Dr. Jo Helmuth. This work was supported in parts by the German Federal Ministry of Research and Education (BMBF) under funding code 031A099.

References

Abramoff MD, Magalhães PJ, Ram SJ (2004) Image processing with ImageJ. Biophoton Int 11(7):36–42

Aigouy B, Farhadifar R, Staple DB, Sagner A, Röper JC, Jülicher F, Eaton S (2010) Cell flow reorients the axis of planar polarity in the wing epithelium of drosophila. Cell 142(5):773–786. doi:10.1016/j.cell.2010.07.042

Ambühl ME, Brepsant C, Meister JJ, Verkhovsky AB, Sbalzarini IF (2012) High-resolution cell outline segmentation and tracking from phase-contrast microscopy images. J Microsc 245(2):161–170

Andres B, Beier T, Kappes JH (2012) OpenGM: a C++ library for discrete graphical models. arXiv preprint arXiv:12060111

Annibale P, Vanni S, Scarselli M, Rothlisberger U, Radenovic A (2011) Quantitative photo activated localization microscopy: unraveling the effects of photoblinking. PLoS One 6(7):e22,678. doi:10.1371/journal.pone.0022678

Avidan S (2004) Support vector tracking. IEEE Trans Pattern Anal Mach Intell 26(8):1064–1072

Bar-Shalom Y, Blair WD (eds) (2000) Multitarget/multisensor tracking: applications and advances, vol III. Artech, Dedham

Batra D, Yadollahpour P, Guzman-Rivera A, Shakhnarovich G (2012) Diverse M-best solutions in Markov random fields. In: Proc. Europ. conf. computer vision (ECCV), Firenze, pp 1–16

Bertelli L, Byun J, Manjunath BS (2007) A variational approach to exploit prior information in object-background segregation: application to retinal images. In: Proc. ICIP, IEEE intl. conf. image processing, vol 6, pp VI–61–VI–64

Berthold MR, Cebron N, Dill F, Gabriel TR, Kötter T, Meinl T, Ohl P, Sieb C, Thiel K, Wiswedel B (2008) KNIME: the Konstanz information miner. Springer, Heidelberg

Bishop CM (2007) Pattern recognition and machine learning, 2nd edn. Springer, Heidelberg

Blake A, Kohli P, Rother C (2011) Markov random fields for vision and image processing. MIT Press, Cambridge

Boykov Y, Veksler O, Zabih R (2001) Fast approximate energy minimization via graph cuts. IEEE Trans Pattern Anal Mach Intell 23:1222–1239. doi:http://doi.ieeecomputersociety.org/10.1109/34.969114

Bradski G, Kaehler A (2008) Learning OpenCV: computer vision with the OpenCV library. O'Reilly, Sebastopol

Breiman L (2001) Random forests. Mach Learn 45:5–32

Brown ES, Chan TF, Bresson X (2011) Completely convex formulation of the Chan-Vese image segmentation model. Int J Comput Vis. doi:10.1007/s11263-011-0499-y

Canny J (1986) A computational approach to edge detection. IEEE Trans Pattern Anal Mach Intell 8(6):679–698

Cardinale J (2013) Unsupervised segmentation and shape posterior estimation under Bayesian image models. PhD thesis, Diss. ETH No. 21026, MOSAIC Group, ETH Zürich

Cardinale J, Rauch A, Barral Y, Székely G, Sbalzarini IF (2009) Bayesian image analysis with on-line confidence estimates and its application to microtubule tracking. In: Proc. IEEE intl. symposium biomedical imaging (ISBI). IEEE, Boston, pp 1091–1094

Cardinale J, Paul G, Sbalzarini IF (2012) Discrete region competition for unknown numbers of connected regions. IEEE Trans Image Process 21(8):3531–3545

Cardona A, Tomancak P (2012) Current challenges in open-source bioimage informatics. Nat Methods 9(7):661–665

Carpenter AE, Jones TR, Lamprecht MR, Clarke C, Kang IH, Friman O, Guertin DA, Chang JH, Lindquist RA, Moffat J, et al (2006) Cellprofiler: image analysis software for identifying and quantifying cell phenotypes. Genome Biol 7(10):R100

Carpenter AE, Kamentsky L, Eliceiri KW (2012) A call for bioimaging software usability. Nat Methods 9(7):666–670

Chan TF, Shen JJ (2005) Image processing and analysis: variational, PDE, wavelet, and stochastic methods. SIAM, Philadelphia

Chang J, Fisher III JW (2011) Efficient MCMC sampling with implicit shape representations. In: IEEE conference on computer vision and pattern recognition (CVPR). IEEE, Washington, pp 2081–2088

de Chaumont F, Dallongeville S, Olivo-Marin JC (2011) ICY: a new open-source community image processing software. In: Proc. IEEE intl. symposium biomedical imaging (ISBI), pp 234–237

de Chaumont F, Dallongeville S, Chenouard N, Hervé N, Pop S, Provoost T, Meas-Yedid V, Pankajakshan P, Lecomte T, Le Montagner Y, Lagache T, Dufour A, Olivo-Marin JC (2012) Icy: an open bioimage informatics platform for extended reproducible research. Nat Methods 9(7):690–696. doi:10.1038/nmeth.2075

Chenouard N, Smal I, de Chaumont F, Maška M, Sbalzarini IF, Gong Y, Cardinale J, Carthel C, Coraluppi S, Winter M, Cohen AR, Godinez WJ, Rohr K, Kalaidzidis Y, Liang L, Duncan J, Shen H, Xu Y, Magnusson KEG, Jaldén J, Blau HM, Paul-Gilloteaux P, Roudot P, Kervrann C, Waharte F, Tinevez JY, Shorte SL, Willemse J, Celler K, van Wezel GP, Dan HW, Tsai YS, de Solórzano CO, Olivo-Marin JC, Meijering E (2014) Objective comparison of particle tracking methods. Nat Methods 11(3):281–289. doi:10.1038/nmeth.2808

Cherkassky VS, Mulier F (1998) Learning from data. Wiley, New York

Chesnaud C, Réfrégier P, Boulet W (1999) Statistical region snake-based segmentation adapted to different physical noise models. IEEE Trans Pattern Anal Mach Intell 21(11):1145–1157

Ciresan D, Meier U, Schmidhuber J (2012) Multi-column deep neural networks for image classification. In: Proc. IEEE intl. conf. computer vision and pattern recognition (CVPR), IEEE, Washington, pp 3642–3649

Clack NG, O'Connor DH, Huber D, Petreanu L, Hires A, Peron S, Svoboda K, Myers EW (2012) Automated tracking of whiskers in videos of head fixed rodents. PLoS Comput Biol 8(7):e1002,591. doi:10.1371/journal.pcbi.1002591

Collinet C, Stöter M, Bradshaw CR, Samusik N, Rink JC, Kenski D, Habermann B, Buchholz F, Henschel R, Mueller MS, Nagel WE, Fava E, Kalaidzidis Y, Zerial M (2010) Systems survey of endocytosis by multiparametric image analysis. Nature 464:243–249

Cooley JW, Tukey JW (1965) An algorithm for the machine calculation of complex Fourier series. Math Comput 19(90):297–301

Couprie C, Grady L, Najman L, Talbot H (2011) Power watershed: A unifying graph-based optimization framework. IEEE Trans Pattern Anal Mach Intell 33(7):1384–1399

Cox IJ, Hingorani SL (1996) An efficient implementation of Reid's multiple hypothesis tracking algorithm and its evaluation for the purpose of visual tracking. IEEE Trans Pattern Anal 18(2):138–150

Crocker JC, Grier DG (1996) Methods of digital video microscopy for colloidal studies. J Colloid Interface Sci 179:298–310

Crosier M, Griffin LD (2010) Using basic image features for texture classification. Int J Comput Vis 88(3):447–460

Danuser G (2011) Computer vision in cell biology. Cell 147(5):973–978. doi:10.1016/j.cell.2011.11.001

Delong A, Osokin A, Isack HN, Boykov Y (2011) Fast approximate energy minimization with label costs. Int J Comput Vis 96(1):1–27

Demirel O, Smal I, Niessen WJ, Meijering E, Sbalzarini IF (2014a) An adaptive distributed resampling algorithm with non-proportional allocation. In: Proc. ICASSP, IEEE intl. conf. acoustics, speech, and signal processing. IEEE, Florence, pp 1635–1639

Demirel O, Smal I, Niessen WJ, Meijering E, Sbalzarini IF (2014b) Piecewise constant sequential importance sampling for fast particle filtering. In: Proc. 10th IET conf. data fusion & target tracking. IET, Liverpool

Demirel O, Smal I, Niessen WJ, Meijering E, Sbalzarini IF (2014c) PPF – a parallel particle filtering library. In: Proc. 10th IET conf. data fusion & target tracking. IET, Liverpool

de Reuille BP, Routier-Kierzkowska AL, Kierzkowski D, Bassel GW, Schüpbach T, Tauriello G, Bajpai N, Strauss S, Weber A, Kiss A, Burian A, Hofhuis H, Sapala A, Lipowczan M, Heimlicher MB, Robinson S, Bayer EM, Basler K, Koumoutsakos P, Roeder AHK, Aegerter-Wilmsen T, Nakayama N, Tsiantis M, Hay A, Kwiatkowska D, Xenarios I, Kuhlemeier C, Smith RS (2015) MorphoGraphX: a platform for quantifying morphogenesis in 4D. Elife 4:e05,864. doi:10.7554/eLife.05864

Dietrich CF (1991) Uncertainty, calibration and probability: the statistics of scientific and industrial measurement. Measurement science and technology, 2nd edn. Adam Hilger, Bristol

Duda RO, Hart PE, Stork DG (2000) Pattern classification, 2nd edn. Wiley, New York

Duncan JS, Papademetris X, Yang J, Jackowski M, Zeng X, Staib LH (2004) Geometric strategies for neuroanatomic analysis from MRI. Neuroimage 23 Suppl. 1:S34–S45. doi:10.1016/j.neuroimage.2004.07.027

Dzyubachyk O, van Cappellen WA, Essers J, Niessen WJ, Meijering E (2010) Advanced level-set-based cell tracking in time-lapse fluorescence microscopy. IEEE Trans Med Imaging 29(3):852–867

Eils R, Athale C (2003) Computational imaging in cell biology. J Cell Biol 161(3):477–481

El-Naqa I, Yang Y, Wernick MN, Galatsanos NP, Nishikawa RM (2002) A support vector machine approach for detection of microcalcifications. IEEE Trans Med Imaging 21(12):1552–1563

El-Zehiry N, Elmaghraby A (2009) An active surface model for volumetric image segmentation. In: Proc. IEEE intl. symposium biomedical imaging (ISBI), pp 1358–1361

Eliceiri KW, Berthold MR, Goldberg IG, Ibáñez L, Manjunath BS, Martone ME, Murphy RF, Peng H, Plant AL, Roysam B, Stuurmann N, Swedlow JR, Tomancak P, Carpenter AE (2012) Biological imaging software tools. Nat Methods 9(7):697–710. doi:10.1038/nmeth.2084

Engelbrecht C, Stelzer E (2006) Resolution enhancement in a light-sheet-based microscope (SPIM). Opt Lett 31:1477–1479

Etyngier P, Ségonne F, Keriven R (2007) Shape priors using manifold learning techniques. In: Proc. IEEE intl. conf. computer vision (ICCV). IEEE, Rio de Janeiro, pp 1–8

Ewers H, Smith AE, Sbalzarini IF, Lilie H, Koumoutsakos P, Helenius A (2005) Single-particle tracking of murine polyoma virus-like particles on live cells and artificial membranes. Proc Natl Acad Sci U S A 102(42):15110–15115

Farhadifar R, Röper JC, Aigouy B, Eaton S, Jülicher F (2007) The influence of cell mechanics, cell-cell interactions, and proliferation on epithelial packing. Curr Biol 17(24):2095–2104. doi:10.1016/j.cub.2007.11.049

Fuchs TJ, Wild PJ, Moch H, Buhmann JM (2008) Computational pathology analysis of tissue microarrays predicts survival of renal clear cell carcinoma patients. In: Medical image computing and computer-assisted intervention – MICCAI 2008. Lecture notes in computer science, vol 5242. Springer, Heidelberg, pp 1–8

Fuchs TJ, Haybaeck J, Wild PJ, Heikenwalder M, Moch H, Aguzzi A, Buhmann JM (2009) Randomized tree ensembles for object detection in computational pathology. In: Proc. intl. symp. visual comput. (ISVC), pp 367–378

Galizia A, D'Agostino D, Clematis A (2015) An MPI–CUDA library for image processing on HPC architectures. J Comput Appl Mech 273:414–427

Geman S, Geman D (1984) Stochastic relaxation, Gibbs distributions, and the Bayesian restoration of images. IEEE Trans Pattern Anal Mach Intell 6(6):721–741

Genovesio A, Olivo-Marin JC (2004) Split and merge data association filter for dense multi-target tracking. In: Proceedings of the 17th international conference on pattern recognition (ICPR'04), vol 4, pp 677–680

Gong Y (2015) Spectrally regularized surfaces. PhD thesis, Diss. ETH No. 22616, MOSAIC Group, ETH Zürich

Gong Y, Sbalzarini IF (2014) Image enhancement by gradient distribution specification. In: Jawahar CV, Shan S (eds) Computer vision – ACCV 2014 workshops, revised selected papers, Part II, Springer, Singapore. Lecture notes in computer science, vol 9009. Springer, Cham, pp 47–62

Hall M, Frank E, Holmes G, Pfahringer B, Reutemann P, Witten IH (2009) The WEKA data mining software: an update. SIGKDD Explor Newsl 11(1):10–18. doi:10.1145/1656274.1656278. http://doi.acm.org/10.1145/1656274.1656278

Halpern JY (2005) Reasoning about uncertainty. MIT Press, Cambridge

Hecht E (2001) Optics, 4th edn. Addison Wesley, Reading

Held M, Schmitz MHA, Fischer B, Walter T, Neumann B, Olma MH, Peter M, Ellenberg J, Gerlich DW (2010) CellCognition: time-resolved phenotype annotation in high-throughput live cell imaging. Nat Methods 7(9):747–754

Helmuth JA, Sbalzarini IF (2009) Deconvolving active contours for fluorescence microscopy images. In: Proc. intl. symp. visual computing (ISVC), Springer, Las Vegas, USA. Lecture notes in computer science, vol 5875. Springer, Heidelberg, pp 544–553

Helmuth JA, Burckhardt CJ, Koumoutsakos P, Greber UF, Sbalzarini IF (2007) A novel supervised trajectory segmentation algorithm identifies distinct types of human adenovirus motion in host cells. J Struct Biol 159(3):347–358

Helmuth JA, Burckhardt CJ, Greber UF, Sbalzarini IF (2009) Shape reconstruction of subcellular structures from live cell fluorescence microscopy images. J Struct Biol 167:1–10

Helmuth JA, Paul G, Sbalzarini IF (2010) Beyond co-localization: inferring spatial interactions between sub-cellular structures from microscopy images. BMC Bioinf 11:372

Hong Y, Kwong S, Chang Y, Ren Q (2008) Consensus unsupervised feature ranking from multiple views. Pattern Recogn Lett 29:595–602

Huang K, Murphy RF (2004) Automated classification of subcellular patterns in multicell images without segmentation into single cells. In: Proc. IEEE intl. symposium biomedical imaging (ISBI), pp 1139–1142

Hue C, Le Cadre JP, Pérez P (2002) Tracking multiple objects with particle filtering. IEEE Trans Aerosp Electron Syst 38(3):791–812

Huisken J, Swoger J, Del Bene F, Wittbrodt J, Stelzer EHK (2004) Optical sectioning deep inside live embryos by selective plane illumination microscopy. Science 305:1007–1009

Ibanez L, Schroeder W, Ng L, Cates J (2005) The ITK software guide. Kitware, Clifton Park. ISBN 1-930934-15-7. http://www.itk.org/ItkSoftwareGuide.pdf, 2nd edn

Jaccard N, Griffin LD, Keser A, Macown RJ, Super A, Veraitch FS, Szita N (2014) Automated method for the rapid and precise estimation of adherent cell culture characteristics from phase contrast microscopy images. Biotechnol Bioeng 111(3):504–517

Jancsary J, Nowozin S, Sharp T, Rother C (2012) Regression tree fields–an efficient, non-parametric approach to image labeling problems. In: Proc. of the 2012 IEEE conference on computer vision and pattern recognition (CVPR). IEEE, Washington, pp 2376–2383

Jaqaman K, Loerke D, Mettlen M, Kuwata H, Grinstein S, Schmid S, Danuser G (2008) Robust single-particle tracking in live-cell time-lapse sequences. Nat Methods 5:695–702. doi:DOI 10.1038/nmeth.1237

Jones TR, Kang IH, Wheeler DB, Lindquist RA, Papallo A, Sabatini DM, Golland P, Carpenter AE (2008) CellProfiler Analyst: data exploration and analysis software for complex image-based screens. BMC Bioinf 9:482. doi:10.1186/1471-2105-9-482

Jung M, Chung G, Sundaramoorthi G, Vese L, Yuille A (2009) Sobolev gradients and joint variational image segmentation, denoising and deblurring. In: SPIE electronic imaging conference proceedings, computational imaging VII, vol 7246

Kalaidzidis Y (2007) Intracellular objects tracking. Eur J Cell Biol 86(9):569–578. doi:10.1016/j.ejcb.2007.05.005

Kalaidzidis Y (2009) Multiple objects tracking in fluorescence microscopy. J Math Biol 58 (1–2):57–80. doi:10.1007/s00285-008-0180-4

Kankaanpää P, Paavolainen L, Tiitta S, Karjalainen M, Päivärinne J, Nieminen J, Marjomäki V, Heino J, White DJ (2012) BioImageXD: an open, general-purpose and high-throughput image-processing platform. Nat Methods 9(7):683–689. doi:10.1038/nmeth.2047

Kannala J, Rahtu E (2012) BSIF: binarized statistical image features. In: Proc. 21st intl. conf. pattern recognition (ICPR). IEEE, Tsukuba, pp 1363–1366

Kass M, Witkin A, Terzopoulos D (1988) Snakes: active contour models. Int J Comput Vis **1**, 321–331

Kaynig V, Fuchs T, Buhmann JM (2010) Neuron geometry extraction by perceptual grouping in ssTEM images. In: Proc. IEEE intl. conf. computer vision and pattern recognition (CVPR), pp 2902–2909

Kohli P, Lempitsky V, Rother C (2010) Uncertainty driven multi-scale optimization. In: Proc. DAGM, pattern recognition. Springer, Darmstadt

Köthe U (1999) Reusable software in computer vision. In: Jähne B, Haußecker H, Geißler P (eds) Handbook on computer vision and applications, vol 3, chap 6. Academic, Boston, pp 105–134

Kvilekval K, Fedorov D, Obara B, Singh A, Manjunath BS (2010) Bisque: a platform for bioimage analysis and management. Bioinformatics 26(4):544–552. doi:10.1093/bioinformatics/btp699

Lagache T, Lang G, Sauvonnet N, Olivo-Marin JC (2013) Analysis of the spatial organization of molecules with robust statistics. PLoS One 8(12):e80,914. doi:10.1371/journal.pone.0080914

Lagache T, Sauvonnet N, Danglot L, Olivo-Marin JC (2015) Statistical analysis of molecule colocalization in bioimaging. Cytometry A 87:568–579

Lamprecht M, Sabatini DM, Carpenter AE (2007) CellProfiler: free, versatile software for automated biological image analysis. Biotechniques 42(1):71–75

Lancaster P, Salkauskas K (1981) Surfaces generated by moving least squares methods. Math Comput 37(155):141–158

Le Maître OP, Knio OM (2010) Spectral methods for uncertainty quantification. Springer, Amsterdam

Li K, Miller ED, Weiss LE, Campbell PG, Kanade T (2006) Online tracking of migrating and proliferating cells imaged with phase-contrast microscopy. In: IEEE proceedings of the 2006 conference on computer vision and pattern recognition workshop (CVPRW). IEEE Computer Society, Washington, pp 65–72
Li K, Chen M, Kanade T (2007) Cell population tracking and lineage construction with spatiotemporal context. Med Image Comput Comput Assist Interv10(Pt 2):295–302
Li S, Kwok JT, Zhu H, Wang Y (2003) Texture classification using the support vector machines. Pattern Recogn 36:2883–2893
Lin G, Adiga U, Olson K, Guzowski JF, Barnes CA, Roysam B (2003) A hybrid 3D watershed algorithm incorporating gradient cues and object models for automatic segmentation of nuclei in confocal image stacks. Cytometry A 56(1):23–36
Linfoot EH, Wolf E (1956) Phase distribution near focus in an aberration-free diffraction image. Proc Phys Soc B 69(8):823–832
Lowe DG (1999) Object recognition from local scale-invariant features. In: Proc. 7th intl. conf. computer vision (ICCV), vol 2. IEEE, Washington, pp 1150–1157
Machacek M, Danuser G (2006) Morphodynamic profiling of protrusion phenotypes. Biophys J 90:1439–1452
Manders EMM, Hoebe R, Strackee J, Vossepoel AM, Aten JA (1996) Largest contour segmentation: a tool for the localization of spots in confocal images. Cytometry 23(1):15–21
Marjoram P, Molitor J, Plagnol V, Tavare S (2003) Markov chain Monte Carlo without likelihoods. Proc Natl Acad Sci U S A 100(26):15,324–15,328
Martin P, Gier PR, Goudail F, Guérault F (2004) Influence of the noise model on level set active contour segmentation. IEEE Trans Pattern Anal Mach Intell 26(6):799–803
Maška M, Ulman V, Svoboda D, Matula P, Matula P, Ederra C, Urbiola A, España T, Venkatesan S, Balak DMW, Karas P, Bolcková T, Štreitová M, Carthel C, Coraluppi S, Harder N, Rohr K, Magnusson KEG, Jaldén J, Blau HM, Dzyubachyk O, Křížek P, Hagen GM, Pastor-Escuredo D, Jimenez-Carretero D, Ledesma-Carbayo MJ, Muñoz Barrutia A, Meijering E, Kozubek M, Ortiz-de Solorzano C (2014) A benchmark for comparison of cell tracking algorithms. Bioinformatics 30(11):1609–1617
McCann MT, Bhagavatula R, Fickus MC, Ozolek JA, Kovacevic J (2012) Automated colitis detection from endoscopic biopsies as a tissue screening tool in diagnostic pathology. In: Proc. of the 2012 19th IEEE international conference on image processing (ICIP). IEEE, Orlando, pp 2809–2812
Meyer F, Vachier C, Oliveras A, Salembier P (1997) Morphological tools for segmentation: Connected filters and watersheds. Annales des télécommunications 52(7–8):367–379
Müller CL (2010) Black-box landscapes: characterization, optimization, sampling, and application to geometric configuration problems. PhD thesis, Diss. ETH No. 19438, ETH Zürich
Murphy RF (2012) CellOrganizer: image-derived models of subcellular organization and protein distribution. Methods Cell Biol 110:179–193. doi:10.1016/B978-0-12-388403-9.00007-2
Myers G (2012) Why bioimage informatics matters. Nat Methods 9(7):659–660
Najman L, Schmitt M (1996) Geodesic saliency of watershed contours and hierarchical segmentation. IEEE Trans Pattern Anal Mach Intell 18(12):1163–1173
Najman L, Talbot H (2010) Mathematical morphology. Wiley, New York
Nandy K (2015) Segmentation and informatics in multidimensional fluorescence optical microscopy images. Ph.D. thesis, University of Maryland
Nicolescu C, Jonker P (2000) Parallel low-level image processing on a distributed-memory system. In: Rolim J (ed) Parallel and distributed processing. Lecture notes in computer science, vol 1800. Springer, Heidelberg, pp 226–233. doi:10.1007/3-540-45591-430. http://dx.doi.org/10.1007/3-540-45591-430
Nilufar S, Perkins TJ (2014) Learning to detect contours with dynamic programming snakes. In: Proc. IEEE intl. conf. pattern recognition (ICPR). IEEE, Stockholm, pp 984–989
North AJ (2006) Seeing is believing? a beginners' guide to practical pitfalls in image acquisition. J Cell Biol 172(1):9–18. doi:10.1083/jcb.200507103

Ober RJ, Tahmasbi A, Ram S, Lin Z, Ward ES (2015) Quantitative aspects of single-molecule microscopy – information-theoretic analysis of single-molecule data. IEEE Signal Proc Mag 32(1):58–69

Olivo-Marin JC (2002) Extraction of spots in biological images using multiscale products. Pattern Recogn 35(9):1989–1996

Orlov N, Shamir L, Macura T, Johnston J, Eckley DM, Goldberg IG (2008) WND-CHARM: multi-purpose image classification using compound image transforms. Pattern Recogn Lett 29(11):1684–1693. doi:10.1016/j.patrec.2008.04.013. http://www.sciencedirect.com/science/article/pii/S0167865508001530

Orlov NV, Chen WW, Eckley DM, Macura TJ, Shamir L, Jaffe ES, Goldberg IG (2010) Automatic classification of lymphoma images with transform-based global features. IEEE Trans Inf Technol Biomed 14(4):1003–1013

Otsu N (1975) A threshold selection method from gray-level histograms. Automatica 11(285–296):23–27

Ovaska K, Laakso M, Haapa-Paananen S, Louhimo R, Chen P, Aittomäki V, Valo E, Núñez-Fontarnau J, Rantanen V, Karinen S, Nousiainen K, Lahesmaa-Korpinen AM, Miettinen M, Saarinen L, Kohonen P, Wu J, Westermarck J, Hautaniemi S (2010) Large-scale data integration framework provides a comprehensive view on glioblastoma multiforme. Genome Med 2(9):65

Papademetris X (2000) Estimation of 3D left ventricular deformation from medical images using biomechanical models. Ph.D. thesis, Yale University

Papademetris X, Sinusas AJ, Dione DP, Duncan JS (1999) 3D cardiac deformation from ultrasound images. In: Proc. MICCAI, medical image computing and computer-assisted intervention, pp 420–429

Paul G, Cardinale J, Sbalzarini IF (2011) An alternating split Bregman algorithm for multi-region segmentation. In: Proc. 45th IEEE Asilomar conf. signals, systems, and computers. IEEE, Asilomar, pp 426–430

Paul G, Cardinale J, Sbalzarini IF (2013) Coupling image restoration and segmentation: a generalized linear model/Bregman perspective. Int J Comput Vis 104(1):69–93. 10.1007/s11263-013-0615-2

Peng H (2008) Bioimage informatics: a new area of engineering biology. Bioinformatics 24(17):1827–1836. doi:10.1093/bioinformatics/btn346

Peng H, Ruan Z, Long F, Simpson JH, Myers EW (2010) V3D enables real-time 3D visualization and quantitative analysis of large-scale biological image data sets. Nat Biotechnol 28(4):348–353. doi:10.1038/nbt.1612

Peng H, Bria A, Zhou Z, Iannello G, Long F (2014) Extensible visualization and analysis for multidimensional images using Vaa3D. Nat Protoc 9(1):193–208

Perona P, Malik J (1990) Scale-space and edge detection using anisotropic diffusion. IEEE Trans Pattern Anal Mach Intell 12(7):629–639

Pock T, Cremers D, Bischof H, Chambolle A (2009) An algorithm for minimizing the Mumford-Shah functional. In: Proc. IEEE intl. conf. computer vision (ICCV), pp 1133–1140. doi:10.1109/ICCV.2009.5459348

Rajaram S, Pavie B, Hac NEF, Altschuler SJ, Wu LF (2012) SimuCell: a flexible framework for creating synthetic microscopy images. Nat Methods 9(7):634–635

Ramakrishna V, Batra D (2012) Mode-marginals: expressing uncertainty via diverse M-best solutions. In: Proc. NIPS, neural information processing systems foundation, Lake Tahoe

Rämö P, Sacher R, Snijder B, Begemann B, Pelkmans L (2009) CellClassifier: supervised learning of cellular phenotypes. Bioinformatics 25(22):3028–3030. doi:10.1093/bioinformatics/btp524

Rantanen V, Valori M, Hautaniemi S (2014) Anima: modular workflow system for comprehensive image data analysis. Front Bioeng Biotechnol 2:25

Reddick WE, Glass JO, Cook EN, Elkin TD, Deaton RJ (1997) Automated segmentation and classification of multispectral magnetic resonance images of brain using artificial neural networks. IEEE Trans Med Imaging 16(6):911–918

Reis Y, Bernardo-Faura M, Richter D, Wolf T, Brors B, Hamacher-Brady A, Eils R, Brady NR (2012) Multi-parametric analysis and modeling of relationships between mitochondrial morphology and apoptosis. PLoS One 7(1):e28,694. doi:10.1371/journal.pone.0028694

Rex DE, Ma JQ, Toga AW (2003) The LONI pipeline processing environment. Neuroimage 19(3):1033–1048

Reynaud EG, Peychl J, Huisken J, Tomancak P (2014) Guide to light-sheet microscopy for adventurous biologists. Nat Methods 12(1):30–34

Rizk A, Paul G, Incardona P, Bugarski M, Mansouri M, Niemann A, Ziegler U, Berger P, Sbalzarini IF (2014) Segmentation and quantification of subcellular structures in fluorescence microscopy images using Squassh. Nat Protoc 9(3):586–596

Royer LA, Weigert M, Günther U, Maghelli N, Jug F, Sbalzarini IF, Myers EW (2015) ClearVolume: open-source live 3D visualization for light-sheet microscopy. Nat Methods 12(6):480–481

Roysam B, Shain W, Robey E, Chen Y, Narayanaswamy A, Tsai CL, Al-Kofahi Y, Bjornsson C, Ladi E, Herzmark P (2008) The FARSIGHT project: associative 4D/5D image analysis methods for quantifying complex and dynamic biological microenvironments. Microsc Microanal 14(S2):60–61

Ruhnow F, Zwicker D, Diez S (2011) Tracking single particles and elongated filaments with nanometer precision. Biophys J 100(11):2820–2828

Ruprecht V, Axmann M, Wieser S, Schütz GJ (2011) What can we learn from single molecule trajectories? Curr Protein Pept Sci 12(8):714–724

Sbalzarini IF (2010) Abstractions and middleware for petascale computing and beyond. Int J Distrib Syst Technol 1(2):40–56

Sbalzarini IF (2013) Modeling and simulation of biological systems from image data. Bioessays 35(5):482–490. doi:10.1002/bies.201200051

Sbalzarini IF, Koumoutsakos P (2005) Feature point tracking and trajectory analysis for video imaging in cell biology. J Struct Biol 151(2):182–195

Schiegg M, Hanslovsky P, Kausler BX, Hufnagel L, Hamprecht F (2013) Conservation tracking. In: Proc. IEEE intl. conf. computer vision (ICCV). IEEE, Sydney, pp 2928–2935

Schindelin J (2008) Fiji is just ImageJ (batteries included). In: ImageJ user and developer conference

Schindelin J, Arganda-Carreras I, Frise E, Kaynig V, Longair M, Pietzsch T, Preibisch S, Rueden C, Saalfeld S, Schmid B, Tinevez JY, White DJ, Hartenstein V, Eliceiri K, Tomancak P, Cardona A (2012) Fiji: an open-source platform for biological-image analysis. Nat Methods 9(7):676–682. doi:10.1038/nmeth.2019

Schneider CA, Rasband WS, Eliceiri KW (2012) NIH Image to ImageJ: 25 years of image analysis. Nat Methods 9(7):671–675

Schölkopf B, Smola AJ (2002) Learning with kernels. Support vector machines, regularization, optimization, and beyond. MIT Press, Cambridge

Seinstra FJ, Koelma D, Geusebroek JM (2002) A software architecture for user transparent parallel image processing. Parallel Comput 28(7–8):967–993. doi:http://dx.doi.org/10.1016/S0167-8191(02)00103-5. http://www.sciencedirect.com/science/article/pii/S0167819102001035

Sethian JA (1999) Level set methods and fast marching methods. Cambridge University Press, Cambridge

Shafer G (1976) A mathematical theory of evidence. Princeton University Press, Princeton

Shamir L, Delaney JD, Orlov N, Eckley DM, Goldberg IG (2010) Pattern recognition software and techniques for biological image analysis. PLoS Comput Biol 6(11):e1000,974. doi:10.1371/journal.pcbi.1000974

Shi Y, Karl W (2005) Real-time tracking using level sets. In: Proc. IEEE conf. CVPR, vol 2, pp 34–41. doi:10.1109/CVPR.2005.294

Shivanandan A, Radenovic A, Sbalzarini IF (2013) MosaicIA: an ImageJ/Fiji plugin for spatial pattern and interaction analysis. BMC Bioinf 14:349

Smal I, Meijering E, Draegestein K, Galjart N, Grigoriev I, Akhmanova A, van Royen ME, Houtsmuller AB, Niessen W (2008) Multiple object tracking in molecular bioimaging by Rao-Blackwellized marginal particle filtering. Med Image Anal 12(6):764–777. doi:10.1016/j.media.2008.03.004

Snijder B, Sacher R, Rämö P, Damm EM, Liberali P, Pelkmans L (2009) Population context determines cell-to-cell variability in endocytosis and virus infection. Nature 461(7263):520–523. doi:10.1038/nature08282

Sommer C, Strähle C, Köthe U, Hamprecht FA (2011) ilastik: interactive learning and segmentation toolkit. In: Proc. IEEE intl. symposium biomedical imaging (ISBI), pp 230–233

Stoeckli M, Chaurand P, Hallahan DE, Caprioli RM (2001) Imaging mass spectrometry: A new technology for the analysis of protein expression in mammalian tissues. Nat Med 7(4):493–496

Swedlow JR, Eliceiri KW (2009) Open source bioimage informatics for cell biology. Trends Cell Biol 19(11):656–660. doi:10.1016/j.tcb.2009.08.007

Swedlow JR, Goldberg I, Brauner E, Sorger PK (2003) Informatics and quantitative analysis in biological imaging. Science 300:100–102

Tomer R, Khairy K, Amat F, Keller PJ (2012) Quantitative high-speed imaging of entire developing embryos with simultaneous multiview light-sheet microscopy. Nat Methods 9(7):755–763. doi:10.1038/nmeth.2062

Vallotton P, Ponti A, Waterman-Storer CM, Salmon ED, Danuser G (2003) Recovery, visualization, and analysis of actin and tubulin polymer flow in live cells: a fluorescent speckle microscopy study. Biophys J 85:1289–1306

Vebjorn L, Sokolnicki KL, Carpenter AE (2012) Annotated high-throughput microscopy image sets for validation. Nat Methods 9(7):637

Wang M, Zhou X, Li F, Huckins J, King RW, Wong STC (2008) Novel cell segmentation and online SVM for cell cycle phase identification in automated microscopy. Bioinformatics 24(1):94–101. doi:10.1093/bioinformatics/btm530

Weber M, Huisken J (2012) Omnidirectional microscopy. Nat Methods 9(7):656–657

Wieser S, Schütz GJ (2008) Tracking single molecules in the live cell plasma membrane–do's and don't's. Methods 46(2):131–140. doi:10.1016/j.ymeth.2008.06.010

Wieser S, Axmann M, Schütz GJ (2008) Versatile analysis of single-molecule tracking data by comprehensive testing against Monte Carlo simulations. Biophys J 95(12):5988–6001. doi:10.1529/biophysj.108.141655

Willert CE, Gharib M (1991) Digital particle image velocimetry. Exp Fluids 10:181–193

Williams O, Blake A, Cipolla R (2005) Sparse Bayesian learning for efficient visual tracking. IEEE Trans Pattern Anal Mach Intell 27(8):1292–1304

Witkin A (1984) Scale-space filtering: a new approach to multi-scale description. In: Proc. of the IEEE international conference on acoustics, speech, and signal processing (ICASSP), vol 9. IEEE, San Diego, pp 150–153. 10.1109/ICASSP.1984.1172729

Witteveen JAS, Iaccarino G (2012) Simplex stochastic collocation with random sampling and extrapolation for nonhypercube probability spaces. SIAM J Sci Comput 34(2):A814–A838

Xiu D (2009) Fast numerical methods for stochastic computations: a review. Commun Comput Phys 5:242–272

Xiu D, Karniadakis GEM (2002) The Wiener-Askey polynomial chaos for stochastic differential equations. SIAM J Sci Comput 24(2):619–644

Xu C, Corso JJ (2012) Evaluation of super-voxel methods for early video processing. In: Proc. of the 2012 IEEE conference on computer vision and pattern recognition (CVPR). IEEE, Providence, pp 1202–1209. 10.1109/CVPR.2012.6247802

Yamauchi Y, Boukari H, Banerjee I, Sbalzarini IF, Horvath P, Helenius A (2011) Histone deacetylase 8 is required for centrosome cohesion and influenza A virus entry. PLoS Pathog 7(10):e1002,316

Yin Z, Li K, Kanade T, Chen M (2010) Understanding the optics to aid microscopy image segmentation. In: Proc. MICCAI, medical image computing and computer-assisted intervention. Springer, Heidelberg, pp 209–217

Yushkevich PA, Piven J, Cody Hazlett H, Gimpel Smith R, Ho S, Gee JC, Gerig G (2006) User-guided 3D active contour segmentation of anatomical structures: Significantly improved efficiency and reliability. NeuroImage 31(3):1116–1128

Zhang B, Zerubia J, Olivo-Marin JC (2007) Gaussian approximations of fluorescence microscope point-spread function models. Appl Opt 46(10):1819–1829

Zhou J, Chan KL, Chong VFH, Krishnan SM (2005) Extraction of brain tumor from MR images using one-class support vector machine. In: Proc. IEEE engineering in medicine and biology, annual conference, Shanghai, pp 6411–6414

Zhu SC, Yuille A (1996) Region competition: Unifying snakes, region growing, and Bayes/MDL for multiband image segmentation. IEEE Trans Pattern Anal Mach Intell 18(9):884–900

Chapter 2
Image Degradation in Microscopic Images: Avoidance, Artifacts, and Solutions

Joris Roels, Jan Aelterman, Jonas De Vylder, Saskia Lippens, Hiêp Q. Luong, Christopher J. Guérin, and Wilfried Philips

Abstract The goal of modern microscopy is to acquire high-quality image based data sets. A typical microscopy workflow is set up in order to address a specific biological question and involves different steps. The first step is to precisely define the biological question, in order to properly come to an experimental design for sample preparation and image acquisition. A better object representation allows biological users to draw more reliable scientific conclusions. Image restoration can manipulate the acquired data in an effort to reduce the impact of artifacts (spurious results) due to physical and technical limitations, resulting in a better representation of the object of interest. However, precise usage of these algorithms is necessary so as to avoid further artifacts that might influence the data analysis and bias the conclusions. It is essential to understand image acquisition, and how it introduces artifacts and degradations in the acquired data, so that their effects on subsequent analysis can be minimized. This paper provides an overview of the fundamental artifacts and degradations that affect many micrographs. We describe why artifacts

J. Roels (✉)
Department of Telecommunications and Information Processing - Image Processing and Interpretation/iMinds, Ghent University, Sint-Pietersnieuwstraat 41, 9000 Gent, Belgium

Inflammation Research Center, Flanders Institute for Biotechnology, Technologiepark 927, 9052 Zwijnaarde, Belgium
e-mail: jbroels@telin.ugent.be

J. Aelterman • J. De Vylder • H.Q. Luong • W. Philips
Department of Telecommunications and Information Processing - Image Processing and Interpretation/iMinds, Ghent University, Sint-Pietersnieuwstraat 41, 9000 Gent, Belgium
e-mail: jonas.devylder@telin.ugent.be

S. Lippens • C.J. Guérin
Bio Imaging Core, Flanders Institute for Biotechnology, Technologiepark 927, 9052 Zwijnaarde, Belgium

Inflammation Research Center, Flanders Institute for Biotechnology, Technologiepark 927, 9052 Zwijnaarde, Belgium

Department of Biomedical Molecular Biology, Ghent University, Technologiepark 927, 9052 Zwijnaarde, Belgium

W.H. De Vos et al. (eds.), *Focus on Bio-Image Informatics*, Advances in Anatomy, Embryology and Cell Biology 219, DOI 10.1007/978-3-319-28549-8_2

appear, in what sense they impact overall image quality, and how to mitigate them by first improving the acquisition parameters and then applying proper image restoration techniques.

2.1 Introduction

Researchers studying life sciences make use of a broad range of technologies, and apply these to an equally broad range of research domains such as: cell biology, biochemistry, genomics, proteomics, transcriptomics, and systems biology. While all of these technologies bring important information regarding the mechanisms underlying health and disease, microscopy is the only one that can show where important cellular processes are occurring.

A biological experiment typically consists of several steps. One has to formulate the specific biological question, perform an imaging experiment, and finally interpret the analyzed image data to form a valid answer to the initial question. More specifically, the crucial part of the imaging experiment is to gather micrographs showing a well-prepared sample at the appropriate magnification and resolution so that the object of interest is clearly represented. The imaging workflow is composed of: sample preparation, image acquisition, image adjustments, and analysis. These steps all contribute to the quality of the image data set and are inevitably connected.

Regardless of what imaging device is chosen, it is critical that the image data sets are a true representation of what the sample looks like. However, due to physical and technical drawbacks of microscopy the acquisition process introduces artifacts or spurious results (indicated by the different blocks in Fig. 2.1) in the image data. This artifact introduction should be minimized by optimizing the acquisition but can also be ameliorated through image adjustment. Although different microscopes have diverse configurations, artifacts will be introduced because of imperfect optical components and inherent problems (largely noise) in digital capture devices, and therefore residual artifacts share many commonalities across different modalities. A generic scheme of an imaging setup is shown in Fig. 2.1, indicating where artifacts are introduced and what the implication is for the captured image.

It is the aim of this paper to discuss the most common artifacts and methods of mitigating them. In the following sections, we will give an overview of significant image artifacts, namely: non-uniform illumination, blur, noise, digitization, and compression artifacts. Note the section ordering of the artifacts is in accordance with their initial appearance in the general image acquisition workflow (Fig. 2.1). We will discuss how they are introduced, propose acquisition adaptations to avoid the specified artifact as much as possible, and discuss image restoration techniques to improve image quality whenever image artifacts are inevitable. For practical application purposes, we refer the reader to freely available software, provided in the footnotes. For the more technically interested reader, we refer to the corresponding cited publications and grayboxes. These grayboxes serve as a more in-depth discussion and are optional reading material in order to follow the complete manuscript.

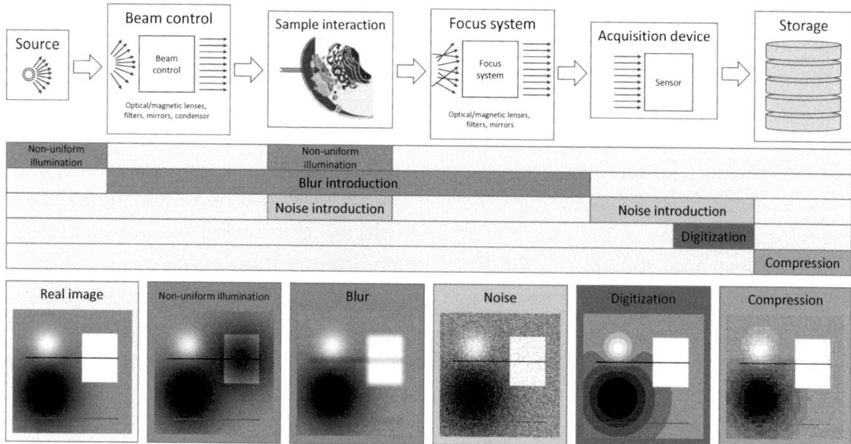

Fig. 2.1 Conceptual workflow of image acquisition in biomedical microscopy-based experiments. Typically an illumination (light or electron) source will illuminate the specimen, but the electromagnetic waves will first pass through optical or magnetic components of the imaging system in order to direct the beam to the specimen (beam control). Light or electron waves coming from the sample will pass through a focus system before they are captured by a sensor. Finally the acquired digital image is stored. These steps typically involve the introduction of several artifacts, specified by the middle timeline. The origin of these artifacts is explained in more detail in the following sections. The *bottom images* illustrate the visual impact of the introduced artifacts compared to the original, artificial image

2.2 Non-uniform Illumination

Non-uniform illumination occurs whenever a sample appears darker in specific regions, compared to others. This is caused by uneven distribution of specimen illumination and can be averted in microscopy using August Köhler's method (Köhler 1893), which is now the recognized standard in widefield microscopy. This method allows alignment of the microscope light path, in which the illumination source is perfectly defocused in the image-forming plane. Even with a perfectly aligned microscope it is possible that a digital image shows areas of non-uniform illumination. In that case an image without specimen will not be entirely even.

The most common solution to non-uniform illumination is flat-field correction. This involves the estimation of the flat-field (the non-uniform illumination pattern) and then corrects the micrograph based on this flat-field.[1] The latter can be achieved by dividing each pixel by its corresponding flat-field coefficient. Different popular strategies are used to estimate the flat-field: based on an image acquired without a

[1]A software implementation for flat-field correction is available at:

- Fiji (Schindelin et al. 2012): http://fiji.sc/Image_Intensity_Processing#Flat-field_correction.

sample (Gareau et al. 2009), based on a time-lapse sequence that is merged into a single intensity image (Douterloigne 2015; Bevilacqua et al. 2011; Piccinini et al. 2012), or based on a single image, e.g., by detecting areas with expected constant intensity and using these areas to model the spatial variation in intensity (De Vylder et al. 2010). Once applied these corrections adjust for the non-uniformity of the original image. Illumination artifacts in three-dimensional image acquisition may occur as well. In point scanning confocal microscopes using visible light lasers and fluorescent probes, the entire sample is illuminated during the acquisition of a single slice; both the area in the focal plane and that above and below it. This leads to photobleaching in the non-imaged areas that can reduce the amount of noise relative to the signal or signal-to-noise ratio (SNR). In addition, optical slices deeper in the sample will exhibit reduced SNR because of signal attenuation due to light scattering in both the illuminating and detected photons.

A simple solution to this problem is to normalize the images by multiplying the intensity of each slice with a certain number such that all slices have the same average intensity. This way there is no need for modeling the difference in intensity with regard to the depth of the slice and potential bleaching of the fluorescence. The downside of this approach is that slices deeper in the sample typically have a lower SNR, which remains unchanged by multiplying their intensity with a constant. As such, while normalizing the data set results in slices with the same average intensity, each slice will have a different SNR. Therefore, this method—which attempts to mitigate for non-uniform illumination—introduces noise that is spatially dependent (see section "Absolute Spatial Dependency"). Another solution is found in single plane illumination microscopy (SPIM) where multiple images are acquired with the light source and objective opposed at 90° (Keller et al. 2010). This way, a point in the sample that has a low SNR in one image might have a better SNR in a different image. Alternatively, there are digital techniques that can deal with low SNR. However, the caveat to using such techniques is that they merely uniformize contrast variations, they do not magically increase SNR in the affected regions. Therefore minimizing these artifacts during acquisition by using very bright and stable fluorochromes, anti-bleaching reagents, and techniques such as multi-photon excitation is considered best practice.

2.3 Blur

Blur is the perceived loss of sharpness in an image. It can have different reasons (resolution limits, out-of-focus regions, etc.—see Fig. 2.1) and the result of this degradation can be modeled as a convolution of the true image data with a point spread function (PSF).

Light waves traveling through a microscope are subject to aberrations. Ideally, a single point light source in a sample is imaged as a single point. However, due to the PSF of the imaging system (characterized by imperfections such as spherical aberrations in the objective lens) a larger spot of light will be visible surrounded by concentric rings of light and dark described by George Airy—the Airy disk (Cole et al. 2011). These rings are caused by diffraction of the light as it images the specimen. This PSF is a 3D function that models the result of a combination of different effects. The lateral component of the blur is related to the in-plane resolution, which is defined as the ability to distinguish two closely spaced point sources. In effect, the optical resolution refers to the number of independently distinguishable points per unit length. Blur implicitly places limits (called the diffraction limit when the blur originates from diffraction) on the resolving power of an imaging system. The optical resolution or the minimal distance between two point sources emitting coherent light that still can be resolved as two individual light sources was calculated by Ernst Abbe in 1873 (Abbe 1873):

$$r = c\frac{\lambda}{NA}, \tag{2.1}$$

with λ the wavelength of the light source, NA the numerical aperture of the objective, and c a modality-specific constant relating to refractive indices. The axial component of the 3D PSF is more blurred than the lateral component because for axial resolution in the above equation c is replaced by $2c$ and NA is replaced by NA^2, which represents a physical property of wavelet interference in diffracted light. Thus even using short wavelength light an optical microscope is limited to approximately 200 nm in lateral and 600 nm in axial resolution. Note that this does not include recent super-resolution techniques, where lateral and axial resolutions of several tens of nanometers are achieved.

2.3.1 Deconvolution

To adjust for the effect of a 3D PSF (which is mathematically the result of a convolution operation) and more accurately localize the point of incoming light, a deconvolution operation can be performed. As can be expected, this requires detailed knowledge of the PSF. Knowledge of the actual PSF determines how accurate the reconstruction can be. Any errors or deviations in the PSF may propagate into huge deviations that render the end result unusable (see Fig. 2.2). Except for along the diagonal of the bottom illustration in this figure, where the correct PSF was used, either false structures appear (below the diagonal) or insufficient sharpening occurs (above the diagonal). The most accurate results are obtained via direct measurement. To do this, a sub-resolution structure of known size and shape (ideally as close to a point singularity as possible) is imaged (Gibson and Lanni 1992). Typically sub-resolution sized beads are used. If direct measurement is impossible, so-called blind deconvolution techniques can be applied. These methods usually jointly estimate the PSF (completely or its parameters) and deconvolve images based on strong assumptions about the shape of the PSF (Sibarita 2005). Regardless of the method of estimating the PSF, the deconvolved image is found

Fig. 2.2 Illustration of the effect of a PSF estimation error in deconvolution (by Tikhonov regularized inverse filtering). Note that the end result is significantly impacted when the PSF used in the deconvolution procedure deviates from the true PSF. Also note that, even with the correct PSF, the restored image quality still depends on the properties of the PSF

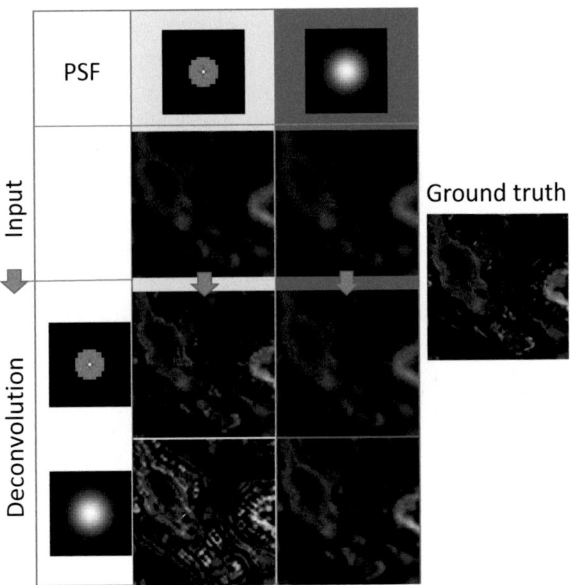

using an estimation algorithm[2] that takes into account the noise processes (to quantify uncertainty), the PSF, as well as prior knowledge about the image.

Popular deconvolution techniques include Richardson–Lucy (Richardson 1972), Maximum Likelihood Expectation Maximization (ML-EM) method (Shepp and Vardi 1982), least-squares solvers (Dougherty 2005), or Tikhonov regularized estimation (Van Kempen et al. 1997; Ramani et al. 2008). A limitation of most deconvolution algorithms is the underlying assumption of the shift-invariance property of the microscope system, which roughly states that a PSF does not change across the image plane. This assumption is not perfectly accurate in practice (Sarder and Nehorai 2006). The use of these algorithms therefore requires an accurate PSF estimation, in order to avoid the artifacts illustrated in Fig. 2.2.

Deconvolution is an inverse problem that becomes highly unstable in the presence of noise. Applying the inverse filter will result in computationally unreliable results and amplification of noise and other undesired artifacts. Figure 2.3 illustrates the effect of the inverse filter at various noise levels. Even barely noticeable levels of noise in blurry image data can lead to large noise levels in the deconvolved image.

[2]Software implementations are available at:

- DeconvolutionLab (Vonesch and Unser 2008) (including Richardson–Lucy, Tikhonov, and wavelet regularizations): http://bigwww.epfl.ch/algorithms/deconvolutionlab/
- TV prior (Tao et al. 2009): http://www.caam.rice.edu/~optimization/L1/ftvd/
- Sparsity prior (Jia and Evans 2011): http://users.ece.utexas.edu/~bevans/papers/2011/sparsity/.

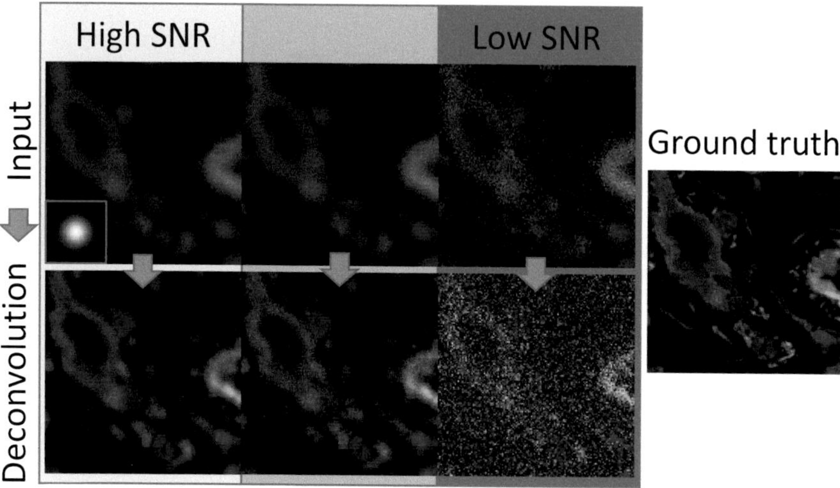

Fig. 2.3 Illustration of the deconvolution (by Tikhonov regularized inverse filtering) performance based on various levels of input noise. Note how even barely noticeable levels of input noise on blurry image data can lead to large noise levels in the output

This problem can be solved by introducing regularization, or even better, using prior knowledge of the image (see Graybox 2).

2.3.2 *Focus Stacking*

Ideally, a lens will focus the light rays of a point light source onto the sensor or eye. However, only point sources at the same distance from the objective can be in focus at the same time. Since microscopic specimens are usually thicker than the focal plane of the objective lens, this means that light source rays at a shorter distance from the objective will converge behind the sensor, whereas light sources further away from the objective will converge in front of the sensor. Both result in a blurred image of the specimen (see Fig. 2.4). In a confocal microscope only in-focus light will pass the pinhole that is placed in front of the photomultiplier tube, while all the out-of-focus light is prevented from reaching the detector. By moving the focal plane over various depths within the sample, optical sectioning can be carried out and z-stacks can be obtained.

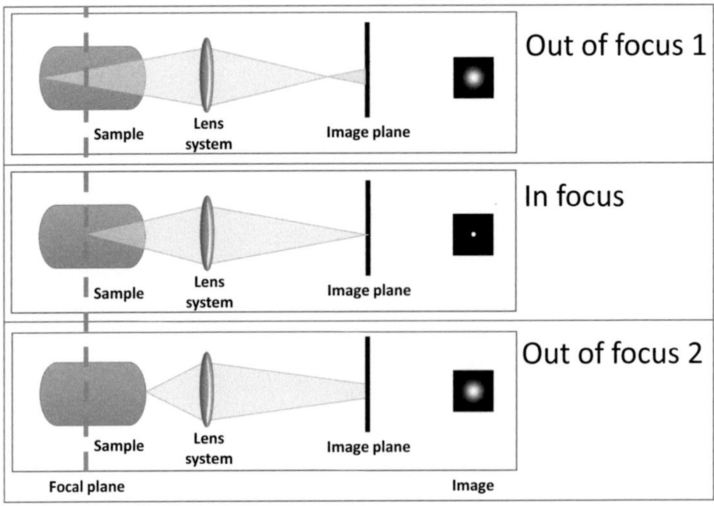

Fig. 2.4 Illustration of out-of-focus light. The *top* example shows the result of light emitted from below the focal plane, the *middle* example shows the focal process for light originating from the focal plane, and the *bottom* example shows light emitted from above the focal plane. Both *top* and *bottom* examples will result in *blurred image* areas

A common image processing approach[3] to get all imaged structures in focus is to extend the depth of field by combining multiple images into a single image (extended focus projections). These methods fuse a stack of optical slices into a single image where everything appears to be in focus. For fluorescence images this often corresponds to a maximum intensity projection (MIP). However, for other microscopic methods such as wide field or differential interference contrast (DIC) microscopy may require more complex methods that ensure good contrast fusion results. Methods fusing information in some feature space generally yield the best results (Tessens et al. 2007; Aguet et al. 2008; Li et al. 2011). Based on the features in a small neighborhood around each pixel a focus-measure is calculated for each slice. Based on these focus-measures the features that are in focus are merged into a single feature image, which then can be transformed into an intensity image.

[3]Software implementations are available at:

- Fiji (Schindelin et al. 2012): http://fiji.sc/Extended_Depth_of_Field
- ImageJ plug-ins (Forster et al. 2004; Aguet et al. 2008): http://bigwww.epfl.ch/demo/edf/.

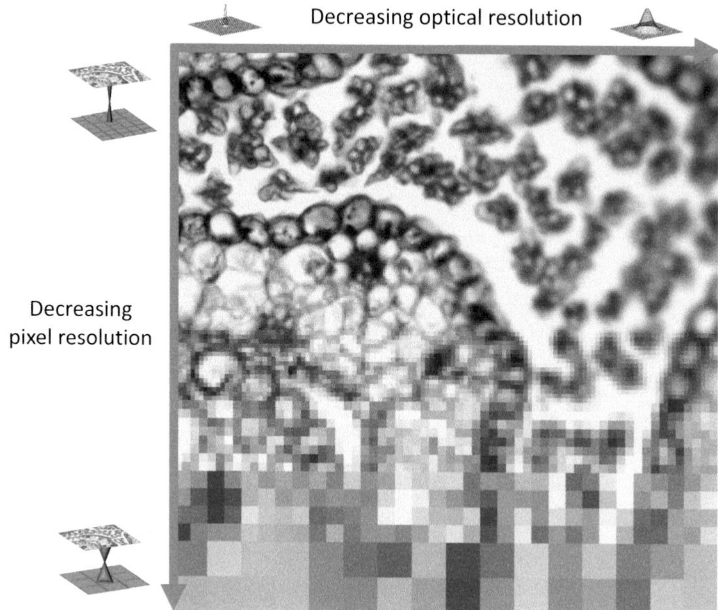

Fig. 2.5 Visualization of different optical and pixel resolutions. While a low pixel resolution results in an image where most details of the image are obscured by the discrete blocks of the pixels, images with a low optical resolution result in smooth intensity changes but lack image details and contrast

Graybox 1: The Nyquist criterion: optical vs pixel resolution

There is a difference between the pixel (or digital) resolution and the optical resolution. The optical resolution is usually fixed by the imaging system and characterized by the PSF (see section "Blur"). Pixel resolution, on the other hand, is defined as the amount of pixels relative to the surface size. Consequently, pixel resolution is inversely proportional to the spatial sampling interval. Figure 2.5 illustrates the impact of both optical and pixel resolution. On the one hand, it does not make sense to use significantly more pixels than the optical resolution of the system would justify: more data needs to be stored, without gaining optical resolving power. On the other hand, information is lost or even distorted when using less pixels, resulting in "blocky" or "pixel" artifacts. Such undersampling artifacts are also known as (spatial) frequency aliasing. Consequently, optical and pixel resolution should be balanced. As higher optical resolution becomes possible, also higher pixel resolution is necessary in order to visualize this high optical resolution detail

(continued)

Graybox 1 (continued)
in specimen of the same size. For this reason, microscopy imaging can be associated with the *big data* problem.

The link between pixel and optical resolution is made formally through the Nyquist criterion (Nyquist 1928). The Nyquist criterion involves the Fourier theorem, which considers (image) signals as being composed of a (possibly infinite) number of sine waves. It states that the spatial sampling interval of a signal should be half of the wavelength of the shortest (highest-frequency) sine wave that makes up the (image) signal. This sampling interval corresponds to a sampling frequency of 2 times the frequency of the highest-frequency sine wave. In microscopy, the common rule-of-thumb is that the pixel size on a digital capture device must be set so that the pixel is roughly 2.3 times smaller than the optical resolution limit of the system (Heintzmann 2006; Pawley 2006). The factor of 2.3 (instead of 2, as suggested by Nyquist) is commonly used to offset the fact that the Airy disk (or system PSF) does not have a literal highest-frequency sine wave. Rather it has an infinite number of sine waves that make up the PSF. Nonetheless, the involved sine waves decay rapidly with increasing frequency, justifying a rule-of-thumb trade-off.

2.4 Noise

By noise we mean statistical noise, i.e. the combination of acquisition factors that introduces uncertainty in the digital data set representing the microscopic image. Visually speaking, such uncertainty manifests as measured image intensity values that differ from their true intensities. An accurate model for noise in microscopy is a combination of a Poisson and a Gaussian-distribution-based model (Luisier et al. 2011). The former is sometimes called shot noise and arises due to the stochastic nature of a photon/electron counting process, which typically follows a Poisson distribution. The latter, often referred to as *dark current*, is caused by complex electronics, where thermal and electromagnetic effects introduce fluctuations that typically follow an additive Gaussian distribution. The defining parameter of Gaussian noise is its variance. The signal intensity relative to its noise variance (SNR) is a good way to express the uncertainty of a measurement. Ideally SNR should be kept as high as possible through a combination of specimen preparation conditions and digital capture optimization. In practice, however, the Poisson distribution is neglected and a more simple assumption of solely Gaussian noise is made, because a Poisson distribution converges to a Gaussian distribution when high numbers of photons/electrons are measured. An illustration comparing the two types of noise is provided in Fig. 2.6.

Most image restoration techniques suppressing noise (or *denoising* algorithms), such as those described below are designed for the additive Gaussian noise model

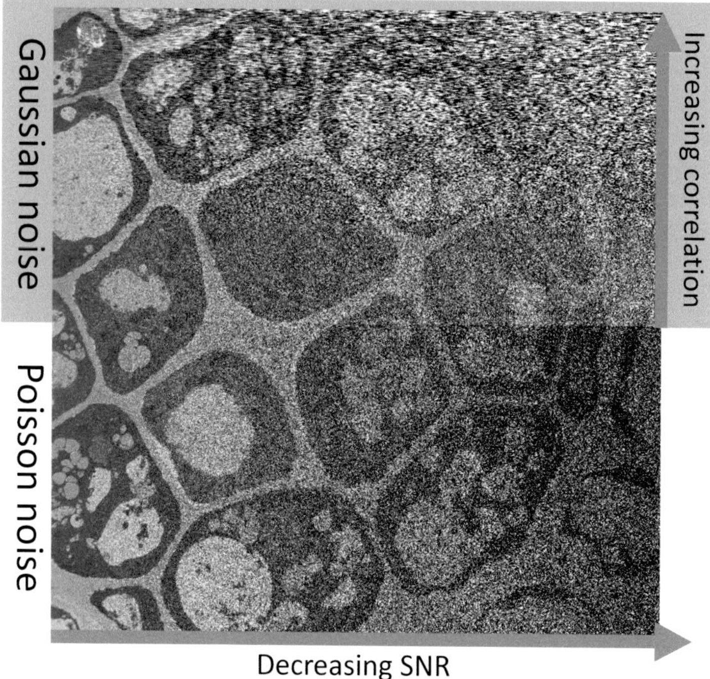

Fig. 2.6 Conventional imaging data is degraded by a combination of Poisson and Gaussian noise. Firstly, note that Poisson noise is indistinguishable from Gaussian noise when the signal-to-noise ratio is high, but deviates from Gaussian noise at low signal-to-noise ratio. The difference is most noticeable by Poisson noise having a higher variance in *dark areas*. Secondly, we made the Gaussian noise exhibit a constant variance, yet increasing horizontal spatial correlation from *bottom* to *top*. Note that noise correlation can severely degrade perceived image quality. This illustration was made by artificially degrading an SBF-SEM image with both types of noise

and exploit *a priori* knowledge about the type of signal that is being imaged by the device (more information about exploiting this *a priori* knowledge can be found in Graybox 2). While these algorithms work well for additive Gaussian noise, their performance is sub-optimal when the encountered noise statistics deviate. This often happens in practice and is discussed in the next sections.

2.4.1 Signal Dependency

In general, noise variance is signal-dependent. Noise is not constant over the complete image but is varying depending on the expected noise-free signal value. A typical scenario for this is shot noise (Poisson noise). Shot noise manifests more in areas with lower signal energies. Because of the lower number of photons or electrons, the Poisson-to-Gaussian simplification, as discussed previously, is not justified anymore. A more accurate approach is to model the acquisition process as a Poisson random arrival process. For a Poisson process, the variance decreases with the increasing signal value. The result is that there is more noise in darker (low-signal) areas than would be the case for pure Gaussian noise (as illustrated in Fig. 2.6). Signal-dependent noise is typical in both fluorescence and electron microscopy (Jezierska et al. 2012; Roels et al. 2014).

State-of-the-art Gaussian denoising algorithms will generally still improve image quality in microscopic images containing signal-dependent noise. However, low-signal areas will still contain noise, or high-signal areas will be denoised too aggressively. As a consequence, blur artifacts are introduced in these areas or even a combination of both effects will occur. This is due to the uneven noise distribution with respect to the underlying signal. A better approach is to correctly estimate the signal dependency relation (Torricelli et al. 2002; Zabrodina et al. 2011) and to improve the underlying noise model of the restoration algorithm taking this extra information into account (Zhang et al. 2008; Roels et al. 2014; Chen et al. 2014). This results in higher noise suppression in low-signal regions compared to the high-signal regions.

2.4.2 Absolute Spatial Dependency

Regardless of the signal value, the amount of noise can also vary across an image depending on its position, i.e. spatially dependent noise. This can happen as a result of inadequate compensation of non-uniform lighting. This compensation for under-exposure is typically done by multiplying the signal but also the noise by a certain value. Although the SNR does not change, the noise variance changes depending on the amplifying value. As a result, the exposure-corrected image suffers from a position-dependent noise variance.

Similar to signal dependency, the performance of denoising algorithms is lower if the spatial noise dependency is not taken into account. In low-noise areas, denoising would be applied too aggressively, in high-noise areas, denoising would be applied insufficiently and noise will remain. One solution is to avoid spatial noise dependency, by postponing non-uniform lighting compensation until after denoising is applied. If this is impractical, an alternative is to modify a denoising algorithm to use locally estimated noise variances so that the appropriate degree of denoising is applied to each area (Goossens et al. 2006).

2.4.3 Relative Spatial Dependency

In pixel-by-pixel scanning microscopes, such as point scanning confocal or scanning electron microscopes, the noise on a measurement can be influenced by measurements on neighboring pixels. When an image is scanned in a certain direction, after the first measurement the next will be influenced by noise of the previous and current measurements. In this case, the directional scanning principle will cause a relative, oriented, spatial dependency between noise intensities (Roels et al. 2014). This is referred to as spatially correlated noise. In this case, noise appears as horizontal stripes, meaning that a positive noise contribution in one pixel is very likely to result in a similar positive noise contribution in the next pixel. This type of noise correlation is illustrated in Fig. 2.6, where in the middle there is hardly any correlation resulting in isolated noise patterns and where in the top the correlation increases into clearly striped patterns.

Apart from resulting in peculiar noise patterns, this is not so much a problem but an opportunity. When properly modeled, the extra knowledge that two neighboring pixels have similar noise contributions reduces uncertainty. Obviously, when subsequent image analysis is oblivious to noise correlation, the resulting quality will be sub-optimal, as the opportunity for reducing measurement uncertainty was not exploited. To make matters worse, denoising methods may mistake the encountered spatial correlation of noise contributions as a property of a biological structure. This significantly hampers some tasks, such as segmentation; e.g., a horizontal stripe due to correlated noise may be erroneously detected as a horizontal segment boundary precisely because of the spatial structure introduced by the noise correlation. Therefore, the proper modeling of noise requires knowledge of how the image was collected and of the analysis of correlation arising from a particular method, such as horizontal scanning (Goossens et al. 2009b; Roels et al. 2014).

2.4.4 *Influencing Noise*

A crucial acquisition parameter influencing noise is the dwell time. Longer dwell times produce less noisy images in general. However, a longer dwell time can introduce additional artifacts due to overexposure. As a consequence, one should be aware of the significance of the dwell time parameter in noise avoidance and the trade-off between noise and overexposure artifacts.

2.4.5 *Denoising*

There are many image processing possibilities in terms of improving the SNR or, equivalently, removing noise while retaining genuine image signal. Generally speaking, these so-called *denoising* algorithms are based on averaging pixel intensities with a similar expected underlying signal and assume white (uncorrelated) Gaussian noise. The main difference between denoising algorithms is therefore in how these algorithms establish which pixel intensities to average.

An elementary class of denoising techniques called low-pass filters assume that pixels in a local neighborhood share similar intensity values. These pixels are averaged out according to a relative weighting function, defined by the low-pass filter, which is characterized by a kernel (a cluster of pixels surrounding

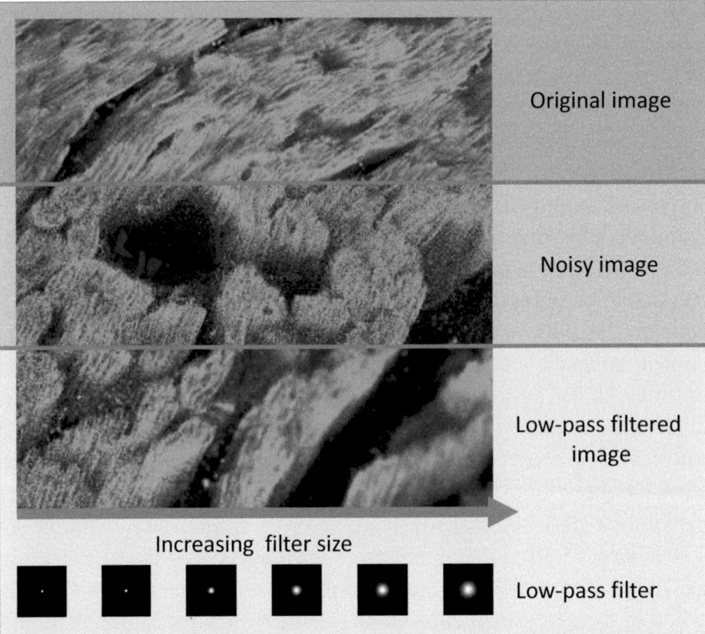

Fig. 2.7 Denoising result using Gaussian low-pass filters of various sizes. A small filter size tends to retain the noise, whereas larger filter sizes result in *blurred edges* and loss of detailed structures

a center point). The size of this kernel determines the user's definition of a "local neighborhood" and significantly influences the denoised image quality (as illustrated in Fig. 2.7).

Because of computational efficiency and simplicity, low-pass filters are widely used in microscopic image denoising. Nevertheless, these filters often fail to restore the true signal in highly textured regions and edges, because the local similarity assumption is not satisfied in general. This problem is solved by anisotropic diffusion (Perona and Malik 1990), in which the low-pass filter kernel is adjusted along edge directions. Although this technique improves restoration quality, it assumes that images consist of constant regions, separated by crisp edges, an assumption that is not always true for biological structures. More specifically, highly detailed structures and smooth gradients cause difficulties for this type of denoising.

An alternative class of denoising algorithms focuses on probability theory as it attempts to determine the underlying true image (Portilla et al. 2003), given the acquired noisy image, a noise model, and optional prior knowledge.[4] The difficulty

[4]Software implementations are available at:

- TV prior (Zhu et al. 2008): http://pages.cs.wisc.edu/~swright/TVdenoising/
- Wavelet prior (Luisier et al. 2007): http://bigwww.epfl.ch/demo/suredenoising/.

with these techniques is the choice of an accurate noise model and prior knowledge (for a more technical discussion we refer to Graybox 2).

Ever since these techniques were proposed, it has become clear that better results can be achieved by averaging more appropriate pixels (i.e., pixels that share local similarity). Such pixels can be found across the entire image and not only in a local neighborhood. This way a new class of algorithms, called non-local denoising algorithms [5] (Buades et al. 2005; Dabov and Foi 2006), have been proposed yielding state-of-the-art performance.

Graybox 2: Prior knowledge in probabilistic image restoration

Image restoration algorithms mitigate many of the artifacts described in this paper. Whether a specific algorithm leads to high-quality restoration results is largely determined by the accuracy of the underlying models. Typically, one may improve the algorithm by incorporating prior knowledge: generic information that is known about the image in advance. Even though images in general tend to be highly variable, they share many commonalities that could be exploited: local and non-local self-similarity, frequency distribution, etc. We briefly describe several common priors that improve image restoration:

- Local smoothness priors: images typically consist of smooth areas, separated by edges. Therefore, a natural prior assumes local smoothness in the image.
  - Edge-stopping prior: edge-stopping priors assume that pixels are locally very similar except on edges (Perona and Malik 1990). In this class, the best-known and most popular variant is based on total variation. The total variation prior models local smoothness by minimizing the amount of edges (i.e., the total variation) in an image. This prior has applications in both denoising (Louchet and Moisan 2014) and deblurring (Oliveira et al. 2009).
  - Sparsity (multi-resolution) prior: alternatively, it is possible to transform the image to a domain wherein edges are compactly represented. Coefficients corresponding to noise are typically easier to identify in such a domain and therefore a prior is more easily defined. There exist

(continued)

[5] Software implementations are available at:

- Non-local means (Buades et al. 2011): http://www.ipol.im/pub/art/2011/bcm_nlm/
- BM3D (Dabov and Foi 2006): http://www.cs.tut.fi/~foi/GCF-BM3D/.

Graybox 2 (continued)

many such transforms [e.g., wavelets (Donoho 1992), curvelets (Candes and Donoho 2000), and shearlets (Guo et al. 2006; Goossens et al. 2009a)] and these transforms differ in the way of representing various types of edges and image structures.

- Self-similarity priors: alternative prior knowledge assumes that natural images consist of similar regions (e.g., repetitive directional edges, texture patterns, etc.).
  - Non-local prior: self-similarity is not always found within close proximity. A non-local prior, implementing a larger search window for locating similarity, is therefore more likely to exploit repetitiveness within images (Aelterman et al. 2012).
- Dictionary-based prior: in some cases, a more specific prior knowledge may lead to higher quality results. Dictionary-based priors specify a limited number of features (a dictionary of features) that can appear in an image. For example, limiting the number of discrete intensities (or colors) that can appear in an image yields excellent restoration results in, e.g., black-and-white or cartoon-like images (Luong et al. 2007) and possibly also in fluorescence microscopy because of the low expected number of possible intensities. Alternatively, dictionaries can consist of a set of (trained) patches, which form the basic elements of the image (Elad and Aharon 2006).

Most image restoration techniques combining artifact models and a priori knowledge are implemented in a probabilistic framework. Assuming the underlying models, the most likely artifact-free signal is estimated from measured data. For example, assuming an image $f(\mathbf{x})$ is degraded by additive, Gaussian noise $n(\mathbf{x})$ and blur (modeled by a PSF $h(\mathbf{x})$), leading to an acquired image $g(\mathbf{x})$, the underlying artifact model would look like

$$g(\mathbf{x}) = \left[h * f + n\right](\mathbf{x}), \tag{2.2}$$

where $\mathbf{x}$ is the spatial coordinate in the image. Equation (2.2) allows us to determine the probability distribution of g, given f, i.e. $p(g|f)$. Usually, a priori knowledge is expressed as a probability distribution $p(f)$. For example, in case of a non-local prior, images with many repeating structures will have relatively large probabilities compared to the ones with many varying structures. We would like to estimate the most likely image f, given the acquired image g, the underlying artifact model, and a priori knowledge. In other words, we have to maximize $p(f|g)$, given g, $p(g|f)$, and $p(f)$, respectively. Bayes' rule is a helpful tool expressing this probability in terms

(continued)

Graybox 2 (continued)
of what is given:

$$\hat{f} = \arg\max_f p(f|g) = \arg\max_f \frac{p(g|f)p(f)}{p(g)} = \arg\max_f p(g|f)p(f). \quad (2.3)$$

Note that $p(g)$ can be removed in the maximization because it is independent of the argument f being maximized. This estimator is usually referred to as a maximum a posteriori (MAP) estimator. The simplification where the prior distribution is uninformative and every image is assumed equally likely ($p(f)$ is constant and disappears in the maximization) leads to a maximum likelihood (ML) estimator. Typically, the extrema in Eq. (2.3) can be found using various optimization solvers: steepest descent, Newton's method, Lagrange multipliers, etc.

2.5 Digitization Artifacts

Mapping measurement values from the analog domain to discrete numerical values in the digital domain (digitization) causes two types of artifacts, called quantization and saturation. The measurement is typically a real number (for example, a voltage), but for the digital image representation, it needs to be rounded to the nearest integer. This rounding is called quantization. As each allowed integer has to be expressed by a unique bit code, the number of bits b of the imaging system restricts the number of different gray values 2^b that a digital image can show. Typically, the number of bits b in microscopic images is between 8 and 16. Furthermore, any measurement that is far lower than the lowest representable integer, or far larger than the largest representable integer, is typically rounded to the closest integer. This effect is called *saturation*. The impact of both quantization and saturation on a distribution (histogram) of measurement values is shown in Fig. 2.8.

Both quantization and saturation involve loss of information. Quantization results in the loss of any contrasts (difference in measurement values) that is smaller than the difference between two representable integers. The smallest desirable contrast to be visible in the resulting digital image determines the number of bits that a digital imaging system can use. Using too few gray values degrades the information available. Saturation results in severe deviations when the difference between a measurement value and the nearest representable integer is large. It is therefore important to match the dynamic range, to the number of bits captured. There is a trade-off between avoiding saturation and avoiding quantization errors. Accommodating to a large dynamic range, to avoid saturation errors, means that the spacing between two representable integers is large, which induces large quantization errors.

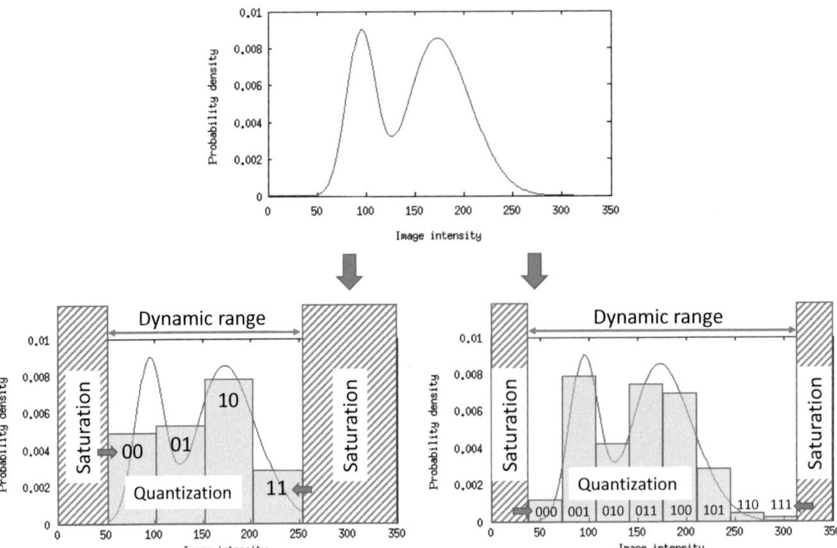

Fig. 2.8 The distribution of measurement values is expressed by a probability density function (*top*). The effect of quantization is that measurement values are grouped into a discrete number of bins that are imposed by the number of bits in the system. Thus each possible measurement value (a real number) is mapped to a bit code of limited length (just a few possible integers). The effect of saturation, indicated by the darker contribution to the "11" and "111" bit code, is that any measurement value too high or too low is mapped to the highest and the lowest bit code, respectively. Note that this may significantly distort the histogram, and subsequently the digital image

An illustration on the visual impact of saturation and quantization is shown in Fig. 2.9.

The impact of using incorrect acquisition parameters is evident. When too few bits are used, subtle contrasts in the image are lost due to quantization errors. When the dynamic range of the digitizer is too large for the input signal, the image becomes dull (not all bit codes are used so only intermediate gray values are displayed instead of white and black). When the dynamic range of the digitizer is too small, saturation causes loss of contrast in areas that are saturated to white or black. Therefore finding the proper bit range for a given sample is an important consideration in acquiring a useful image.

Quantization and saturation are consequences of working in the digital world. These processes result in image contrast to be distorted from their original (analog) values, sometimes to the extent that objects can no longer be resolved. Luckily, these effects can be mitigated by using sufficient bits in the digitization process and respecting the dynamic range of the signal, at the cost of using more memory and storage.

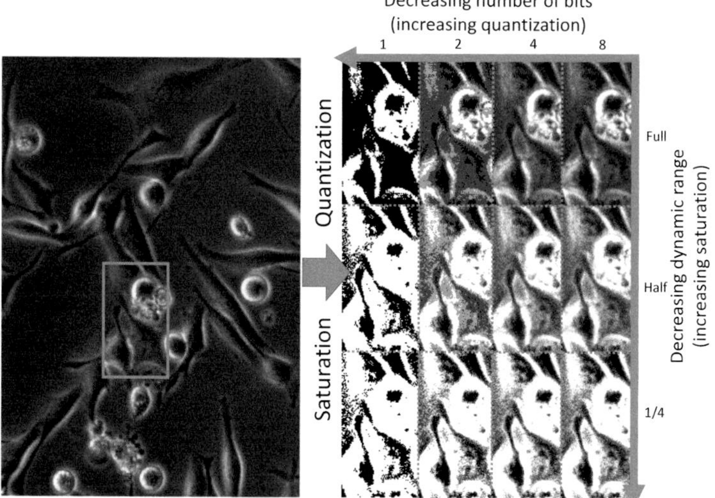

Fig. 2.9 Visualization of different levels of quantization and different levels of saturation artifacts. Decreasing the number of bits leads to quantization artifacts such as posterization along soft edges. Decreasing the dynamic range leads to more saturation or complete loss of image information in the high- and low-intensity (saturated) regions

2.6 Compression Artifacts

Modern day microscopes produce huge data sets: a single experiment can easily result in a data set of a few to hundreds of gigabytes, making efficient storage an important issue for many biologists. One way to cope with such large data sets is by using image compression, the practice of modifying the representation of an image with the aim of reducing the number of bytes. This helps by minimizing storage capacities and increasing bandwidth-efficient transmission. There are two fundamental categories of compression techniques: lossless compression, the techniques that are exactly reversible (thus resulting in no information loss), and lossy compression, the techniques that are not perfectly reversible (where information is lost).

The compression ratio (i.e., the ratio of the file size of the compressed image to the uncompressed image) for lossless compression generally ranges from 30 to 70 %, but obviously depends on the image content. These lossless compression methods do not impact the resulting image quality. The only reasons not to use such methods are because of software compatibility or because of memory, storage, or bandwidth limitations. However, distinguishing lossless from lossy compression can be a challenge as some file formats allow both forms of compression. Table 2.1 shows a non-exhaustive list of common image formats (Cox 2006).

The compression ratio of lossy compression can be higher than that of lossless compression (compression ratios of 10 % or lower are frequently encountered). This

Table 2.1 An overview of properties of common bio-image formats

	Principle	Lossy vs. lossless	Compression ratio[a]
JPEG	Block-based DCT	Lossy (with a lossless extension)	Good
JPEG2000	Wavelet-based	Usually lossy (has lossless option)	Very good
PNG	LZ77 + Huffman	Lossless	Neutral
TIFF	Container for other formats	Can be both	Neutral to bad
Proprietary formats	Often based on TIFF	Typically lossless	Neutral to bad

[a]An indication of how good a compression ratio can be achieved without introducing visible artifacts

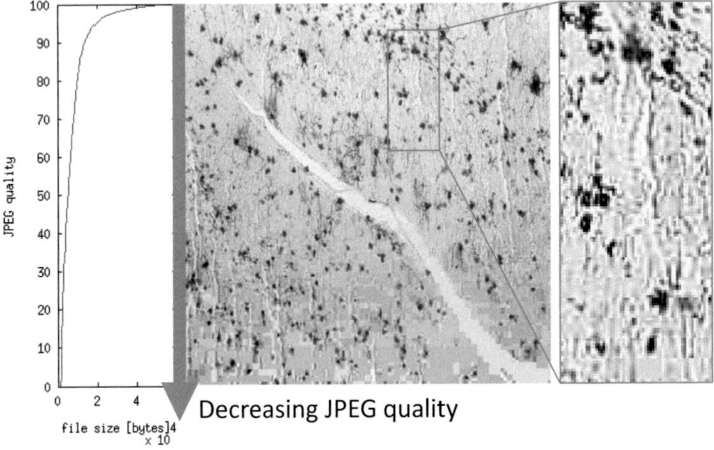

Fig. 2.10 Visualization of increasing levels of lossy (JPEG) compression. Note how an inflection point is reached at a compression level before visually disturbing errors start to appear and how a large decrease in visual quality at very high compression rates (*right part of the image*) only results in slightly smaller file sizes

comes at a cost; lossy compression is not reversible because information is lost. The vast majority of such algorithms are designed to make this lost information imperceptible to the human observer. The aim is solely to "trick" the human observer into thinking that the image is still the original image. Nevertheless, lossy compression can severely impact any image analysis and inevitably degrades the scientific information to some extent. Note that lossy image compression algorithms typically offer a parameter to trade off how aggressively information is removed from the image. Therefore lossy compression is not a binary process, there exist various degrees of "heavy" or "light" compression. Figure 2.10 illustrates how JPEG compression, still the most popular form of image compression, causes severe artifacts when carried to extremes. A tell-tale sign for JPEG compression

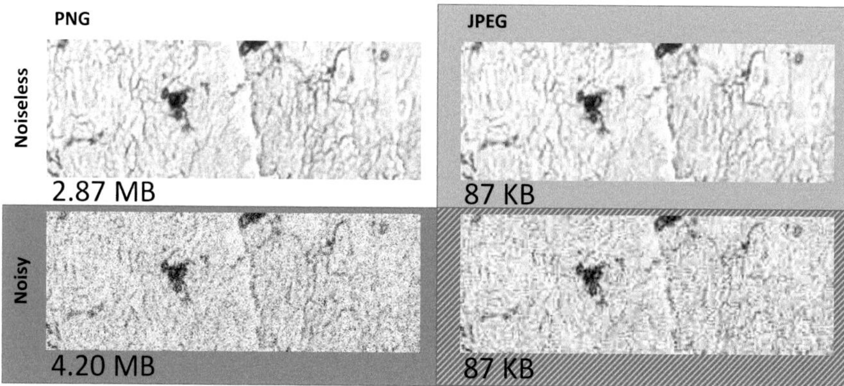

Fig. 2.11 Comparison of lossless compression (PNG) and lossy compression (JPG) in noisy and noiseless condition. Note that, despite the fact that noise causes the lossless representation to increase in size, the JPG (as well as the PNG) version of the noisy version shows reduced image quality. Also note that the JPEG compression process even results in false structures, especially in the presence of noise

is the *block artifact* where unwanted block structures become visible in the image as the compression factor increases. If the goal is to present an esthetically pleasing representation of a micrographic image that accurately reflects the data collected, then lossy compression can be acceptable. However, if the end goal is analyzing the scientific content accurately, then lossy compression merely for the sake of convenience is unacceptable and the use of lossless compression should be considered.[6]

If the input data set contains useful signal, as well as noise, then the compression algorithm will "waste" bytes on accurately representing the noise. This results in significantly larger file sizes for lossless compression, despite the fact that noise actually reduces the useful information. Depending on the compression ratio, the impact of noise on lossy compression can be more complex. A lossy compression algorithm may significantly change the structure of the noise, in an effort to maintain the overall look of the image to the human observer while using less bytes. This effect as well as a compression ratio comparison is illustrated in Fig. 2.11.

[6]Software that allows for lossless/lossy compression and compression quality adjustment:

- IrfanView: http://www.irfanview.com/
- GIMP: http://www.gimp.org/.

Graybox 3: Why is image restoration such a challenging problem?
The goal of image restoration is to retrieve the ideal image, i.e. the structure of a microscopic sample, based on the observed image, as already mentioned in Graybox 2. A first step in image restoration is to correctly model the imaging system. For example, taking into account the imaging artifacts discussed in this paper, an underlying true image $f(\mathbf{x})$ would result in an observed degraded image $g(\mathbf{x})$ according to the following model:

$$g(\mathbf{x}) = c\left[q_i\Big(u\big(s(h * f + b) + k * n\big)\Big)\right](\mathbf{x}). \tag{2.4}$$

where $\mathbf{x}$ denotes the spatial position in the image. In this model, the true image is first blurred by the PSF $h(\mathbf{x})$ of the microscopic setup, leading to $(h * f)(\mathbf{x})$. To this blurred image a background signal, $b(\mathbf{x})$, is added. The resulting image is further corrupted by shot noise, which we model here as $s(\lambda)$, which corresponds to a Poisson random variable of intensity λ. After this shot noise, the image is further corrupted by additional sensor and electronic circuit noise, which can be accurately modeled by additive white Gaussian noise, $n(\mathbf{x})$. This noise can be correlated, e.g., due to the scanning process (see section "Relative Spatial Dependency"). This correlation can be expressed by another convolution step, $(k * n)(\mathbf{x})$. Next, the effect of non-uniform illumination (see section "Non-uniform Illumination") is modeled using a point-wise multiplication with a flat-field $u(.)$. Finally, the image is quantized and compressed using the quantization operator $q_i(.)$, which quantizes the image using i bits, and the compression operator $c(.)$. Note that Eq. (2.3) is a special case of Eq. (2.4): background signal, shot noise, noise correlation, flat-field correction, quantization, and compression artifacts are ignored in this simplified model.

It is clear that realistic degradation consists of a cascade of several operators that jointly transform an ideal image to an observed image. Most of these operators are not invertible. For example, different input images can result in the same quantized output image and as such it is impossible to ascertain what the exact original input image was. One can only attempt to approximate the true image using advanced image restoration techniques while minimizing the number of additional artifacts the algorithms might introduce. Figure 2.2 illustrates this effect: Each deconvolution result is equally consistent with the observed image, but the choice of PSF determines the quality of the result. An accurate estimation of the PSF is therefore crucial in this case.

The best approach is to try to minimize the number of operators in the model without dissatisfying the physical conditions, e.g., by using Köhler

(continued)

Graybox 3 (continued)
illumination, non-uniform light is not an issue anymore and the corresponding operator can be discarded from the equation. Optimizing the imaging setup to avoid most artifacts is best practice. However, for many applications it is impossible to avoid all artifacts, e.g., by increasing the dwell time one improves the SNR but increases bleaching in a fluorescent image. The optimal trade-off between the artifacts depends on the application and should be carefully defined within the protocol development. In order to remove the remaining artifacts dedicated image processing tools can be used. These algorithms are typically hampered by a number of challenges:

- Most advanced methods work iteratively: an initial estimate of the image is gradually improved until the expected optimal image is found. For many of these methods, the result depends on the initialization. If this is the case, the method will usually only converge to the optimal solution if the initialization is close to the optimum. More often it is found to converge to less desirable but equally data-consistent solutions. Recently many image degradation problems are reformulated such that finding the solution is unhampered by the initialization (Goldstein and Osher 2009). These new problems can then be solved using efficient optimization algorithms that find the global optimum (Bertsekas 2015).
- Another major challenge is maintaining scalability. Many advanced image restoration algorithms are computationally intensive. As data volume increases, higher spatiotemporal resolution can be achieved, at the cost of an increased computational burden. For the challenges related to data-storage we refer to section "Compression Artifacts." However, these huge data sets also pose severe problems related to image processing and data mining. In order to cope with this big data, efficient implementations using the massive parallel power of computer clusters and graphics cards (GPUs) are currently being investigated. However, GPU software is notoriously complex to program effectively. Fortunately, the advent of new programming languages such as Quasar seems to facilitate and speed up parallel implementations both on multi-core CPUs and on GPUs (Goossens et al. 2014).
- The model proposed in Eq. (2.4) uses many operators, which each are based on a number of parameters, e.g., the Gaussian noise depends on its variance. In order to get a reliable and optimal image restoration, all operators and parameters should be accurately tuned according to the microscopic setup and sample. The restoration can be further improved by exploiting prior knowledge of the type of images expected (see Graybox 2). While there exist fixed protocols for the estimation of certain parameters, e.g., the PSF can be measured using small beads or the flat-field can be

(continued)

Graybox 3 (continued)
estimated by acquiring an image without a sample, these protocols are not always followed. For other parameters proper estimation protocols seem to be lacking. Little work has been done on the accurate modeling and estimation of noise parameters in microscopy. For noise in scanning electron microscopic images (Roels et al. 2014) is one of the few works thoroughly analyzing the noise characteristics but has not led to a fixed procedure yet that can be used in combination with image processing software.

2.7 Concluding Remarks

It is a basic principle that most artifacts can be avoided before or during the acquisition process. The impact of artifacts can be mitigated through a proper protocol development (choice of microscope and detection system, illumination methods, and optical setup) and acquisition parameters (choice of dwell time and pixel size). It is important to realize that acquisition parameters can generally only be chosen as pareto-optimal values as these involve a trade-off between artifacts. This is because artifacts are not independent of each other. Since trade-offs between different types of artifacts are ubiquitous in image formation, artifacts will never be completely absent from an image. In the end, modern microscopy is a series of choices between what is possible and what is ideal. The need for a carefully considered imaging chain, from experimental setup to acquisition parameters to image analysis, all are crucial in the final quality and scientific validity of the image.

Acknowledgements This research has been made possible by the Agency for Innovation by Science and Technology in Flanders (IWT) and the iMinds BAHAMAS project (http://www.iminds.be/en/projects/2015/03/11/bahamas). We would like to thank Evelien Van Hamme (VIB—Bio Imaging Core/IRC/DMBR) for the microscopy images.

References

Abbe E (1873) Beiträge zur Theorie des Mikroskops und der mikroskopischen Wahrnehmung. Arch Mikrosk Anat 9(1):413–418. doi:10.1007/BF02956173. http://dx.doi.org/10.1007/BF02956173

Aelterman J, Goossens B, Luong H, De Vylder J, Pizurica A, Philips W (2012) Combined non-local and multi-resolution sparsity prior in image restoration. In: 2012 19th IEEE international conference on image processing (ICIP), pp 3049–3052. doi:10.1109/ICIP.2012.6467543

Aguet F, Van De Ville D, Unser M (2008) Model-based 2.5-D deconvolution for extended depth of field in brightfield microscopy. IEEE Trans Image Process 17(7):1144–1153

Bertsekas DP (2015) Convex optimization algorithms. Athena Scientific, Belmont

Bevilacqua A, Piccinini F, Gherardi A (2011) Vignetting correction by exploiting an optical microscopy image sequence. In: 2011 annual international conference of the IEEE engineering in medicine and biology society, EMBC, pp 6166–6169. doi:10.1109/IEMBS.2011.6091523

Buades A, Coll B, Morel JM (2005) A non-local algorithm for image denoising. In: IEEE computer society conference on computer vision and pattern recognition, 2005. CVPR 2005, vol 2, pp 60–65. doi:10.1109/CVPR.2005.38

Buades A, Coll B, Morel J-M (2011) Non-local means denoising. Image Process On Line 4(2):490–530. doi:10.5201/ipol.2011.bcm_nlm. arXiv:1112.0311

Candes E, Donoho D (2000) Curvelets: a surprisingly effective nonadaptive representation of objects with edges. Curves and surface fitting. http://citeseerx.ist.psu.edu/viewdoc/summary?doi=10.1.1.43.6593

Chen G, Xie W, Dai S (2014) Image denoising with signal dependent noise using block matching and 3D filtering. In: Zeng Z, Li Y, King I (eds) Advances in neural networks – ISNN 2014. Lecture notes in computer science, vol 8866. Springer, Heidelberg, pp 423–430. doi:10.1007/978-3-319-12436-0_47. http://dx.doi.org/10.1007/978-3-319-12436-0_47

Cole RW, Jinadasa T, Brown CM (2011) Measuring and interpreting point spread functions to determine confocal microscope resolution and ensure quality control. Nat Protoc 6:1929–1941

Cox G (2006) Mass storage, display, and hard copy. In: Pawley JB (ed) Handbook of biological confocal microscopy. Springer, New York, pp 580–594. doi:10.1007/978-0-387-45524-2_32. http://dx.doi.org/10.1007/978-0-387-45524-2_32

Dabov K, Foi A (2006) Image denoising with block-matching and 3D filtering. Electron Imaging 6064:1–12. doi:10.1117/12.643267. http://proceedings.spiedigitallibrary.org/proceeding.aspx?articleid=728740

De Vylder J, Rooms F, Philips W (2010) Automatic detection of cell nuclei in fluorescence micrographs. In: Focus on microscopy 2010, Shanghai

Donoho DL (1992) Nonlinear solution of linear inverse problems by wavelet-vaguelette decomposition. Appl Comput Harmon Anal 2:101–126

Dougherty R (2005) Extensions of DAMAS and benefits and limitations of deconvolution in beamforming. In: AIAA/CEAS aeroacoustics conference, proceedings

Douterloigne K (2015) Camera calibration and intensity based image registration for mapping and other applications. PhD thesis, Ghent University

Elad M, Aharon M (2006) Image denoising via sparse and redundant representations over learned dictionaries. IEEE Trans Image Process 15(12):3736–3745. doi:10.1109/TIP.2006.881969. http://dx.doi.org/10.1109/TIP.2006.881969

Forster B, Van De Ville D, Berent J, Sage D, Unser M (2004) Complex wavelets for extended depth-of-field: a new method for the fusion of multichannel microscopy images. Microsc Res Tech 65(1–2):33–42

Gareau DS, Patel YG, LI Y, Aranda I, Halpern AC, Nehal KS, Rajadhyaksha M (2009) Confocal mosaicing microscopy in skin excisions: a demonstration of rapid surgical pathology. J Micros 233(1):149–159. doi:10.1111/j.1365-2818.2008.03105.x. http://dx.doi.org/10.1111/j.1365-2818.2008.03105.x

Gibson SF, Lanni F (1992) Experimental test of an analytical model of aberration in an oil-immersion objective lens used in three-dimensional light microscopy. J Opt Soc Am 9(1):154–166. doi:10.1364/JOSAA.9.000154. http://josaa.osa.org/abstract.cfm?URI=josaa-9-1-154

Goldstein T, Osher S (2009) The split Bregman method for L1-regularized problems. SIAM J Imaging Sci 2(2):323–343. doi:10.1137/080725891. http://dx.doi.org/10.1137/080725891, http://dx.doi.org/10.1137/080725891

Goossens B, Pizurica A, Philips W (2006) Wavelet domain image denoising for non-stationary noise and signal-dependent noise. In: 2006 IEEE international conference on image processing, pp 1425–1428. doi:10.1109/ICIP.2006.312694

Goossens B, Aelterman J, Luong Q, Pizurica A, Philips W (2009a) Efficient design of a low redundant discrete shearlet transform. In: LNLA: 2009 international workshop on local and non-local approximation in image processing. IEEE, Tuusula, pp 112–124. http://dx.doi.org/10.1109/LNLA.2009.5278394

Goossens B, Pizurica A, Philips W (2009b) Removal of correlated noise by modeling the signal of interest in the wavelet domain. IEEE Trans Image Process 18(6):1153–1165. doi:10.1109/TIP.2009.2017169

Goossens B, De Vylder J, Philips W (2014) Quasar: a new heterogeneous programming framework for image and video processing algorithms on CPU and GPU. In: IEEE international conference on image processing proceedings. IEEE, Paris, pp 2183–2185

Guo K, Kutyniok G, Labate D (2006) Sparse multidimensional representations using anisotropic dilation and shear operators. In: Wavelets and splines, pp 189–201. http://www.home.uos.de/kutyniok/papers/ShearletsContDiscr.pdf; nhttp://www3.math.tu-berlin.de/numerik/mt/mt/www.shearlet.org/papers/SMRuADaSO.pdf

Heintzmann R (2006) Band-limit and appropriate sampling in microscopy. In: Cell biology. A laboratory handbook, vol 3. Academic, Burlington

Jezierska A, Talbot H, Chaux C, Pesquet JC, Engler G (2012) Poisson-Gaussian noise parameter estimation in fluorescence microscopy imaging. In: Proc. IEEE int. symp. biomedical imaging (ISBI), Barcelona

Jia C, Evans B (2011) Patch-based image deconvolution via joint modeling of sparse priors. In: 2011 18th IEEE international conference on image processing, pp 681–684. doi:10.1109/ICIP.2011.6116644

Keller PJ, Schmidt AD, Santella A, Khairy K, Bao ZR, Wittbrodt J, Stelzer EHK (2010) Fast, high-contrast imaging of animal development with scanned light sheet-based structured-illumination microscopy. Nat Methods 7(8):637–U55. <GotoISI>://000280500000009

Köhler A (1893) Ein neues Beleuchtungsverfahren für mikrophotographische Zwecke. Wissenschaftliche Mikroskopie und für Mikroskopische Technik 10 (4), Zeitschrift für

Li S, Yang B, Hu J (2011) Performance comparison of different multi-resolution transforms for image fusion. Inf Fusion 12(2):74–84

Louchet C, Moisan L (2014) Total variation denoising using iterated conditional expectation. In: 2014 Proceedings of the 22nd European signal processing conference (EUSIPCO), pp 1592–1596

Luisier F, Blu T, Unser M (2007) A new SURE approach to image denoising: interscale orthonormal wavelet thresholding. IEEE Trans Image Process 16:593–606

Luisier F, Blu T, Unser M (2011) Image denoising in mixed Poisson-Gaussian noise. IEEE Trans Image Process 20:696–708

Luong H, Goossens B, Philips W (2007) Image upscaling using global multimodal priors. In: Blanc-Talon J, Philips W, Popescu D, Scheunders P (eds) Advanced concepts for intelligent vision systems. Lecture notes in computer science, vol 4678. Springer, Berlin/Heidelberg, pp 473–484. doi:10.1007/978-3-540-74607-2_43. http://dx.doi.org/10.1007/978-3-540-74607-2_43

Nyquist H (1928) Certain topics in telegraph transmission theory. Trans Am Inst Electr Eng 47(2):617–644. doi:10.1109/T-AIEE.1928.5055024

Oliveira JP, Bioucas-Dias JM, Figueiredo MA (2009) Adaptive total variation image deblurring: a majorization–minimization approach. Signal Process 89(9):1683–1693. doi:http://dx.doi.org/10.1016/j.sigpro.2009.03.018. http://www.sciencedirect.com/science/article/pii/S0165168409001224

Pawley JB (2006) Points, pixels and gray levels: digitizing image data. Handbook of biology confocal microscopy, 3rd edn. Springer, New York

Perona P, Malik J (1990) Scale-space and edge detection using anisotropic diffusion. IEEE Trans Pattern Anal Mach Intell 12(7):629–639. doi:10.1109/34.56205

Piccinini F, Lucarelli E, Gherardi A, Bevilacqua A (2012) Multi-image based method to correct vignetting effect in light microscopy images. J Microsc 248(1):6–22. doi:10.1111/j.1365-2818.2012.03645.x. http://dx.doi.org/10.1111/j.1365-2818.2012.03645.x

Portilla J, Strela V, Wainwright MJ, Simoncelli EP (2003) Image denoising using scale mixtures of Gaussians in the wavelet domain. IEEE Trans Image Process 12(11):1338–1351. doi:10.1109/TIP.2003.818640

Ramani S, Vonesch C, Unser M (2008) Deconvolution of 3D fluorescence micrographs with automatic risk minimization. In: 5th IEEE international symposium on biomedical imaging: from nano to macro, 2008. ISBI 2008, pp 732–735. doi:10.1109/ISBI.2008.4541100

Richardson WH (1972) Bayesian-based iterative method of image restoration. J Opt Soc Am 62(1):55–59. doi:10.1364/JOSA.62.000055. http://www.osapublishing.org/abstract.cfm?URI=josa-62-1-55

Roels J, Aelterman J, De Vylder J, Luong H, Saeys Y, Lippens S, Philips W (2014) Noise analysis and removal in 3D electron microscopy. In: Advances in visual computing. Lecture notes in computer science, vol 8887. Springer, Heidelberg, pp 31–40. doi:10.1007/978-3-319-14249-4_4. http://dx.doi.org/10.1007/978-3-319-14249-4_4

Sarder P, Nehorai A (2006) Deconvolution methods for 3-D fluorescence microscopy images. IEEE Signal Process Mag 23(3):32–45. doi:10.1109/MSP.2006.1628876

Schindelin J, Arganda-Carreras I, Frise E, Kaynig V, Longair M, Pietzsch T, Preibisch S, Rueden C, Saalfeld S, Schmid B, Tinevez JY, White DJ, Hartenstein V, Eliceiri K, Tomancak P, Cardona A (2012) Fiji: an open-source platform for biological-image analysis. Nat Methods 9:676–682

Shepp L, Vardi M (1982) Maximum likelihood reconstruction for emission tomography. IEEE Trans Med Imaging 1:113–122

Sibarita JB (2005) Deconvolution microscopy. Adv Biochem Eng Biotechnol 95:201–243

Tao M, Yang J, He B (2009) Alternating direction algorithms for total variation deconvolution in image reconstruction. Research report, Rice University

Tessens L, Ledda A, Pizurica A, Philips W (2007) Extending the depth of field in microscopy through curvelet-based frequency-adaptive image fusion. IEEE international conference on acoustics, speech, and signal processing, pp 861–864

Torricelli G, Argenti F, Alparone L (2002) Modelling and assessment of signal-dependent noise for image de-noising. In: 2002 11th European signal processing conference, pp 1–4

Van Kempen GMP, Van Vliet LJ, Verveer PJ, Van Der Voort HTM (1997) A quantitative comparison of image restoration methods for confocal microscopy. J Microsc 185:354–365

Vonesch C, Unser M (2008) A fast thresholded landweber algorithm for wavelet-regularized multidimensional deconvolution. IEEE Trans Image Process 17(4):539–549

Zabrodina V, Abramov S, Lukin V, Astola J, Vozel B, Chehdi K (2011) Blind estimation of mixed noise parameters in images using robust regression curve fitting. In: 2011 19th European signal processing conference, pp 1135–1139

Zhang B, Fadili J, Starck JL (2008) Wavelets, ridgelets, and curvelets for Poisson noise removal. IEEE Trans Image Process 17(7):1093–1108. doi:10.1109/TIP.2008.924386

Zhu M, Wright SJ, Chan TF (2008) Duality-based algorithms for total variation image restoration. Tech. rep., University of California, Los Angeles

Chapter 3
Transforms and Operators for Directional Bioimage Analysis: A Survey

Zsuzsanna Püspöki, Martin Storath, Daniel Sage, and Michael Unser

Abstract We give a methodology-oriented perspective on directional image analysis and rotation-invariant processing. We review the state of the art in the field and make connections with recent mathematical developments in functional analysis and wavelet theory. We unify our perspective within a common framework using operators. The intent is to provide image-processing methods that can be deployed in algorithms that analyze biomedical images with improved rotation invariance and high directional sensitivity. We start our survey with classical methods such as directional-gradient and the structure tensor. Then, we discuss how these methods can be improved with respect to robustness, invariance to geometric transformations (with a particular interest in scaling), and computation cost. To address robustness against noise, we move forward to higher degrees of directional selectivity and discuss Hessian-based detection schemes. To present multiscale approaches, we explain the differences between Fourier filters, directional wavelets, curvelets, and shearlets. To reduce the computational cost, we address the problem of matching directional patterns by proposing steerable filters, where one might perform arbitrary rotations and optimizations without discretizing the orientation. We define the property of steerability and give an introduction to the design of steerable filters. We cover the spectrum from simple steerable filters through pyramid schemes up to steerable wavelets. We also present illustrations on the design of steerable wavelets and their application to pattern recognition.

3.1 Introduction

Directionality and orientation information is very useful for the quantitative analysis of images. By those terms, we refer to local directional cues and features that one can identify in natural images. The area of applications based on the detection of orientation is continuously growing as the importance of directionality is becoming

Z. Püspöki (✉) • M. Storath • D. Sage • M. Unser
Biomedical Imaging Group, École polytechnique fédérale de Lausanne (EPFL), Station 17, 1015 Lausanne VD, Switzerland
e-mail: zsuzsanna.puspoki@epfl.ch

W.H. De Vos et al. (eds.), *Focus on Bio-Image Informatics*,
Advances in Anatomy, Embryology and Cell Biology 219,
DOI 10.1007/978-3-319-28549-8_3

more and more relevant in image processing. The range of applications spans topics from astronomy (Bernasconi et al. 2005; Yuan et al. 2011; Schuh et al. 2014), aerial and satellite imagery (Tupin et al. 1998; Jiuxiang et al. 2007), material sciences (Dan et al. 2012) to biological and medical applications. Focusing on the last two categories, the palette is quite broad: detection of nodules in the lungs (Agam et al. 2005) and vessels in retinal fundus images (Lam et al. 2010; Patton et al. 2006), bioimaging (Honnorat et al. 2011), neuroimaging (Meijering et al. 2004; Gonzalez et al. 2009). Investigations of collagen in the arterial adventitia also rely on directional analysis (Rezakhaniha et al. 2012). Neuron tracking is of primal importance to understand the development of the brain and requires robust directional image-analysis tools to capture dendrites in 2D and 3D (Meijering 2010). In Jacob et al. (2006), the authors used steerable ridge detector [based on Canny (1986)] to study the aging of elastin in human cerebral arteries. In Aguet et al. (2009), 3D steerable filters were applied to the estimation of orientation and localization of fluorescent dipoles.

Researchers in image analysis are getting inspiration from the human visual system. In the early 1960s, it was demonstrated that directionality plays a key role in visual perception: The neurophysiological findings of Huber and Wiesel initiated a field of research for decades to come (Hubel and Wiesel 1962). Follow-up studies confirmed that the organization of the primary visual cortex makes our visual perception particularly sensitive to the directional clues carried by edges, ridges, and corners (Olshausen and Field 1996; Marr and Hildreth 1980). Our visual system is able to efficiently capture and summarize this information using a small number of neuronal cells.

Based on these structures, many image-analysis methods have been proposed, but they face several challenges. One of them is efficiency with respect to computational resources, because real-time applications and the processing of large multidimensional data (e.g., multichannel time-lapse sequences of images or volumes) demand fast algorithms. Another challenge is to design algorithmic detectors of orientation that acknowledge that patterns in natural images usually have an unknown size and location. Robustness to noise is another desirable trait.

This survey aims at providing the reader with a broad overview of techniques for the directional analysis of images. It is intended to be used as a guide to state-of-the-art methods and techniques in the field. In this paper, we focus on the applications in bioimaging, presenting and comparing the described methods on experimental data.

We focus on the continuous domain setup for explaining the relevant concepts because it allows for convenient, compact, and intuitive formulation. It primarily involves differential and convolution operators (smoothing filters and wavelets) that are acting on continuously defined images, $f(\boldsymbol{x}), \boldsymbol{x} = (x_1, x_2) \in \mathbb{R}^2$. The final transcription of a continuous domain formula into an algorithm requires the discretization of the underlying filters which can be achieved using standard techniques. For instance, partial derivatives can be closely approximated using finite differences, while there are well-established techniques for computing wavelets using digital filters. For further implementation details, we are giving pointers to the specialized literature.

3.2 Derivative-Based Approaches

3.2.1 *Gradient Information and Directional Derivatives*

Some of the earliest and simplest techniques in image analysis to account for orientation rely on gradient information. Intuitively, the direction of the gradient corresponds to the direction of steepest ascent. The local direction of an image f at $\boldsymbol{x}_0$ can be estimated in terms of the direction orthogonal to its gradient. A direction is specified in $\mathbb{R}^2$ by a unit vector $\boldsymbol{u} = (u_1, u_2) \in \mathbb{R}^2$ with $\|\boldsymbol{u}\| = 1$. The first-order directional derivative $\mathrm{D}_{\boldsymbol{u}} f$ along the direction $\boldsymbol{u}$ can be expressed in terms of the gradient

$$\mathrm{D}_{\boldsymbol{u}} f(\boldsymbol{x}_0) = \lim_{h \to 0} \frac{f(\boldsymbol{x}_0) - f(\boldsymbol{x}_0 - h\boldsymbol{u})}{h} = \langle \boldsymbol{u}, \nabla f(\boldsymbol{x}_0) \rangle, \tag{3.1}$$

where the right-hand side is the inner product between $\boldsymbol{u}$ and the gradient vector $\nabla f(\boldsymbol{x}_0)$ evaluated at x_0. We note that (3.1) is maximum when $\boldsymbol{u}$ is collinear to $\nabla f(\boldsymbol{x}_0)$ (by the Cauchy–Schwartz inequality). Conversely, $\mathrm{D}_{\boldsymbol{u}_0} f(\boldsymbol{x}_0)$ vanishes when $\boldsymbol{u}_0 \perp \nabla f$, so that $\boldsymbol{u}_0$ provides us with a local estimate of the directionality of the image. Figure 3.1 illustrates the application of the gradient operators.

Gradient-based orientation estimators are frequently used as they can be discretized and implemented easily. However, the gradient-based estimation of the orientations is sensitive to noise. The robustness can be improved by smoothing the image by a Gaussian kernel before taking the derivative. A still very popular method based on gradients is Canny's classical edge detector (Canny 1986).

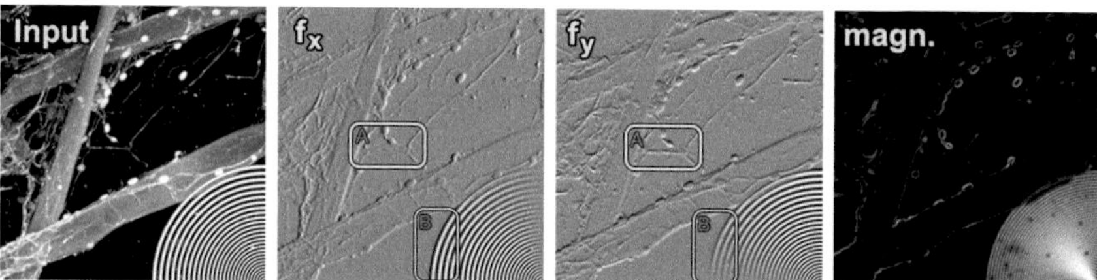

Fig. 3.1 Illustration of the use of gradient operators, from left to right: (1) Input image: confocal micrograph showing nerve cells growing along fibers made from a specially modified silk that is similar to that made by spiders and silkworms. This input image is from the Cell Image Library (http://www.cellimagelibrary.org/images/38921). *Right lower corner*: bright arcs with different scales, artificially added. (2) the x_1 (or horizontal) component of the gradient is the directional derivative along $\boldsymbol{u} = (1, 0)$. (3) the x_2 (or vertical) component of the gradient is the directional derivative along $\boldsymbol{u} = (0, 1)$. (4) Magnitude of the gradient vector. *Highlighted window A*: Horizontal edges are attenuated in case of directional derivative along $\boldsymbol{u} = (1, 0)$ and enhanced/kept in case of directional derivative along $\boldsymbol{u} = (0, 1)$. *Highlighted window B*: Vertical edges are attenuated in case of directional derivative along $\boldsymbol{u} = (0, 1)$ and enhanced/kept in case of directional derivative along $\boldsymbol{u} = (1, 0)$. All the images were produced by the ImageJ/Fiji plugin OrientationJ

3.2.2 Improving Robustness by the Structure Tensor

The estimation of the local orientation using derivatives can be made more robust by using the structure tensor (Jahne 1997). The structure tensor is a matrix derived from the gradient of the image and can be interpreted as a localized covariance matrix of the gradient. Since the pioneering work of Förstner (1986), Bigun (1987), and Harris and Stephens (1988), the structure tensor has become a tool for the analysis of low-level features, in particular for corner and edge detection as well as texture analysis. In 2D, the structure tensor at location $\boldsymbol{x}_0$ is defined by

$$\boldsymbol{J}(\boldsymbol{x}_0) = \int_{\mathbb{R}^2} w(\boldsymbol{x} - \boldsymbol{x}_0)\,(\nabla f(\boldsymbol{x}))\,\nabla^T f(\boldsymbol{x})\mathrm{d}x_1\mathrm{d}x_2, \tag{3.2}$$

where w is a nonnegative isotropic observation window (e.g., a Gaussian) centered at $\boldsymbol{x}_0$. More explicitly, the (2×2) matrix $\boldsymbol{J}(\boldsymbol{x}_0)$ reads

$$\boldsymbol{J}(\boldsymbol{x}_0) = \int_{\mathbb{R}^2} w(\boldsymbol{x} - \boldsymbol{x}_0) \begin{pmatrix} f_{x_1}^2(\boldsymbol{x}) & f_{x_1}(\boldsymbol{x})f_{x_2}(\boldsymbol{x}) \\ f_{x_2}(\boldsymbol{x})f_{x_1}(\boldsymbol{x}) & f_{x_2}^2(\boldsymbol{x}) \end{pmatrix} \mathrm{d}x_1\mathrm{d}x_2 \tag{3.3}$$

$$= \begin{pmatrix} (w * f_{x_1}^2)(\boldsymbol{x}_0) & (w * f_{x_1}f_{x_2})(\boldsymbol{x}_0) \\ (w * f_{x_2}f_{x_1})(\boldsymbol{x}_0) & (w * f_{x_2}^2)(\boldsymbol{x}_0) \end{pmatrix}, \tag{3.4}$$

where $w * f$ denotes the convolution of w and f. The partial derivative of f with respect to some variable x_i is denoted by f_{x_i}. This reveals that $\boldsymbol{J}$ is a smoothed version of

$$\begin{pmatrix} f_{x_1}^2(\boldsymbol{x}) & f_{x_1}(\boldsymbol{x})f_{x_2}(\boldsymbol{x}) \\ f_{x_2}(\boldsymbol{x})f_{x_1}(\boldsymbol{x}) & f_{x_2}^2(\boldsymbol{x}) \end{pmatrix}. \tag{3.5}$$

The eigenvalues of the structure tensor are noted $\lambda_{\max}$ and $\lambda_{\min}$, with $\lambda_{\min}, \lambda_{\max} \in \mathbb{R}$. They carry information about the distribution of the gradient within the window w. Depending on the eigenvalues, one can discriminate between homogenous regions, rotational symmetric regions without predominant direction, regions where the eigenvector is well aligned with one of the gradient directions, or regions where the dominant direction lies in between the gradient directions. For such purpose, two measures are defined, the so-called energy E and the coherence C. The energy is defined based on the eigenvalues of the structure tensor as $E = |\lambda_1| + |\lambda_2|$. If $E \approx 0$, which corresponds to $\lambda_{\max} = \lambda_{\min} \approx 0$, then the region is homogenous. If $E \gg 0$, then the characteristic of the structure is determined by the coherency information. The coherency information C is a measure of confidence, defined as

$$0 \leq C = \frac{\lambda_{\max} - \lambda_{\min}}{\lambda_{\max} + \lambda_{\min}} = \frac{\sqrt{(J_{22} - J_{11})^2 + 4J_{12}^2}}{J_{22} + J_{11}} \leq 1, \tag{3.6}$$

where J_{ij} denotes an element of the structure tensor. If $C \approx 0$, which corresponds to $\lambda_{\max} \approx \lambda_{\min}$, then the region is rotational symmetric without predominant direction, the structure has no orientation. If $C \approx 1$, which corresponds to $\lambda_{\max} > 0, \lambda_{\min} \approx 0$ or $\lambda_{\max} \gg \lambda_{\min}$, the eigenvector is well aligned with one of the gradient directions. For $0 < C < 1$, the predominant orientation lies between the gradient directions. In general, a coherency close to 1 indicates that the structure in the image is locally 1D, a coherency close to 0 indicates that there is no preferred direction.

The energy of the derivative in the direction $\boldsymbol{u}$ can be expressed as

$$\|\mathrm{D}_{\boldsymbol{u}} f\|_w^2 = \langle \boldsymbol{u}^T \nabla f, \boldsymbol{u}^T \nabla f \rangle_w = \boldsymbol{u}^T \langle \nabla f, \nabla f \rangle_w \boldsymbol{u} = \boldsymbol{u}^T \mathbf{J} \boldsymbol{u}. \tag{3.7}$$

This means that, in the window centered around $\boldsymbol{x}_0$, the dominant orientation of the neighborhood can be computed by

$$\boldsymbol{u}_1 = \arg \max_{\|\boldsymbol{u}\|=1} \|\mathrm{D}_{\boldsymbol{u}} f\|_w^2. \tag{3.8}$$

We interpret $\|\mathrm{D}_{\boldsymbol{u}} f\|_w^2$ as the average energy in the window defined by w and centered at $\boldsymbol{x}_0$. Moreover, $\mathrm{D}_{\boldsymbol{u}} f = \langle \nabla f, \boldsymbol{u} \rangle$ is the derivative in the direction of $\boldsymbol{u}$. The maximizing argument corresponds to the eigenvector with the largest eigenvalue of the structure tensor at $\boldsymbol{x}_0$. The dominant orientation of the pattern in the local window w is computed as $\boldsymbol{u}_1 = (\cos\theta, \sin\theta)$, with $\theta = \frac{1}{2} \arctan\left(\frac{2J_{12}}{J_{22} - J_{11}}\right)$.

Figure 3.2 illustrates the improved robustness of the structure tensor in terms of the estimation of the orientation. Figure 3.3 provides another concrete example on the structure-tensor analysis produced by the freely available OrientationJ plugin for Fiji/ImageJ.[1] We have chosen a HSB (hue, saturation, and brightness) cylindrical-coordinate color representation to visualize the results. The HSB components

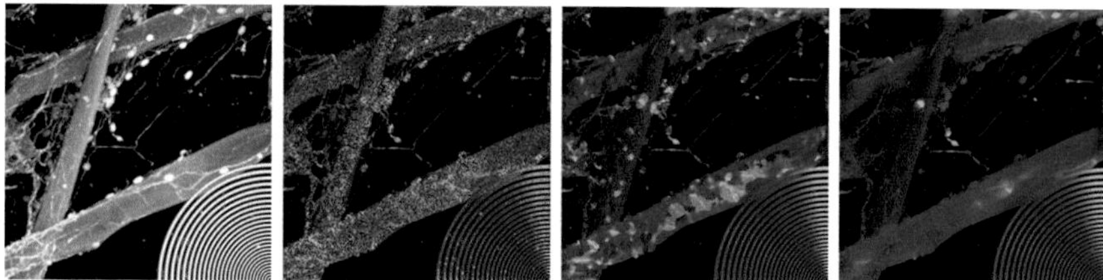

Fig. 3.2 Illustration of the robustness of the structure tensor in terms of estimation of the orientation, from *left to right*: (1) Input image: confocal micrograph, same as the original image in Fig. 3.1. (2) Local dominant orientation, color-coded, no filtering applied. (3) Orientation given by the structure tensor with a small window size (standard deviation of the Gaussian window = 1). (4) Orientation given by the structure tensor large window size (standard deviation of the Gaussian window = 1). All the images were produced by the ImageJ/Fiji plugin OrientationJ

[1] Software available at http://bigwww.epfl.ch/demo/orientation/.

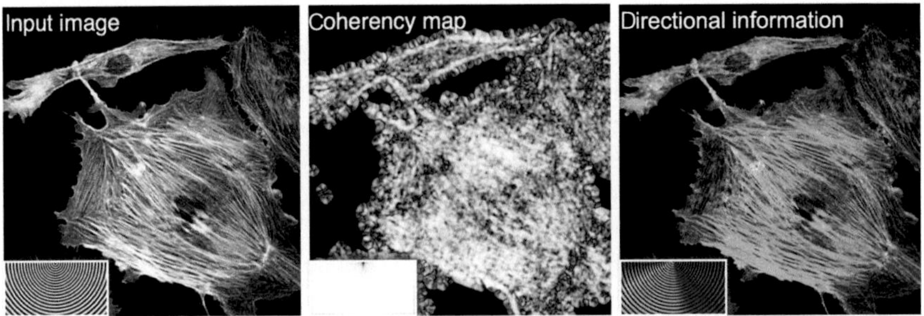

Fig. 3.3 Illustration of the use of structure tensors. Large images, from *left to right*: (1) Input image (800 × 800 pixels): immunofluorescence cytoskeleton (actin fibers), courtesy of Caroline Aemisegger, University of Zürich. (2) Coherency map: coherency values close to 1.0 are represented in white, coherency values close to 0.0 are represented in *black*. (3) Construction of color representation in HSB, H: angle of the orientation, S: coherency, B: input image. Small images in the *left bottom corners*, from *left to right*: (1) Input image: wave pattern with constant wavelength. (2) Coherency map: coherency values are close to 1.0 as expected. (3) The *color representation* reflects the different orientations. All the images were produced by the ImageJ/Fiji plugin OrientationJ

correspond to the following values: angle of the orientation, coherency, and input image, respectively. The advantage of the proposed model is that it gives a direct link between the quantities to display and the color coding. In the cylindrical-coordinate color representation, the angle around the central vertical axis corresponds to hue. The distance along the axis corresponds to brightness, thus we preserve the visibility of the original structures. The distance from the axis corresponds to saturation: the higher the coherency is, the more saturated the corresponding colors are.

In the 3D shape estimation of DNA molecules from stereo cryo-electron micrographs (Fonck et al. 2008), the authors took advantage of its structure-tensor method. Other applications can be found in Köthe (2003) and Bigun et al. (2004).

While simple and computationally efficient, the structure-tensor method has drawbacks: it only takes into account one specific scale, the localization accuracy for corners is low, and the integration of edge and corner detection is ad hoc (e.g., Harris' corner detector).

3.2.3 Higher-Order Directional Structures and the Hessian

To capture higher-order directional structures, the gradient information is replaced by higher-order derivatives. In general, an nth-order directional derivative is associated with n directions. Taking all of these to be the same, the directional derivative of order n in $\mathbb{R}^2$ is defined as

$$\mathrm{D}_{\boldsymbol{u}}^n f(\boldsymbol{x}) = \sum_{k=0}^{n} \binom{n}{k} u_1^k u_2^{n-k} \partial_{x_1}^k \partial_{x_2}^{n-k} f(\boldsymbol{x}), \tag{3.9}$$

which is a linear combination of partial derivatives of order n. More specifically, if we fix $n = 2$ and the unit vector $\boldsymbol{u}_\theta = (\cos\theta, \sin\theta)$, we obtain

$$\mathrm{D}^2_{\boldsymbol{u}_\theta} f(\boldsymbol{x}) = \cos^2(\theta)\,\partial^2_{x_1} f(\boldsymbol{x}) + 2\cos(\theta)\sin(\theta)\,\partial_{x_1}\partial_{x_2} f(\boldsymbol{x}) + \sin^2(\theta)\,\partial^2_{x_2} f(\boldsymbol{x}). \tag{3.10}$$

The Hessian filter is a square matrix of second-order partial derivatives of a function. For example, in 2D, the smoothed Hessian matrix, useful for ridge detection at location $\boldsymbol{x}_0$, can be written as

$$\mathbf{H}(x_0) = \begin{pmatrix} (w_{11} * f)(\boldsymbol{x}_0) & (w_{12} * f)(\boldsymbol{x}_0) \\ (w_{21} * f)(\boldsymbol{x}_0) & (w_{22} * f)(\boldsymbol{x}_0) \end{pmatrix}, \tag{3.11}$$

where w is a smoothing kernel and $w_{\mathrm{ij}} = \partial_{x_{\mathrm{i}}}\partial_{x_{\mathrm{j}}} w$ denotes its derivatives with respect to the coordinates x_{i} and x_{j}. In the window centered around $\boldsymbol{x}_0$, the dominant orientation of the ridge is

$$\boldsymbol{u}_2 = \arg\max_{\|\boldsymbol{u}\|=1} \left(\boldsymbol{u}^T \mathbf{H} \boldsymbol{u}\right). \tag{3.12}$$

The maximizing argument corresponds to the eigenvector with the largest eigenvalue of the Hessian at $\boldsymbol{x}_0$. The eigenvectors of the Hessian are orthogonal to each other, so the eigenvector with the smallest eigenvalue corresponds to the direction orthogonal to the ridge.

A sample application of the Hessian filter is vessel enhancement (Frangi et al. 1998). There, the authors define a measure called vesselness which corresponds to the likeliness of an image region to contain vessels or other image ridges. The vesselness measure is derived based on the eigenvalues of the steerable Hessian filter. In 2D, a vessel is detected when one of the eigenvalues is close to zero ($\lambda_1 \approx 0$) and the other one is much larger $|\lambda_2| \gg |\lambda_1|$. The direction of the ridge is given by the eigenvector of the Hessian filter output corresponding to λ_1. In (Frangi et al. 1998), the authors define the measure of vesselness as

$$V(\boldsymbol{x}) = \begin{cases} 0, & \text{if } \lambda_1 > 0 \\ \exp\left(-\frac{(\lambda_1/\lambda_2)^2}{2\beta_1}\right)\left(1 - \exp\left(-\frac{\lambda_1^2+\lambda_2^2}{2\beta_2}\right)\right), & \text{otherwise,} \end{cases} \tag{3.13}$$

where β_1 and β_2 control the sensitivity of the filter.[2] A particular application of the vesselness index on filament enhancement is shown in Fig. 3.4. Alternative vesselness measures based on the Hessian have been proposed by Lorenz et al. (1997) and Sato et al. (1998).

[2]Plugin available at http://fiji.sc/Frangi/.

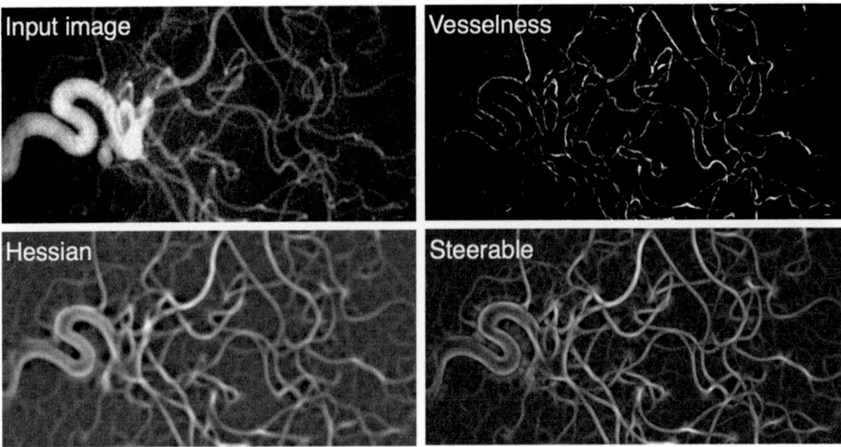

Fig. 3.4 Rotation-invariant enhancement of filaments. From *top to bottom*, *left to right*: (1) Input image (512 × 256 pixels) with neuron, cell body, and dendrites (maximum-intensity projection of a z-stack, fluorescence microscopy, inverted scale). (2) Output of the Hessian filter. The largest eigenvalue of the Hessian matrix was obtained after a Gaussian smoothing (standard deviation = 5). The image was produced using the ImageJ/Fiji plugin FeatureJ available at: http://www.imagescience.org/meijering/software/featurej/. (3) Output of the vesselness index obtained by the Fiji plugin Frangi-Vesselness. (4) Output of the steerable filters (Gaussian-based, 4th order). The image was produced using the ImageJ/Fiji plugin SteerableJ

3.3 Directional Multiscale Approaches

In natural images, oriented patterns are typically living on different scales, for example, thin and thick blood vessels. To analyze them, methods that extract oriented structures separately at different scales are required. The classical tools for a multiscale analysis are wavelets. In a nutshell, a wavelet is a bandpass filter that responds almost exclusively to features of a certain scale. The separable wavelet transform that is commonly used is computationally very efficient but provides only limited directional information. Its operation consists of filtering with 1D wavelets with respect to the horizontal and vertical directions. As a result, two pure orientations (vertical and horizontal) and a mixed channel of diagonal directions are extracted. Using the dual-tree complex wavelet transform (Kingsbury 1998),[3] one can increase the number of directions to six while retaining the computational efficiency of the separable wavelet transform. [We refer to Selesnick et al. (2005) for a detailed treatment of this transform.] Next, we describe how to achieve wavelets with an even higher orientational selectivity at the price of higher computational costs.

[3] Available at http://eeweb.poly.edu/iselesni/WaveletSoftware/.

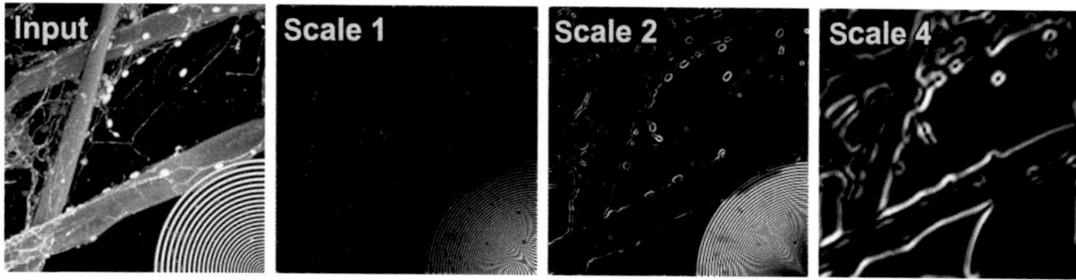

Fig. 3.5 Illustration of the gradient at different scales, from *left to right*: (1) Input image: confocal micrograph, same as the original image in Fig. 3.1. (2) Magnitude of the gradient at scale 1. (3) Magnitude of the gradient at scale 2. (4) Magnitude of the gradient at scale 4. All the images were produced by the ImageJ/Fiji plugin OrientationJ

Figure 3.5 illustrates the gradient at different scales. We can observe that different features are kept at different scales.

3.3.1 Construction of Directional Filters in the Fourier Domain

In order to construct orientation-selective filters, methods based on the Fourier transform are powerful. From now on, we denote the Cartesian and polar representations of the same 2D function f by $f(\boldsymbol{x})$ with $\boldsymbol{x} \in \mathbb{R}^2$ and $f_{\text{pol}}(r,\theta)$ with $r \in \mathbb{R}^+$, $\theta \in [0, 2\pi)$ [similarly in the Fourier domain: $\hat{f}(\boldsymbol{\omega})$ and $\hat{f}_{\text{pol}}(\rho,\phi)$]. The key property for directional analysis is that rotations in the spatial domain propagate as rotations to the Fourier domain. Formally, we write that

$$f(\mathbf{R}_\theta \boldsymbol{x}) \overset{\mathscr{F}}{\longleftrightarrow} \hat{f}(\mathbf{R}_\theta \boldsymbol{\omega}), \tag{3.14}$$

where $\mathbf{R}_\theta$ denotes a rotation by the angle θ. The construction is based on a filter ψ whose Fourier transform $\hat{\psi}$ is supported on a wedge around the ω_1 axis; see Fig. 3.6. In order to avoid favoring special orientations, one typically requires that $\hat{\psi}$ be nonnegative and that it forms (at least approximately) a partition of unity of the Fourier plane under rotation, like

$$\sum_{\theta_i} |\hat{\psi}(\mathbf{R}_{\theta_i}\boldsymbol{\omega})|^2 = 1, \text{ for all } \boldsymbol{\omega} \in \mathbb{R}^2 \setminus \{\mathbf{0}\}. \tag{3.15}$$

Here, $\theta_1, \ldots, \theta_n$ are arbitrary orientations which are typically selected to be equidistant, with $\theta_i = (i-1)\pi/n$. To get filters that are well localized in the spatial domain, one chooses $\hat{\psi}$ to be a smooth function; for example, the Meyer window function (Daubechies 1992; Ma and Plonka 2010). A directionally filtered image f_{θ_i} can be easily computed by rotating the window $\hat{\psi}$ by θ_i and multiplying it with the

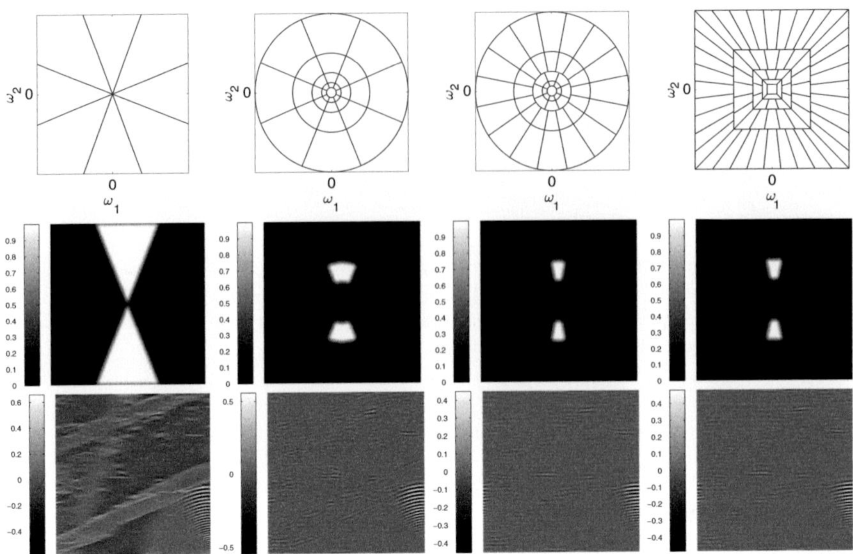

Fig. 3.6 *Top row (from left to right):* schematic tilings of the frequency plane by Fourier filters, directional wavelets, curvelets, and shearlets (the origin of the Fourier domain lies in the center of the images); *Middle row:* a representative Fourier multiplier. *Bottom row:* Corresponding filtering result for the image of Fig. 3.1. The Fourier filter extracts oriented patterns at all scales whereas the wavelet-type approaches are sensitive to oriented patterns of a specific scale. Curvelets and shearlets additionally increase the directional selectivity at the finer scales

Fourier transform $\hat{f}$ of the image, and by transforming back to the spatial domain. This is written

$$f_{\theta_i}(\boldsymbol{x}) = \mathscr{F}^{-1}\{\hat{\psi}(\mathbf{R}_{\theta_i}\cdot)\hat{f}\}(\boldsymbol{x}). \tag{3.16}$$

[We refer to Chaudhury et al. (2010) for filterings based on convolutions in the spatial domain.] The resulting image f_{θ_i} contains structures that are oriented along the direction θ_i. The local orientation θ is given by the orientation of the maximum filter response

$$\theta(\boldsymbol{x}) = \arg\max_{\theta_i} |f_{\theta_i}(\boldsymbol{x})|. \tag{3.17}$$

Such directional filters have been used in fingerprint enhancement (Sherlock et al. 1994) and in crossing-preserving smoothing of images (Franken et al. 2007; Franken and Duits 2009).

3.3.2 Directional Wavelets with a Fixed Number of Directions

Now we augment the directional filters by scale-selectivity. Our starting point is the radial windowing function of (3.15). The simplest way to construct a directional wavelet transform is to partition the Fourier domain into dyadic frequency bands ("octaves"). To ensure a complete covering of the frequency plane, we postulate again nonnegativity and a partition-of-unity property of the form

$$\sum_{s\in\mathbb{Z}}\sum_{\theta_i}|\hat{\psi}(2^{-s}\mathbf{R}_{\theta_i}\boldsymbol{\omega})|^2 = 1, \text{ for } \boldsymbol{\omega}\in\mathbb{R}^2\setminus\{\mathbf{0}\}. \tag{3.18}$$

Classical examples of this type are the Gabor wavelets that cover the frequency plane using Gaussian windows which approximate (rescaled) partition-of-unity (Mallat 2008; Lee 1996). These serve as model for the filters in the mammalian visual system (Daugman 1985, 1988). Alternative constructions are Cauchy wavelets (Antoine et al. 1999) or constructions based on the Meyer window functions (Daubechies 1992; Ma and Plonka 2010). We refer to Vandergheynst and Gobbers (2002) and Jacques et al. (2011) for further information on the design of directional wavelets. In particular, sharply direction-selective Cauchy wavelets have been used for symmetry detection (Antoine et al. 1999).

3.3.3 Curvelets, Shearlets, Contourlets, and Related Transforms

Over the past decade, curvelets (Candès and Donoho 2004), shearlets (Labate et al. 2005; Yi et al. 2009; Kutyniok and Labate 2012), and contourlets (Do and Vetterli 2005) have attracted a lot of interest. They are constructed similarly to the directional wavelets. The relevant difference in this context is that they increase the directional selectivity on the finer scales according to a parabolic scaling law. This means that the number of orientations is increased by a factor of about $\sqrt{2}$ at every scale or by 2 at every other scale; see Fig. 3.6. Therefore, they are collectively called parabolic molecules (Grohs and Kutyniok 2014). Curvelets are created by using a set of basis functions from a series of rotated and dilated versions of an anisotropic mother wavelet to approximate rotation and dilation invariance. Contourlets use a tree-structured filterbank to reproduce the same frequency partitioning as curvelets. Their structure is more flexible, enabling different subsampling rates. To overcome the limitations of the Cartesian grid (i.e., exact rotation invariance is not achievable on it), shearlets are designed in the discrete Fourier domain with constraints on exact sheer invariance.

These transforms are well suited to the analysis and synthesis of images with highly directional features. Applications include texture classification of tissues in computed tomography (Semler and Dettori 2006), texture analysis (Dong et al.

2015), image denoising (Starck et al. 2002), contrast enhancement (Starck et al. 2003), and reconstruction in limited-angle tomography (Frikel 2013). Furthermore, they are closely related to a mathematically rigorous notion of the orientation of image features, the so-called wavefront set (Candès and Donoho 2005; Kutyniok and Labate 2009). Loosely speaking, the wavefront set is the collection of all edges along with their normal directions. This property is used for the geometric separation of points from curvilinear structures, for instance, to separate spines and dendrites (Kutyniok 2012) and for edge detection with resolution of overlaying edges (Yi et al. 2009; Guo et al. 2009; Storath 2011b). We show in Fig. 3.7 the result of the curvelet/shearlet-based edge-detection scheme of Storath (2011b) which is obtained as follows: For every location (pixel) $\boldsymbol{b}$ and every available orientation θ, the rate of decay $d_{\boldsymbol{b},\theta}$ of the absolute values of the curvelet/shearlet coefficients over the scale is computed. The reason for computing the rate of decay of the coefficients is their connection to the local regularity: the faster the decay rate, the smoother the image at location $\boldsymbol{b}$ and orientation θ (see Candès and Donoho 2005; Kutyniok and Labate 2009; Guo et al. 2009). We denote the curvelet/shearlet coefficients at scale a, location $\boldsymbol{b}$, and orientation θ by $c_{a,\boldsymbol{b},\theta}$. Then, $d_{\boldsymbol{b},\theta}$ corresponds to the least-squares fit to the set of constraints $|c_{a,\boldsymbol{b},\theta}| = C'_{\boldsymbol{b},\theta} a^{d_{\boldsymbol{b},\theta}}$, where a runs over all available scales (in the example of Fig. 3.7, $a = 2^{-s/3}$ with $s = 0, \ldots, 15$). Note that this reduces to solving a system of linear equations in terms of $\log C'_{\boldsymbol{b},\theta}$ and $d_{\boldsymbol{b},\theta}$, after having taken a logarithm on both sides. Having computed d, we perform for each orientation θ a non-maximum suppression on d; that is, we set to $(-\infty)$ all pixels that are not a local maximum of the image $d_{\cdot,\theta}$ with respect to the direction θ. Finally, a threshold is applied and the connected components of the (3D-array) d are determined (and colored). The image displayed in Fig. 3.7 is the maximum-intensity projection of the three-dimensional image d with respect to the θ component.

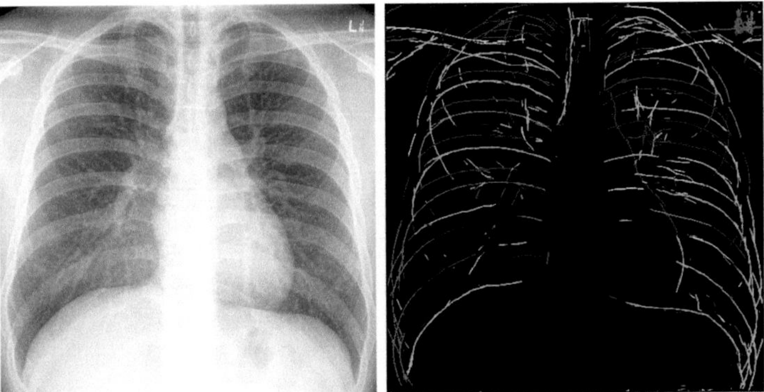

Fig. 3.7 Edge detection with resolution of crossing edges using the curvelet transform. The *colors* correspond to connected edge segments. Note that crossing edges are resolved, for instance near the shoulder bones. [Original image courtesy of Dr. Jeremy Jones, Radiopaedia.org]

Relevant software packages implementing these transforms are the Matlab toolboxes CurveLab,[4] ShearLab,[5] FFST,[6] and the 2D Shearlet Toolbox.[7]

3.4 Steerable Filters

For the purpose of detecting or enhancing a given type of directional pattern (edge, line, ridge, corner), a natural inclination is to try to match directional patterns. The simplest way to do that is to construct a template and try to align it with the pattern of interest. Usually, such algorithms rely on the discretization of the orientation. To obtain accurate results, a fine discretization is required. In general, Fourier filters and wavelet transforms are computationally expensive in this role because a full 2D filter operation has to be computed for each discretized direction. However, an important exception is provided by steerable filters, where one may perform arbitrary (continuous) rotations and optimizations with a substantially reduced computational overhead. The basics of steerability were formulated by Freeman and Adelson in the early 1990s (Freeman and Adelson 1990; Freeman 1992; Freeman and Adelson 1991) and developed further by Perona (1992), Simoncelli and Farid (1996), Unser and Chenouard (2013), Unser and Van De Ville (2010), Ward et al. (2013), and Ward and Unser (2014). We now explain the property of steerability and show the development of steerable wavelets.

A function f on the plane is steerable in the finite basis $\{f_1, \ldots, f_N\}$ if, for any rotation matrix $\mathbf{R}_{\theta_0}$, we can find coefficients $c_1(\theta_0), \ldots, c_N(\theta_0)$ such that

$$f(\mathbf{R}_{\theta_0}\boldsymbol{x}) = \sum_{n=1}^{N} c_n(\theta_0) f_n(\boldsymbol{x}). \tag{3.19}$$

It means that a function f in $\mathbb{R}^2$ is steerable if all of its rotations can be expressed in the same finite basis as the function itself. Thus, any rotation of f can be computed with a systematic modification (i.e., matrix multiplication) of the coefficients. The importance of this property is that, when doing pattern matching, it is enough to compute the coefficients only once, for one particular angle. Based on that, one can then easily determine the coefficients for arbitrary angles. A simple example of steerable functions are the polar-separable functions, whose amplitude (radial part) is 1 and angular part is $\cos(\theta)$ or $\sin(\theta)$. All rotations of these functions can be

[4] Available at http://www.curvelet.org/.

[5] Available at http://www.shearlab.org/.

[6] Available at http://www.mathematik.uni-kl.de/imagepro/members/haeuser/ffst/.

[7] Available at http://www.math.uh.edu/~dlabate/software.html.

expressed in the basis $\{\sin(\theta), \cos(\theta)\}$. The rotations of $\cos(\theta)$ and $\sin(\theta)$ by θ_0 are determined by

$$\begin{pmatrix} \cos(\theta + \theta_0) \\ \sin(\theta + \theta_0) \end{pmatrix} = \begin{pmatrix} \cos(\theta_0) & -\sin(\theta_0) \\ \sin(\theta_0) & \cos(\theta_0) \end{pmatrix} \begin{pmatrix} \cos(\theta) \\ \sin(\theta) \end{pmatrix}. \tag{3.20}$$

Instead of setting the amplitude to 1, one can choose any nonvanishing isotropic function for the radial part. Also, replacing sin and cos with exponentials will preserve the property (since $e^{j(\theta+\theta_0)} = e^{j\theta_0}e^{j\theta}$).

The simplest examples of steerable filters are the ones that are based on the gradient or the Hessian. Starting from an isotropic lowpass function $\varphi(x_1, x_2)$, one can create a subspace of steerable derivative-based templates which can serve as basic edge or ridge detectors. In 2D, let $\varphi_{k,l} = \partial^k_{x_1}\partial^l_{x_2}\varphi$ be anisotropic derivatives of the isotropic function φ. By the chain rule of differentiation, for any rotation matrix $\mathbf{R}_{\theta_0}$, the function $\partial^k_{x_1}\partial^l_{x_2}\varphi(\mathbf{R}_{\theta_0}\cdot)$ can be written as a linear combination of $\varphi_{i,j}$ with $i + j = k + l$. Therefore, any anisotropic filter of the form

$$h(x_1, x_2) = \sum_{m=1}^{M} \sum_{k+l=m} \alpha_{k,l}\varphi_{k,l}(x_1, x_2) \tag{3.21}$$

is steerable. Consequently, for any rotation matrix $\mathbf{R}_{\theta_0}$, an application of the rotated filter to an image f yields

$$(f * h(\mathbf{R}_\theta\cdot))(\boldsymbol{x}) = \sum_{m=1}^{M} \sum_{k+l=m} \alpha_{k,l}(\theta) f_{k,l}(\boldsymbol{x}), \tag{3.22}$$

where $f_{k,l} = f * \partial^k_{x_1}\partial^l_{x_2}\varphi$ and $\alpha_{k,l}(\theta)$ is a trigonometric polynomial in $\cos(\theta)$ and $\sin(\theta)$. Once every $f_{k,l}$ is precomputed, the linear combination (3.22) allows us to quickly evaluate the filtering of the image by the anisotropic filter rotated by any angle. We can then „steer" h by manipulating θ, typically to determine the direction along which the response is maximized and across which is minimized.

Figure 3.4 contains the outcome of three derivative-based methods used to enhance filaments in a rotation-invariant way. The computation complexity is the same for the different methods, with approximatively the same number of filters with the same computation time. The directionality is best captured in the steerable case.

Jacob and Unser (2004) improved the basic steerable filters by imposing Canny-like criteria of optimality (Canny 1986) on the properties of the detector: reasonable approximation of the ideal detector, maximum signal-to-noise ratio, good spatial localization, and reduced oscillations. Their formalism boils down to a constrained optimization of the expansion coefficients $\alpha_{k,i}$ using Lagrange multipliers.

3.5 Steerable Multiscale Approaches

In Simoncelli and Freeman (1995), the authors proposed a new take on steerable filters: the steerable pyramid. The goal of their design was to combine steerability with a multiscale detection scheme. His pioneering work had many successful applications: contour detection (Perona 1992), image filtering and denoising (Bharath and Ng 2005), orientation analysis (Simoncelli and Farid 1996), and texture analysis and synthesis (Portilla and Simoncelli 2000). In Karssemeijer and te Brake (1996) multiscale steerable filters were involved in the detection of stellate distortions in mammograms. Classical multiresolution steerable methods use a purely discrete framework with no functional analytic counterpart. They are not amenable to extensions to dimensions higher than two. Fortunately, it is possible to address these limitations. In this section we overview a continuous-domain formulation that extends the technique proposed by Simoncelli and Freeman (1995). Multiresolution directional techniques were motivated by their invariance with respect to primary geometric transformations: translation, dilation, and rotation. Translation and dilation invariance is satisfied by the application of the wavelet transform. Rotation invariance is achieved by the Riesz transform, which also gives a connection to gradient-like signal analysis (Held et al. 2010).

3.5.1 *The Riesz Transform*

The complex Riesz transform was introduced to the literature by Larkin et al. (2001) and Larkin (2001) as a multidimensional extension of the Hilbert transform. The Hilbert transform is a 1D shift-invariant operator that maps all cosine functions into sine functions without affecting their amplitude (allpass filter). Expressed in the Fourier domain, the Hilbert transform of a function f is

$$\mathscr{F}\{\mathscr{H}\{f\}\}(\omega) = -\frac{\mathrm{j}\omega}{|\omega|}\hat{f}(\omega) = -\mathrm{j}\,\mathrm{sgn}(\omega)\hat{f}(\omega). \tag{3.23}$$

Similarly to the Hilbert transform, the Riesz transform is defined in the Fourier domain as

$$\mathscr{F}\{\mathscr{R}\{f\}\}(\boldsymbol{\omega}) = \frac{(\omega_x + \mathrm{j}\omega_y)}{\|\boldsymbol{\omega}\|}\hat{f}(\boldsymbol{\omega}) = \mathrm{e}^{\mathrm{j}\phi}\hat{f}_{\mathrm{pol}}(\rho, \phi), \tag{3.24}$$

where the subscript "pol" denotes the polar representation of the function. The transform is a convolution-type operator that also acts as an allpass filter, with a phase response that is completely encoded in the orientation.

The Riesz transform is translation- and scale-invariant since

$$\forall \boldsymbol{x}_0 \in \mathbb{R}^d, \quad \mathscr{R}\{f(\cdot - \boldsymbol{x}_0)\}(\boldsymbol{x}) = \mathscr{R}f(\boldsymbol{x} - \boldsymbol{x}_0) \tag{3.25}$$

$$\forall a \in \mathbb{R}^+ \setminus \{0\}, \quad \mathscr{R}\left\{f\left(\frac{\cdot}{a}\right)\right\}(\boldsymbol{x}) = \mathscr{R}f\left(\frac{\boldsymbol{x}}{a}\right). \tag{3.26}$$

The Riesz transform is also rotation-invariant.

The nth-order complex 2D Riesz transform $\mathscr{R}^n$ represents the n-fold iterate of $\mathscr{R}$, defined in the Fourier domain as

$$\mathscr{F}\{\mathscr{R}^n\{f\}\}(\rho\cos\phi, \rho\sin\phi) = \mathrm{e}^{\mathrm{j}n\phi}\hat{f}_{\mathrm{pol}}(\rho, \phi). \tag{3.27}$$

The nth order Riesz transform decomposes a 2D signal into $n + 1$ distinct components. It inherits the invariance properties of the Riesz transform since they are preserved through iteration. This means that we can use the Riesz transform to map a set of primary wavelets into an augmented one while preserving the scale- and shift-invariant structure.

3.5.2 Connection to the Gradient and Directional Derivatives

In this section, we describe the connection between the Riesz transform, the directional Hilbert transform, the gradient, and the directional derivatives. Assuming a zero-mean function f, the Riesz transform is related to the complex gradient operator as

$$\mathscr{R}f(x_1, x_2) = -\mathrm{j}\left(\frac{\partial}{\partial x_1} + \mathrm{j}\frac{\partial}{\partial x_2}\right)(-\Delta)^{-1/2}f(x_1, x_2). \tag{3.28}$$

Here, $(-\Delta)^{\alpha}, \alpha \in \mathbb{R}^+$ is the isotropic fractional differential operator of order 2α. Conversely, the corresponding fractional integrator of order 2α is $(-\Delta)^{-\alpha}, \alpha \in \mathbb{R}^+$. The value $\alpha = 1/2$ is of special interest, providing the link between the Riesz transform and the complex gradient operator. The integral operator acts on all derivative components and has an isotropic smoothing effect, thus, the Riesz transform acts as the smoothed version of the image gradient.

Assuming a zero-mean function f, the high-order Riesz transform is related to the partial derivatives of f by

$$\mathscr{R}^n f(x_1, x_2) = (-\Delta)^{-\frac{n}{2}} \sum_{n_1=0}^{n} \binom{n}{n_1} (-\mathrm{j})^{n_1} \partial_{x_1}^{n_1} \partial_{x_2}^{n-n_1} f(x_1, x_2). \tag{3.29}$$

The fractional integrator acts as an isotropic lowpass filter whose smoothing strength increases with n. The Riesz transform captures the same directional information as derivatives. However, it has the advantage of being better conditioned since, unlike them, it does not amplify the high frequencies.

The directional Hilbert transform is the Hilbert transform along a direction $\boldsymbol{u}$. It is related to the Riesz transform by

$$\mathscr{H}_{\boldsymbol{u}_\theta} f(\boldsymbol{x}) = \cos\theta f_1(\boldsymbol{x}) + \sin\theta f_2(\boldsymbol{x}), \tag{3.30}$$

where $f_1 = \mathrm{Re}(\mathscr{R}f)$ and $f_2 = \mathrm{Im}(\mathscr{R}f)$ are the real and imaginary parts of $\mathscr{R}f$. Assuming again a zero-mean function f, the directional Hilbert transform is related to the derivative in the direction $\boldsymbol{u}$ by

$$\mathscr{H}_{\boldsymbol{u}} f(\boldsymbol{x}) = -(-\Delta)^{-\frac{1}{2}} \mathrm{D}_{\boldsymbol{u}} f(\boldsymbol{x}). \tag{3.31}$$

Here, the operator $\mathrm{D}_{\boldsymbol{u}}$ is the one defined in (3.1). This result corresponds to the interpretation that the Hilbert transform acts as a lowpass-filtered version of the derivative operator. The n-fold version of the directional Hilbert transform acting on a zero-mean function f along the direction specified by $\boldsymbol{u}$ can be expressed in terms of the partial derivatives of f as

$$\mathscr{H}_{\boldsymbol{u}}^n f(\boldsymbol{x}) = (-1)^n (-\Delta)^{-\frac{n}{2}} \mathrm{D}_{\boldsymbol{u}}^n f(\boldsymbol{x}). \tag{3.32}$$

3.5.3 *Steerable Wavelets*

In this section, we present the construction of steerable wavelet frames that are shaped to capture the local orientation of features in images within a multiresolution hierarchy. Their construction has two main parts: first, generation of circular harmonic wavelet frames by applying the multiorder complex Riesz transform on a bandlimited isotropic mother wavelet; second, shaping of the wavelet frames to a particular desired profile with an orthogonal transform.

We start our construction of steerable wavelet frames from a bandlimited isotropic mother wavelet in $L_2(\mathbb{R}^2)$, denoted by ψ, whose shifts and dilations form a wavelet frame (e.g., Simoncelli's wavelet). This isotropic wavelet at scale s and grid point (location) $\boldsymbol{x}_0 = 2^s\boldsymbol{k}$, $\boldsymbol{k} \in \mathbb{Z}^2$ (in 2D) takes the form of

$$\psi_s(\boldsymbol{x} - \boldsymbol{x}_0) = \frac{1}{2^s}\psi\left(\tfrac{\boldsymbol{x}-\boldsymbol{x}_0}{2^s}\right) = \frac{1}{2^s}\psi\left(\tfrac{\boldsymbol{x}}{2^s} - \boldsymbol{k}\right). \tag{3.33}$$

We then apply the multiorder complex Riesz transform on $\psi_s(\cdot - \boldsymbol{x}_0)$. The transform preserves the frame properties. Thus, by choosing N distinct values for the integer n (distinct set of harmonics), one can form a frame of steerable wavelets, referred to as circular harmonic wavelets (Jacovitti and Neri 2000). An nth-order harmonic wavelet has a rotational symmetry of order n around its center, corresponding to the nth-order rotational symmetry of $\mathrm{e}^{\mathrm{j}n\phi}$. The illustration of

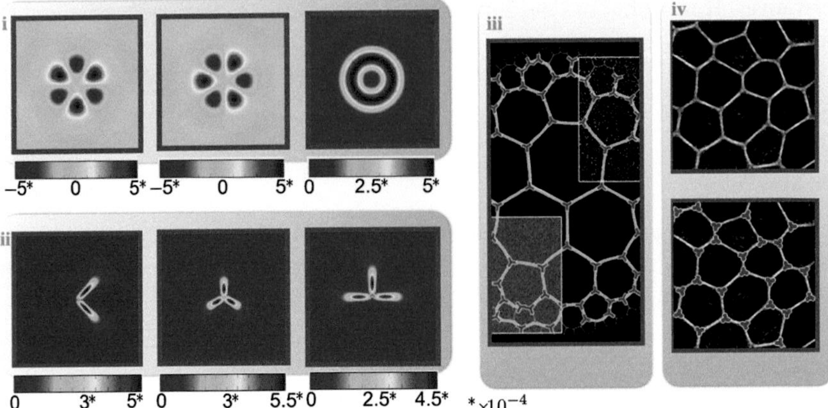

Fig. 3.8 From *left to right*: (1) Circular harmonic wavelets of order three, real part, imaginary part, and magnitude; (2) steerable wavelets (different shapes, magnitude); (3) detections in synthetic data exhibiting multiple scales; (4) original image and detections in a micrograph of embryonic stem cells

circular harmonics for order three is presented in Fig. 3.8 (i). An application of circular harmonic wavelets on local symmetry detection is presented in Püspöki and Unser (2015).

3.5.4 Detection of Junctions

An important step in the analysis of microscopic images is the detection of key points, or junctions of coinciding branches. The automatic detection of these junctions can facilitate further image-processing tasks such as cell segmentation, counting of cells, or image statistics. M-fold symmetric structures (including the case of ridges, assimilated to twofold symmetries) are present in filaments, fibers, membranes, or endothelial cells (e.g., in the eyes). The difficulty in the detection of these junctions is twofold. First, they can appear at arbitrary orientation and scale; second, biological micrographs are frequently contaminated by local variations in intensity and measurement noise. With the modified wavelet schemes presented above, one can design an efficient detector of the location and orientation of local symmetries.

From the circular harmonic wavelet representation, one constructs new steerable representations by using an orthogonal shaping matrix $\mathbf{U}$ to define new steerable-frame functions (Unser and Chenouard 2013). The role of the shaping matrix $\mathbf{U}$

is to give the wavelet functions a desired angular profile. The transform can be formulated as

$$\begin{bmatrix} \xi_{s,\boldsymbol{k}}^{(n_1)} \\ \vdots \\ \xi_{s,\boldsymbol{k}}^{(n_N)} \end{bmatrix} = \mathbf{U} \begin{bmatrix} \psi_{s,\boldsymbol{k}}^{(1)} \\ \vdots \\ \psi_{s,\boldsymbol{k}}^{(N)} \end{bmatrix}, \tag{3.34}$$

where $\psi_{s,\boldsymbol{k}}^{(n)} = \mathscr{R}^n \psi_{s,\boldsymbol{k}}$ is the nth-order circular harmonic wavelet at scale s and location $\boldsymbol{k}$, $\mathbf{U}$ is an orthonormal transformation matrix, and $\{\xi_{s,\boldsymbol{k}}^{(n)}\}$ are the new wavelet channels at scale s and location $\boldsymbol{k}$. The number of channels corresponds to the number of harmonics. The new wavelets span the same space as the wavelet frame $\psi_{s,\boldsymbol{k}}^{(n)}$.

Specific detectors are designed by defining the right weights for $\mathbf{U}$. Typically, the process involves an optimization problem, either in the space or in the Fourier domain. The list of coefficients to optimize can be expressed as a vector that takes the form $\boldsymbol{u} = (u_1, \ldots, u_N)$, with $\boldsymbol{u}\boldsymbol{u}^{\mathbf{H}} = 1$. One can specify a quadratic energy term to minimize in the space domain like

$$E[\xi] = \tfrac{1}{2\pi} \int_0^\infty \int_{-\pi}^{\pi} |\xi(r,\theta)|^2 v(\theta) r \mathrm{d}\theta \mathrm{d}r, \tag{3.35}$$

or (for symmetric patterns) in the Fourier domain like

$$E[\hat{\xi}] = \frac{1}{2\pi} \int_0^\infty \int_{-\pi}^{+\pi} |\hat{\xi}(\rho,\phi)|^2 v(\phi) \rho \mathrm{d}\phi \mathrm{d}\rho. \tag{3.36}$$

The angular weighting function $v(\theta) \geq 0$ or $v(\phi) \geq 0$ should have minima on the unit circle that enforces the concentration of the energy along the desired pattern (for instance, symmetric shape or T-shape). Minimizing E will thus force the solution ξ to be localized at the corresponding angles. Once the mother wavelet ξ is found, its translates and dilates naturally share the same optimal angular profile around their center. By expanding ξ as $\sum_n u_n \psi^{(n)}$ and imposing unit norm on $\boldsymbol{u}$, this formalism leads to a quadratic optimization problem with quadratic constraints that can be solved through eigen decomposition. In Fig. 3.8 (ii) are shown the amplitude of three different detectors that one can design with the proposed method: corner, symmetric threefold junction, and T-junction detector. Key points in the image correspond to maxima in the response of the wavelet detector. The optimal steering angle can be determined by root finding, as presented in (Püspöki et al. 2013, Sec. 4.1). The rest of the detection scheme is achieved by traditional techniques which may combine the results at different scales, local maxima search, thresholding, among others. Detections in synthetic and microscopic data is visualized are Fig. 3.8 (iii), and (iv), respectively.

The construction presented here makes it possible to capture the local orientation of features in an image within a multiresolution hierarchy Püspöki et al. (2016). The relation between the Riesz transform and steerable filters is studied in (Felsberg and Sommer 2001). The properties of steerable filters using low-order harmonics are analyzed in Koethe (2006). The extension of the steerable wavelet design based on the Riesz transform for higher dimensions, along with potential biomedical applications, are presented in Chenouard and Unser (2012). Application of steerable wavelets in texture learning for the analysis of CT images are presented in Depeursinge et al. (2015, 2014b).

3.6 Conclusion and Outlook

We have presented a survey on the directional analysis of bioimages. We have discussed the benefits and drawbacks of classical gradient-based methods, directional multiscale transforms, and multiscale steerable filters. From a user perspective, we have identified steerable wavelets and shearlets as the most attractive methods. They unify high directional selectivity and multiscale analysis, which allows the processing of oriented patterns at different scales. There exist computationally efficient implementations of such schemes that are available for the public. Finally, they are still an active field of research, see the recent papers (Kutyniok 2014; Bodmann et al. 2015; Duval-Poo et al. 2015; Kutyniok et al. 2014) for shearlets and (Ward and Unser 2014; Pad et al. 2014; Depeursinge et al. 2014a; Dumic et al. 2014; Schmitter et al. 2014) for steerable filters and wavelets. Furthermore, the corresponding user packages and plugins are maintained and continuously extended.

The state of the art in the field will need to be adjusted to fulfill the upcoming needs of biomedical and biological imaging. Advances in microscopy and in some other measurement systems (CT, X-ray) will shape the future of research. Currently, microscopes are already routinely producing and storing large datasets (often several GBs per measurement) that have to be handled in a fast and efficient way. Moreover, the need to process spatio-temporal data (2D/3D images over time) is becoming unavoidable and is going to require the proper extension of current filter-based schemes. Along with efficiency, the robustness, the precision, and the depth of the extracted information can be improved. Another promising direction of future research is the recovery of directional phase information using complex-valued wavelet transforms such as the monogenic wavelets (Felsberg and Sommer 2001; Olhede and Metikas 2009; Held et al. 2010; Unser et al. 2009; Storath 2011a; Soulard et al. 2013; Häuser et al. 2014; Heise et al. 2014). Preliminary applications include equalization of brightness (Held et al. 2010), detection of salient points (Storath et al. 2015), enhancement of anisotropic structures in fluorescence microscopy (Chenouard and Unser 2012), and texture segmentation (Storath et al. 2014). Image-analysis tools based on monogenic wavelets are provided by the

ImageJ/Fiji plugins MonogenicJ[8] and Monogenic Wavelet Toolbox.[9] A further possible direction is the extension of directional wavelet transforms to nonuniform lattices such as polar grids or general graphs (Hammond et al. 2011; Shuman et al. 2013; Sandryhaila and Moura 2013).

Acknowledgements The research leading to these results has received funding from the European Research Council under the European Union's Seventh Framework Programme (FP7/2007-2013) / ERC grant agreement n° 267439. The work was also supported by the Hasler Foundation, and the Swiss National Foundation (grant number: 200020-162343).

References

Agam G, Armato S, Wu C (2005) Vessel tree reconstruction in thoracic CT scans with application to nodule detection. IEEE Trans Med Imaging 24(4):486–499

Aguet F, Geissbühler S, Märki I, Lasser T, Unser M (2009) Super-resolution orientation estimation and localization of fluorescent dipoles using 3-D steerable filters. Opt Express 17(8):6829–6848

Antoine JP, Murenzi R, Vandergheynst P (1999) Directional wavelets revisited: Cauchy wavelets and symmetry detection in patterns. Appl Comput Harmon Anal 6(3):314–345

Bernasconi P, Rust D, Hakim D (2005) Advanced automated solar filament detection and characterization code: description, performance, and results. Sol Phys 228(1–2):97–117

Bharath A, Ng J (2005) A steerable complex wavelet construction and its application to image denoising. IEEE Trans Image Process 14(7):948–959

Bigun J (1987) G.H.: Optimal orientation detection of linear symmetry. In: Proceedings of the first IEEE international conference on computer vision, London, pp 433–438

Bigun J, Bigun T, Nilsson K (2004) Recognition by symmetry derivatives and the generalized structure tensor. IEEE Trans Pattern Anal Mach Intell 26(12):1590–1605

Bodmann B, Kutyniok G, Zhuang X (2015) Gabor shearlets. Appl Comput Harmon Anal 38(1):87–114

Candès E, Donoho D (2004) New tight frames of curvelets and optimal representations of objects with piecewise C2 singularities. Commun Pure Appl Math 57(2):219–266

Candès E, Donoho D (2005) Continuous curvelet transform: I. Resolution of the wavefront set. Appl Comput Harmon Anal 19(2):162–197

Canny J (1986) A computational approach to edge detection. IEEE Trans Pattern Anal Mach Intell 8(6):679–698

Chaudhury K, Muñoz Barrutia A, Unser M (2010) Fast space-variant elliptical filtering using box splines. IEEE Trans Image Process 19(9):2290–2306

Chenouard N, Unser M (2012) 3D steerable wavelets in practice. IEEE Trans Image Process 21(11):4522–4533

Dan B, Ma AWK, Hároz EH, Kono J, Pasquali M (2012) Nematic-like alignment in SWNT thin films from aqueous colloidal suspensions. Ind Eng Chem Res 51(30):10232–10237

Daubechies I (1992) Ten lectures on wavelets, vol 61. SIAM, Philadelphia

Daugman J (1985) Uncertainty relation for resolution in space, spatial frequency, and orientation optimized by two-dimensional visual cortical filters. J Opt Soc Am A 2(7):1160–1169

Daugman J (1988) Complete discrete 2D Gabor transforms by neural networks for image analysis and compression. IEEE Trans Acoust Speech Signal Process 36(7):1169–1179

[8] Available at http://bigwww.epfl.ch/demo/monogenic/.

[9] Available at http://www-m6.ma.tum.de/Mamebia/MonogenicWaveletToolbox/.

Depeursinge A, Foncubierta-Rodriguez A, Van De Ville D, Müller H (2014a) Rotation-covariant texture learning using steerable Riesz wavelets. IEEE Trans Image Process 23(2):898–908

Depeursinge A, Kurtz C, Beaulieu C, Napel S, Rubin D (2014b) Predicting visual semantic descriptive terms from radiological image data: preliminary results with liver lesions in CT. IEEE Trans Med Imaging 33(8):1669–1676

Depeursinge A, Yanagawa M, Leung A, Rubin D (2015) Predicting adenocarcinoma recurrence using computational texture models of nodule components in lung ct. Med Phys 42(4):2054–2063

Do M, Vetterli M (2005) The contourlet transform: an efficient directional multiresolution image representation. IEEE Trans Image Process 14(12):2091–2106

Dong Y, Tao D, Li X, Ma J, Pu J (2015) Texture classification and retrieval using shearlets and linear regression. IEEE Trans Med Imaging 45(3):358–369

Dumic E, Grgic S, Grgic M (2014) IQM2: new image quality measure based on steerable pyramid wavelet transform and structural similarity index. Signal Image Video Process 8(6):1159–1168

Duval-Poo M, Odone F, De Vito E (2015) Edges and corners with shearlets. IEEE Trans Image Process 24(11):3768–3780. doi:10.1109/TIP.2015.2451175

Felsberg M, Sommer G (2001) The monogenic signal. IEEE Trans Signal Process 49(12):3136–3144

Fonck E, Feigl G, Fasel J, Sage D, Unser M, Rüfenacht D, Stergiopulos N (2008) Effect of ageing on elastin functionality in human cerebral arteries. In: Proceedings of the ASME 2008 summer bioengineering conference (SBC'08), Marco Island, FL, pp SBC2008–192,727–1/2

Förstner W (1986) A feature based correspondence algorithm for image matching. Int Arch Photogramm Remote Sens 26(3):150–166

Frangi A, Niessen W, Vincken K, Viergever M (1998) Multiscale vessel enhancement filtering. In: Medical image computing and computer-assisted intervention. Lecture notes in computer science, vol 1496. Springer, Berlin, pp 130–137

Franken E, Duits R (2009) Crossing-preserving coherence-enhancing diffusion on invertible orientation scores. Int J Comput Vis 85(3):253–278

Franken E, Duits R, ter Haar Romeny B (2007) Nonlinear diffusion on the 2D Euclidean motion group. In: Scale space and variational methods in computer vision. Springer, Berlin, pp 461–472

Freeman W, Adelson E (1990) Steerable filters for early vision, image analysis, and wavelet decomposition. In: Proceedings of the third international conference on computer vision. IEEE, New York, pp 406–415

Freeman W, Adelson E (1991) The design and use of steerable filters. IEEE Trans Pattern Anal Mach Intell 13(9):891–906

Freeman WT (1992) Steerable filters and local analysis of image structure. PhD thesis, Massachusetts Institute of Technology

Frikel J (2013) Sparse regularization in limited angle tomography. Appl Comput Harmon Anal 34(1):117–141

Gonzalez G, Fleurety F, Fua P (2009) Learning rotational features for filament detection. In: Proceedings of the IEEE computer society conference on computer vision and pattern recognition (CVPR'09). IEEE, New York, pp 1582–1589

Grohs P, Kutyniok G (2014) Parabolic molecules. Found Comput Math 14(2):299–337

Guo K, Labate D, Lim WQ (2009) Edge analysis and identification using the continuous shearlet transform. Appl Comput Harmon Anal 27(1):24–46

Hammond D, Vandergheynst P, Gribonval R (2011) Wavelets on graphs via spectral graph theory. Appl Comput Harmon Anal 30(2):129–150

Harris C, Stephens M (1988) A combined corner and edge detector. In: Proceedings of the fourth alvey vision conference, pp 147–151

Häuser S, Heise B, Steidl G (2014) Linearized Riesz transform and quasi-monogenic shearlets. Int J Wavelets Multiresolution Inf Process 12(03):1450027. doi:10.1142/S0219691314500271. http://www.worldscientific.com/doi/abs/10.1142/S0219691314500271

Heise B, Reinhardt M, Schausberger S, Häuser S, Bernstein S, Stifter D (2014) Fourier plane filtering revisited – analogies in optics and mathematics. Sampl Theory Signal Image Process 13(3):231–248

Held S, Storath M, Massopust P, Forster B (2010) Steerable wavelet frames based on the Riesz transform. IEEE Trans Image Process 19(3):653–667

Honnorat N, Vaillant R, Duncan J, Paragios N (2011) Curvilinear structures extraction in cluttered bioimaging data with discrete optimization methods. In: Proceedings of the eighth IEEE international symposium on biomedical imaging: from nano to macro (ISBI'11), IEEE, New York, pp 1353–1357

Hubel D, Wiesel T (1962) Receptive fields, binocular interaction and functional architecture in the cat's visual cortex. J Physiol 160(1):106–154

Jacob M, Unser M (2004) Design of steerable filters for feature detection using Canny-like criteria. IEEE Trans Pattern Anal Mach Intell 26(8):1007–1019

Jacob M, Blu T, Vaillant C, Maddocks J, Unser M (2006) 3-D shape estimation of DNA molecules from stereo cryo-electron micro-graphs using a projection-steerable snake. IEEE Trans Image Process 15(1):214–227

Jacovitti G, Neri A (2000) Multiresolution circular harmonic decomposition. IEEE Trans Signal Process 48(11):3242–3247

Jacques L, Duval L, Chaux C, Peyré G (2011) A panorama on multiscale geometric representations, intertwining spatial, directional and frequency selectivity. Signal Process 91(12):2699–2730

Jahne B (1997) Digital image processing: concepts, algorithms, and scientific applications, 4th edn. Springer, New York, Secaucus, NJ

Jiuxiang H, Razdan A, Femiani J, Ming C, Wonka P (2007) Road network extraction and intersection detection from aerial images by tracking road footprints. IEEE Trans Geosci Remote Sens 45(12):4144–4157

Karssemeijer N, te Brake G (1996) Detection of stellate distortions in mammograms. IEEE Trans Med Imaging 15(5):611–619

Kingsbury NG (1998) The dual-tree complex wavelet transform: a new technique for shift invariance and directional filters. In: Proceedings of 8th IEEE DSP workshop, Bryce Canyon, August 1998

Köthe U (2003) Edge and junction detection with an improved structure tensor. In: Pattern recognition. Lecture notes in computer science, vol 2781. Springer, Berlin, pp 25–32

Koethe U (2006) Low-level feature detection using the boundary tensor. In: Weickert J, Hagen H (eds) Visualization and processing of tensor fields, mathematics and visualization. Springer, Berlin, Heidelberg, pp 63–79

Kutyniok G (2012) Data separation by sparse representations. In: Compressed sensing: theory and applications. Cambridge University Press, Cambridge

Kutyniok G (2014) Geometric separation by single-pass alternating thresholding. Appl Comput Harmon Anal 36(1):23–50

Kutyniok G, Labate D (2009) Resolution of the wavefront set using continuous shearlets. Trans Am Math Soc 361(5):2719–2754

Kutyniok G, Labate D (2012) Shearlets: multiscale analysis for multivariate data. Birkhauser, Basel

Kutyniok G, Lim WQ, Reisenhofer R (2014) Shearlab 3D: faithful digital shearlet transforms based on compactly supported shearlets. arXiv:14025670

Labate D, Lim W, Kutyniok G, Weiss G (2005) Sparse multidimensional representation using shearlets. In: Proceedings of SPIE. Wavelets XI, San Diego, vol 5914. SPIE, Bellingham, pp 254–262

Lam B, Gao Y, Liew AC (2010) General retinal vessel segmentation using regularization-based multiconcavity modeling. IEEE Trans Med Imaging 29(7):1369–1381

Larkin K (2001) Natural demodulation of two-dimensional fringe patterns. II. stationary phase analysis of the spiral phase quadrature transform. J Opt Soc Am A 18:1871–1881

Larkin K, Bone DJ, Oldfield MA (2001) Natural demodulation of two-dimensional fringe patterns. I. general background of the spiral phase quadrature transform. J Opt Soc Am A 18:1862–1870

Lee T (1996) Image representation using 2D Gabor wavelets. IEEE Trans Pattern Anal Mach Intell 18(10):959–971

Lorenz C, Carlsen IC, Buzug T, Fassnacht C, Weese J (1997) Multi-scale line segmentation with automatic estimation of width, contrast and tangential direction in 2D and 3D medical images. In: Proceedings of the first joint conference computer vision, virtual reality and robotics in medicine and medical robotics and computer-assisted surgery CVRMed - MRCAS'97. Springer, Berlin, pp 233–242

Ma J, Plonka G (2010) The curvelet transform. IEEE Signal Process Mag 27(2):118–133

Mallat S (2008) A wavelet tour of signal processing: the sparse way. Academic Press, New York

Marr D, Hildreth E (1980) Theory of edge detection. Proc R Soc London Ser B 207(1167):187–217

Meijering E (2010) Neuron tracing in perspective. Cytometry A 77(7):693–704

Meijering E, Jacob M, Sarria JC, Steiner P, Hirling H, Unser M (2004) Design and validation of a tool for neurite tracing and analysis in fluorescence microscopy images. Cytometry Part A 58A(2):167–176

Olhede S, Metikas G (2009) The monogenic wavelet transform. IEEE Trans Signal Process 57(9):3426–3441

Olshausen B, Field D (1996) Emergence of simple-cell receptive field properties by learning a sparse code for natural images. Nature 381:607–609

Pad P, Uhlmann V, Unser M (2014) VOW: Variance-optimal wavelets for the steerable pyramid. In: IEEE international conference on image processing (ICIP), pp 2973–2977

Patton N, Aslam T, MacGillivray T, Deary I, Dhillon B, Eikelboom R, Yogesan K, Constable I (2006) Retinal image analysis: concepts, applications and potential. Prog Retin Eye Res 25(1):99–127

Perona P (1992) Steerable-scalable kernels for edge detection and junction analysis. Image Vis Comput 10(10):663–672

Portilla J, Simoncelli E (2000) A parametric texture model based on joint statistics of complex wavelet coefficients. Int J Comput Vis 40(1):49–70

Püspöki Z, Unser M (2015) Template-free wavelet-based detection of local symmetries. IEEE Trans Image Process 24(10):3009–3018

Püspöki Z, Vonesch C, Unser M (2013) Detection of symmetric junctions in biological images using 2-D steerable wavelet transforms. In: Proceedings of the tenth IEEE international symposium on biomedical imaging: from nano to macro (ISBI'13), San Francisco, CA, pp 1488–1491

Püspöki Z, Uhlmann V, Vonesch C, Unser M (2016) Design of steerable wavelets to detect multifold junctions. IEEE Trans Image Process 25(2):643–657. http://bigwww.epfl.ch/publications/puespoeki1601.html

Rezakhaniha R, Agianniotis A, Schrauwen J, Griffa A, Sage D, Bouten C, van de Vosse F, Unser M, Stergiopulos N (2012) Experimental investigation of collagen waviness and orientation in the arterial adventitia using confocal laser scanning microscopy. Biomech Model Mechanobiol 11(3–4):461–473

Sandryhaila A, Moura J (2013) Discrete signal processing on graphs. IEEE Trans Signal Process 61(7):1644–1656

Sato Y, Nakajima S, Shiraga N, Atsumi H, Yoshida S, Koller T, Gerig G, Kikinis R (1998) Three-dimensional multi-scale line filter for segmentation and visualization of curvilinear structures in medical images. Med Image Anal 2(2):143–168

Schmitter D, Delgado-Gonzalo R, Krueger G, Unser M (2014) Atlas-free brain segmentation in 3D proton-density-like MRI images. In: Proceedings of the IEEE international symposium on biomedical imaging: from nano to macro (ISBI), pp 629–632

Schuh M, Banda J, Bernasconi P, Angryk R, Martens P (2014) A comparative evaluation of automated solar filament detection. Sol Phys 289(7):2503–2524

Selesnick I, Baraniuk R, Kingsbury N (2005) The dual-tree complex wavelet transform. IEEE Signal Process Mag 22(6):123–151

Semler L, Dettori L (2006) Curvelet-based texture classification of tissues in computed tomography. In: Proceedings of the thirteenth IEEE international conference on image processing (ICIP'06), IEEE, New York, pp 2165–2168

Sherlock B, Monro D, Millard K (1994) Fingerprint enhancement by directional Fourier filtering. In: IEE proceedings on vision, image and signal processing, vol 141, pp 87–94

Shuman D, Narang S, Frossard P, Ortega A, Vandergheynst P (2013) The emerging field of signal processing on graphs: extending high-dimensional data analysis to networks and other irregular domains. IEEE Signal Process Mag 30(3):83–98

Simoncelli E, Farid H (1996) Steerable wedge filters for local orientation analysis. IEEE Trans Image Process 5(9):1377–1382

Simoncelli E, Freeman WT (1995) The steerable pyramid: a flexible architecture for multi-scale derivative computation. In: Proceedings of the second IEEE international conference on image processing (ICIP'95), vol 3. IEEE, New York, pp 444–447

Soulard R, Carre P, Fernandez-Maloigne C (2013) Vector extension of monogenic wavelets for geometric representation of color images. IEEE Trans Image Process 22(3):1070–1083

Starck JL, Candès E, Donoho D (2002) The curvelet transform for image denoising. IEEE Trans Image Process 11(6):670–684

Starck JL, Murtagh F, Candes E, Donoho D (2003) Gray and color image contrast enhancement by the curvelet transform. IEEE Trans Image Process 12(6):706–717

Storath M (2011a) Directional multiscale amplitude and phase decomposition by the monogenic curvelet transform. SIAM J Imaging Sci 4(1):57–78

Storath M (2011b) Separation of edges in X-Ray images by microlocal analysis. Proc Appl Math Mech 11(1):867–868

Storath M, Weinmann A, Unser M (2014) Unsupervised texture segmentation using monogenic curvelets and the Potts model. In: IEEE international conference on image processing (ICIP), pp 4348–4352

Storath M, Demaret L, Massopust P (2015) Signal analysis based on complex wavelet signs. Appl Comput Harmon Anal. doi:http://dx.doi.org/10.1016/j.acha.2015.08.005. http://www.sciencedirect.com/science/article/pii/S1063520315001086

Tupin F, Maitre H, Mangin J-F, Nicolas J-M, Pechersky E (1998) Detection of linear features in SAR images: application to road network extraction. IEEE Trans Geosci Remote Sens 36(2):434–453. doi:10.1109/36.662728

Unser M, Chenouard N (2013) A unifying parametric framework for 2D steerable wavelet transforms. SIAM J Imaging Sci 6(1):102–135

Unser M, Van De Ville D (2010) Wavelet steerability and the higher-order Riesz transform. IEEE Trans Image Process 19(3):636–652

Unser M, Sage D, Van De Ville D (2009) Multiresolution monogenic signal analysis using the Riesz-Laplace wavelet transform. IEEE Trans Image Process 18(11):2402–2418

Vandergheynst P, Gobbers JF (2002) Directional dyadic wavelet transforms: design and algorithms. IEEE Trans Image Process 11(4):363–372

Ward J, Unser M (2014) Harmonic singular integrals and steerable wavelets in $l_2(\mathbb{R}^d)$. Appl Comput Harmon Anal 36(2):183–197

Ward J, Chaudhury K, Unser M (2013) Decay properties of Riesz transforms and steerable wavelets. SIAM J Imaging Sci 6(2):984–998

Yi S, Labate D, Easley G, Krim H (2009) A shearlet approach to edge analysis and detection. IEEE Trans Image Process 18(5):929–941

Yuan Y, Shih F, Jing J, Wang H, Chae J (2011) Automatic solar filament segmentation and characterization. Sol Phys 272(1):101–117

Chapter 4
Analyzing Protein Clusters on the Plasma Membrane: Application of Spatial Statistical Analysis Methods on Super-Resolution Microscopy Images

Laura Paparelli, Nikky Corthout, Benjamin Pavie, Wim Annaert, and Sebastian Munck

Abstract The spatial distribution of proteins within the cell affects their capability to interact with other molecules and directly influences cellular processes and signaling. At the plasma membrane, multiple factors drive protein compartmentalization into specialized functional domains, leading to the formation of clusters in which intermolecule interactions are facilitated. Therefore, quantifying protein distributions is a necessity for understanding their regulation and function. The

L. Paparelli
VIB Bio Imaging Core, Herestraat 49, Box 602, Leuven 3000, Belgium

Laboratory of Membrane Trafficking, Department of Human Genetics, KU Leuven, Herestraat 49, Box 602, Leuven 3000, Belgium

VIB Center for the Biology of Disease, KU Leuven, Herestraat 49, Box 602, Leuven 3000, Belgium

N. Corthout • S. Munck (✉)
VIB Bio Imaging Core, Herestraat 49, Box 602, Leuven 3000, Belgium

VIB Center for the Biology of Disease, KU Leuven, Department of Human Genetics, Herestraat 49, Box 602, Leuven 3000, Belgium

VIB, LiMoNe, Herestraat 49, Box 602, Leuven 3000, Belgium
e-mail: sebastian.munck@cme.vib-kuleuven.be

B. Pavie
VIB Bio Imaging Core, Herestraat 49, Box 602, Leuven 3000, Belgium

VIB Center for the Biology of Disease, KU Leuven, Department of Human Genetics, Herestraat 49, Box 602, Leuven 3000, Belgium

W. Annaert
Laboratory of Membrane Trafficking, Department of Human Genetics, KU Leuven, Herestraat 49, Box 602, Leuven 3000, Belgium

VIB Center for the Biology of Disease, KU Leuven, Herestraat 49, Box 602, Leuven 3000, Belgium

W.H. De Vos et al. (eds.), *Focus on Bio-Image Informatics*,
Advances in Anatomy, Embryology and Cell Biology 219,
DOI 10.1007/978-3-319-28549-8_4

recent advent of super-resolution microscopy has opened up the possibility of imaging protein distributions at the nanometer scale. In parallel, new spatial analysis methods have been developed to quantify distribution patterns in super-resolution images. In this chapter, we provide an overview of super-resolution microscopy and summarize the factors influencing protein arrangements on the plasma membrane. Finally, we highlight methods for analyzing clusterization of plasma membrane proteins, including examples of their applications.

4.1 Introduction

In the last decades, several technological breakthroughs have been published that allow retrieving information from beyond the diffraction limit (Schermelleh et al. 2010). These super-resolution techniques offer the possibility to observe structures in biological samples at improved resolution compared to conventional microscopy, providing novel insights into cell organization and cellular signaling (Lang and Rizzoli 2010). For cellular signal transduction, the distribution of proteins on the plasma membrane plays a crucial role. To efficiently exploit their function, the proteins of the plasma membrane are often organized in clusters. Their spatiotemporal localization on the plasma membrane is, in fact, one of the elements regulating their interactions (Simons and Toomre 2000).

Even if super-resolution imaging enables the visualization of the organization of plasma membrane proteins at the nanometer scale, the high level of detailed information present in the resulting images is often difficult to interpret. Quantitative approaches, optimized from spatial statistics, have therefore been employed to decipher if a protein distribution is organized in clusters or if it is random. In this context, membrane biology, super-resolution microscopy, and novel analysis schemes lead to a positive feedback loop pushing the boundaries of science and have contributed to the emerging field of BioImage Informatics, which combines computational image analysis and life sciences. The developments in the fields mentioned are the major topic of discussion in this review.

Starting from an overview of the different super-resolution microscopy techniques, we next describe the factors affecting the distribution of plasma membrane proteins. Finally, we discuss the statistical techniques available for assessing clusterization, focusing on nearest neighbor distance, K-Ripley function, and pair correlation approaches. In this context, we provide examples of applications to emphasize the transformative role of these techniques for quantitative biology, with particular focus on the neuronal and the immunological synapse, where protein distribution analysis has provided novel insights into the biology of signaling within these systems.

4.2 Super-Resolved Fluorescence Microscopy

The recent past has seen the emergence of several super-resolution techniques that allow to observe structures in the sample beyond the diffraction limit (Abbe 1873; Rayleigh 1903; Sparrow 1916) (see Box 4.1, Fig. 4.1) of a conventional fluorescence microscope (which is about 200 nm). Among these are near-field scanning optical microscopy (NSOM) (Betzig and Trautman 1992; Betzig and Chichester 1993), stimulated emission depletion (STED) (Hell and Wichmann 1994; Hell and Kroug 1995), structured illumination microscopy (SIM) (Heintzmann and Cremer 1999; Gustafsson 2000), and single-molecule detection methods like photoactivated localization microscopy (PALM) (Betzig et al. 2006), fluorescence photoactivation localization microscopy (FPALM) (Hess et al. 2006), and stochastic optical reconstruction microscopy (STORM) (Rust et al. 2006) (see Table 4.1).

In NSOM (Betzig and Trautman 1992; Betzig and Chichester 1993), the surface of an object is scanned by a probe with nanosized aperture located at subwavelength distance to the sample. The size of the aperture of the probe imposes the resolution, which is typically in the range of 20–120 nm. STED (Hell and Wichmann 1994; Hell and Kroug 1995) uses the combination of excitation and stimulated emission of fluorophores to create an effective point spread function (PSF) (see Box 4.1, Fig. 4.1) smaller than the diffraction limit. NSOM and STED share the ability to achieve super-resolution without processing the images after the acquisition. Structured illumination microscopy is a super-resolution technique that uses an illumination pattern (Heintzmann and Cremer 1999; Gustafsson 2000) for achieving optical sectioning. The interference pattern (moiré fringes) generated by the interaction of the grid with the sample is used to back-calculate the structure of the sample, gaining spatial information that is otherwise not accessible. This approach can achieve a lateral resolution of about 100 nm (Fig. 4.2a–c), which can be further enhanced by using saturated structured illumination microscopy (SSIM) (Heintzmann et al. 2002; Heintzmann 2003) to <50 nm (Gustafsson 2005).

Single-molecule localization techniques are based on the localization of a molecule by determining the center of a single fluorescence emitting object. Once the single localizations are determined, they can be assembled in a single super-resolution image, as initially proposed by pointillism (Lidke et al. 2005). Since this calculation is possible only if the PSFs from different fluorophores are at a distance higher than the diffraction limit, diverse image acquisition approaches were proposed to avoid signal overlapping. These approaches rely on resolving fluorescent molecules in time by stochastically exciting a subset of fluorophores in the sample and by repeating the acquisition several times. In photoactivated localization microscopy (PALM) (Betzig et al. 2006) and in fluorescence photoactivation localization microscopy (FPALM) (Hess et al. 2006), this is achieved by the use of photoactivatable or photoconvertible dyes, while in stochastic optical reconstruction microscopy (STORM), pairs of photoswitchable dyes (like Cy3-Cy5) are adopted (Rust et al. 2006) (Fig. 4.2d). These techniques have been described to localize a protein of interest to approximately 20 nm. Localization precision,

Table 4.1 Characteristics of super-resolution microscopy techniques

Technique	Principle	Typical *xy*-resolution (nm)	Typical *z*-resolution (nm)	Required label	Image processing	References
NSOM	A probe, used as near-field excitation source, scans the surface of the sample to generate an image. The probe is typically an optical fiber with a nanosized aperture placed in close proximity to the sample	20–120	<10	Conventional dyes	No	Betzig and Trautman (1992), Betzig and Chichester (1993), Lange et al. (2001)
STED	A small excitation spot is generated by the overlapping of two beams, one that excites the sample and the other that depletes the fluorescence at the periphery of the excitation beam	~50	~80[a]	Conventional dyes	No	Hell and Wichmann (1994), Hell and Kroug (1995), Willig et al. (2007), Donnert et al. (2007), Hein et al. (2008)
SIM	The sample is excited with multiple structured illumination patterns to extract high-spatial frequency information and reconstruct the super-resolved image	100–130	250–360	Conventional dyes	Yes	Heintzmann and Cremer (1999), Gustafsson (2000), Gustafsson et al. (2008), Schermelleh et al. (2010), Shao et al. (2011)
PALMFPALMSTORM	Stochastic excitation of a portion of fluorophores in the sample is used to resolve in time their fluorescence emission. The multiple images acquired are then processed to reconstruct a final super-resolved image	10–50	20–100	Photoactivatable or photoconvertible fluorescent proteins (PALM/FPALM) Organic dyes (STORM)	Yes	Betzig et al. (2006) Hess et al. (2006), Rust et al. (2006) (Huang et al. (2008) Juette et al. (2008) Pavani et al. (2009) Baddeley et al. (2011)

[a]Resolution achieved by commercial devices (Leica TCS SP8 STED 3× super-resolution microscope)

labeling density and other factors define the final resolution. (Deschout et al. 2014a) (see Box 4.1). Improved versions of the mentioned techniques include PALM with independently running acquisition (PALMIRA) (Geisler et al. 2007), stroboscopic PALM (sPALM) (Flors et al. 2007), direct STORM (dSTORM) (Heilemann et al. 2008; Heilemann et al. 2009), and ground-state depletion microscopy (GSDIM) (Fölling et al. 2008). Super-resolution has also been achieved by the use of other approaches like point accumulation for imaging in nanoscale topography (PAINT) (Sharonov and Hochstrasser 2006), blink microscopy (Steinhauer et al. 2008), super-resolution optical fluctuation imaging (SOFI) (Dertinger et al. 2009), focal modulation microscopy with annular aperture (AFMM) (Gong et al. 2010), Bayesian analysis of the blinking and bleaching (3B analysis) (Cox et al. 2012), photobleaching microscopy with nonlinear processing (PiMP) (Munck et al. 2012), nonnegative matrix factorization with iterative restarts (iNMF) (Mandula et al. 2014), and nanometer accuracy by stochastic chemical reactions (NASCA) (Ristanović et al. 2015).

Overall these new approaches have greatly enhanced the quality and the quantity of information gained on the organization of the cell and its compartments, including the plasma membrane and its complex architecture.

Box 4.1. Microscope Resolution

The resolution of a microscope is its ability to discriminate the features of the observed sample. The optical resolution is limited because of the diffraction of the light. Light diffraction is a phenomenon occurring when light interacts with a physical barrier or has to pass through a circular aperture with a size comparable to light wavelength. In a microscope, the diffraction occurs at the aperture of the objective or at the specimen (Fornasiero and Rizzoli 2014). When light passes through the circular aperture, light waves are spread out behind the obstacle, generating a diffraction pattern called Airy disk pattern, whose intensity distribution is called point spread function (PSF) (see Fig. 4.1). For describing the resolution of the microscope, several definitions are used. The Rayleigh criterion (Rayleigh 1903) states that two objects with overlapping diffraction patterns can be resolved if they are separated by a distance greater than the distance at which the maximum of the Airy disk (central diffraction disk) coincides with the first minimum of the other diffraction pattern. Sparrow (Sparrow 1916) used as definition the distance at which the PSFs of two objects are so close that their peaks exhibit constant brightness and therefore cannot be discerned. Both measures are based on the PSF, and consequently the full width at half maximum of the PSF is equally used as a resolution measure. Apart from these straightforward measures, Ernst Abbe (Abbe 1873) described the resolution limit as a fundamental barrier. Abbe used the cutoff frequency of the optical transfer function as

(continued)

Box 4.1 (continued)
the limit of diffraction. Abbes' diffraction limit would correspond in real space to a grid of objects at a distance that cannot be resolved resulting in an image of homogeneous intensities. In the context of single-molecule localization microscopy, an important parameter that has an influence on the achieved resolution is the localization precision. This indicates the level of accuracy with which the position of a single fluorescent emitter can be detected, and it is inversely proportional to the square root of the number of detected photons. In addition, in single-molecule localization microscopy, the resolution is influenced by the labeling density, by the photoactivation and photoswitching rate, and by the spatial structure of the sample (Deschout et al. 2014a). In order to take these factors into account, recently, an image-based measure of resolution for localization microscopy images has been proposed (Nieuwenhuizen et al. 2013); this measurement takes advantage of the Fourier ring correlation which is directly computed on the image.

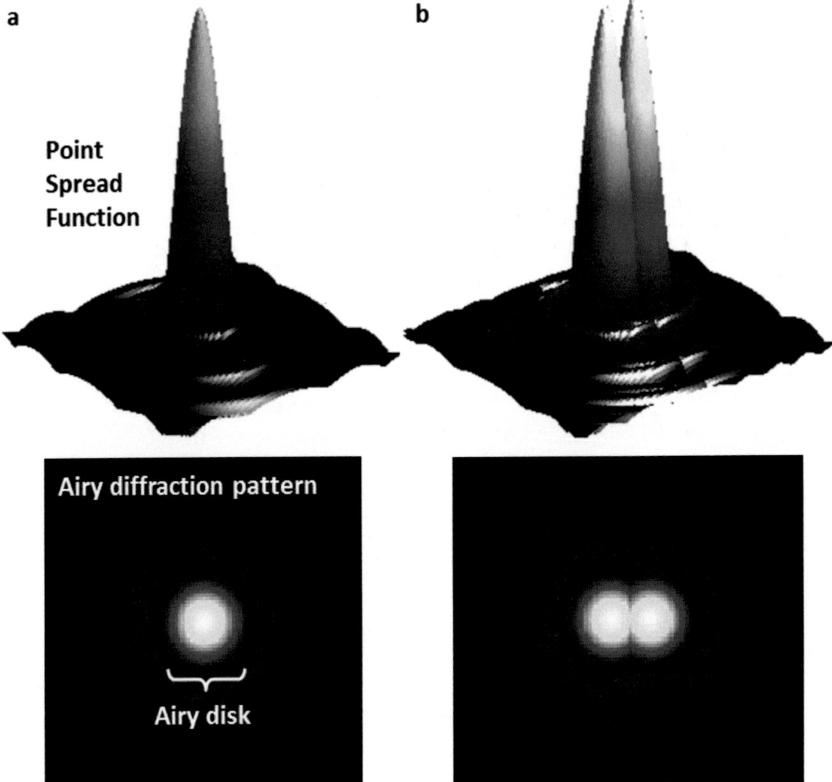

Fig. 4.1 Sketch of the point spread function. (**a**) Representation of the point spread function (PSF) (*top image*) and the Airy pattern (*bottom figure*) generated by a single source of light. (**b**) Overlapping PSFs and Airy patterns generated by two different sources of light

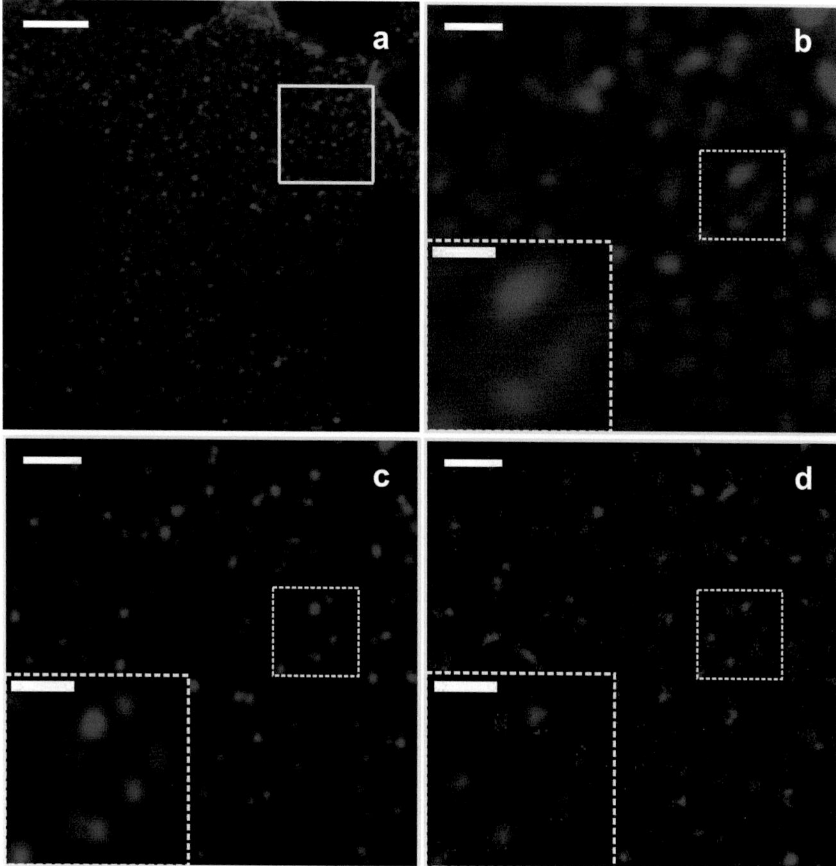

Fig. 4.2 (**a**) Widefield image of immunostained Na^+/K^+ ATPase on an isolated plasma membrane (according to (Chaney and Jacobson 1983) (scale bar: 5 μm). (**b–d**) Zoom of the *yellow* selected region in figure (**a**) including a detailed area; the images were acquired by different types of microscopy techniques (similar as in Hamel et al. (2014): (**b**) widefield, (**c**) SIM, and (**d**) dSTORM. Scale bars: (**b–d**) 1 μm. Scale bars of the inserts in (**b–d**) 0.5 μm. The sample was stained for Na^+/K^+ATPase using standard immunochemistry procedures with a mouse monoclonal primary antibody (#NB300-146, Novus Biologicals) followed by the secondary antibody staining with goat anti-mouse Alexa Fluor 647 (A-21241, Invitrogen, polyclonal). Widefield and SIM images were acquired on an inverted Zeiss Elyra S.1 microscope using a Plan-APOCHROMAT 63X 1.4 oil objective lens. dSTORM images were acquired using a breadboard setup as described before (Adam et al. 2011; Munck et al. 2012). dSTORM images were reconstructed using QuickPALM (Henriques et al. 2010). The widefield, SIM, and dSTORM images were aligned using TrakEM2 (Cardona et al. 2010)

4.3 Overview on Plasma Membrane Architecture Distance

The essential structure of cellular membranes is a lipid bilayer (Gorter and Grendel 1925) in which proteins are embedded. Even though early thermodynamic studies suggested the free lateral diffusion of membrane proteins in this bilayer (leading to the proposition of the fluid mosaic model (Singer and Nicolson 1972)), our current perception of the plasma membrane organization has been integrated by novel discoveries suggesting the presence of constraints to protein movement (Nicolson 2014). During the past decades, in fact, a more complex architecture of the plasma membrane has been revealed, demonstrating how the compartmentalization of proteins and lipids is directed by several factors and events.

4.3.1 Lipid Domains

One of these factors is the existence of highly ordered lipid domains called lipid rafts (Simons and Van Meer 1988; Simons and Ikonen 1997). These are small (10–200 nm), dynamic domains enriched in cholesterol, sphingolipids, and saturated phospholipids that interact with specific proteins to compartmentalize signaling processes (Pike 2006; Simons and Sampaio 2011). Model membranes and biophysical studies showed lipid rafts to be in a liquid-ordered state, where saturated lipids and cholesterol are tightly packed together (as in a solid-ordered phase) but show rapid lateral mobility (as in a liquid disorder state) (Ipsen et al. 1987; London 2002). In the plasma membrane, this state coexists with the liquid disorder phase characteristic of the unsaturated lipid-rich membrane surrounding lipid rafts (Veatch and Keller 2003; Feigenson 2007). These states regulate the activity of membrane proteins, like receptors, mediating the signaling output (Sezgin et al. 2015). Membrane trafficking (Diaz-Rohrer et al. 2014), cancer regulation (Mollinedo and Gajate 2015), lymphocyte activation (Horejsi and Hrdinka 2014), and neuropathogenesis (Marin et al. 2013) are some of the processes known to be influenced by lipid rafts through the partition or exclusion of signaling proteins within these domains. Protein affinity to raft domains is often regulated by lipid modifications that allow anchoring the lipidated proteins to the membranes. Particularly, the interaction with lipid raft domains is facilitated by the addition of sterols and saturated fatty acids, for instance, through palmitoylation (Levental et al. 2010). Alternatively, protein partition into lipid rafts occurs via glycophosphatidylinositol (GPI) anchors (Brown and Rose 1992; Schroeder et al. 1994).

A specialized type of cholesterol-rich domains is present at the cell surface in the form of omega-shaped invaginations called caveolae. These small pits of the plasma membrane, 60–70 nm in diameter, were first observed through electron microscopy in the middle of the last century (Palade 1953; Yamada 1955). Caveolae present a characteristic coat (Peters et al. 1985) consisting of different proteins including caveolins, cavins, and the recently discovered components pacsin2 and

EHD2 (Shvets et al. 2014). The assembly of these components results in a tightly organized ultrastructure (Ludwig et al. 2013; Gambin et al. 2014) that contributes to the spatial organization of signaling molecules (Parton and del Pozo 2013). Caveolae form a "sink" for molecules involved in cellular signaling, restricting the environment in which they can function, as in the case of endothelial nitric oxide synthase (eNOS) (Ramadoss et al. 2013). In addition to other functions, such as cholesterol homeostasis, mechanosensing, and cell proliferation (Parton and Simons 2007), caveolae also play a role in internalization and transport of proteins (Pelkmans et al. 2004; Chaudhary et al. 2014).

4.3.2 Protein Domains

The organization of plasma membrane constituents is additionally directed by scaffolding proteins that promote the formation of other specialized domains. Tetraspanins, for instance, are a large group of proteins (Boucheix and Rubinstein 2001) characterized by four transmembrane domains (Wright and Tomlinson 1994; Stipp et al. 2003). The association of tetraspanins with other tetraspanins and their interaction with various transmembrane receptors (i.e., adhesion receptors, growth factors receptors, immunoglobulin-domain containing factors, and cytokine receptors) leads to the formation of an extended network, which is at the base of tetraspanin-enriched microdomains (TEMs) (Hemler 2003; Hemler 2005). TEMs regulate the spatial distribution of the associated molecules by packing them in large clusters (~300 nm) (Barreiro et al. 2008; Espenel et al. 2008) that have roles in various cell functions spanning cell adhesion, motility, differentiation, and protein trafficking (Yáñez-Mó et al. 2009). Flotillins are other scaffolding proteins known to form microdomains at the plasma membrane through the interaction of the two homologues flotillin 1 and flotillin 2 (Solis et al. 2007). These domains have been shown to act as platforms for signaling processes and endocytosis (Otto and Nichols 2011; Meister and Tikkanen 2014).

4.3.3 Cytoskeleton

Besides interactions among plasma membrane constituents, other cellular components like cytoskeletal proteins have an influence on the distribution of the plasma membrane proteins and lipids. In 1993, Kusumi et al. proposed the membrane skeleton fence model to explain the nonhomogeneous distribution of plasma membrane proteins and their confined diffusion (Kusumi et al. 1993). According to this model, the lateral diffusion of integral plasma membrane proteins is regulated by the subcortical skeleton network, which represents a barrier ("fence") to the free diffusion of proteins due to steric interactions between cytoplasmatic domains of transmembrane proteins and the actin meshwork in proximity to the plasma membrane (Kusumi and Sako 1996). This obstacle causes a transient confinement of the proteins in domains; the movement from one domain to another ("hop diffusion") is possible thanks to the fluctuation of the skeleton network position. Hop diffusion has also been associated with membrane lipids (Fujiwara et al. 2002; Ehrig et al. 2011) and is explained by the presence of transmembrane proteins anchored to the cytoskeleton that slows down lipid diffusion. This slower diffusion is due to hydrodynamic friction between the lipids and the immobilized proteins. These findings have led to the definition of a more complete model that explains the influence of the cytoskeleton on the organization of membrane constituents, the so-called picket-fence model (Ritchie et al. 2003).

4.3.4 Other Factors

An extracellular mechanism for microdomain organization at the cell surface is represented by secreted glycan-binding proteins, such as galectins, which interact with specific glycan structures of plasma membrane glycoproteins and glycolipids. Galectins, like other glycan-binding proteins, are multivalent and, therefore, able to cross-link different glycans promoting the formation of a lattice (Brewer et al. 2002). Galectin lattice directs glycoprotein clustering in specialized domains regulating the associated signaling events (Boscher et al. 2011; Belardi et al. 2012). In addition, galectin lattice has been shown to play a role in both clathrin-mediated (Torreno-Pina et al. 2014) and clathrin-independent endocytosis (Lakshminarayan et al. 2014).

These different levels of compartmentalization (Kusumi et al. 2011) provide the plasma membrane with the ability to increase the frequency of specific interactions by spatiotemporal confinement of the signaling events (Saka et al. 2014). Beyond the compartmentalization of plasma membrane components, cellular polarity, which is the asymmetric organization of cellular organelles and plasma membrane constituents, imposes additional complexity to the plasma membrane of polarized cells (for details on cell polarity, see review (Li and Gundersen 2008)).

4.4 Plasma Membrane Protein Distribution Analysis

The strong relationship between the localization of plasma membrane proteins and cell signaling (Simons and Toomre 2000) has stimulated the development of analytical tools to quantify protein localization and distribution. Beyond the information on proteins subcellular localization, which can be determined by colocalization analysis (Bolte and Cordelières 2006), and beyond protein movements, which can be followed by molecules tracking (Kusumi et al. 2014), it is of increasing interest in biology to assess the plasma membrane protein organization in terms of clusterization, cluster size, and number of protein per clusters. Understanding whether a protein clusters or redistributes is, in fact, relevant to define its interactions with partner molecules and the spatiotemporal features of the events in which it is involved. To determine the distribution patterns of proteins, spatial statistical methods are employed. These approaches are the focus of the following description.

4.4.1 Point Processes Analysis

For analyzing the distribution pattern of plasma membrane proteins, the latter are often regarded as point processes in a bidimensional space. A point process can adopt various patterns including complete spatial randomness (CSR), cluster, and regular distributions. In CSR points are randomly distributed (Fig. 4.3a), the distribution in clusters consists of points that tend to aggregate (Fig. 4.3b), while the regular distribution is characterized by points more ordered than in the random distribution (Fig. 4.3c) (Baddeley 2007). To determine the deviation from CSR, in terms of clustering or regularity, various statistical approaches can be exploited. These approaches can be divided into two main categories (Cressie 1993): (1) analyses that explore the variation of point density in the space and (2) analyses that explore the relationship between the points of the pattern by measuring the interpoint distances. The methods belonging to the first category are based on the subdivision of the space in different areas, according to certain rules, and on counting the number of points in each area to determine the level of aggregation or sparsity in comparison with CSR. Quadrant count (Greig-Smith 1952) is an example of a method of the first category; in this case, the space is divided into a number of nonoverlapping quadrants, usually with squared or rectangular shape. These methods are limited by the dependency of the results to the size of the areas, and they have not found yet large application in the analysis of plasma membrane protein distributions. Differently, methods of the second category, which explore the spatial dependence between points of the process, have been extensively optimized for their application on protein pattern analysis in microscopy images. Therefore, methods of the second category are the main focus of the following paragraphs. These include nearest neighbor distances analysis, the *K*-Ripley function, and the pair correlation function (see Box 4.2) (Cressie 1993). Nearest neighbor distances (Fig. 4.3d) allow determining if a pattern is random, clusterized, or regular, based on the mean distance between the points of the process and their closest neighbor (Cressie 1993). *K*-Ripley (Fig. 4.3e) and pair correlation functions (Fig. 4.3f) both measure the number of neighbors within a certain radius to a point (Ripley 1977; Cressie 1993), with the difference that *K*-Ripley function analysis considers all the points within a circle, while pair correlation function analysis considers only the points within a ring. Therefore, pair correlation function analysis is not influenced by the points at shorter distances to the center of the circle (Wiegand and Moloney 2013). These techniques allow, in addition to pattern recognition, the determination of cluster characteristics, such as size, and therefore are broadly used and optimized for protein distribution analysis (Lagache et al. 2013; Shivanandan et al. 2015).

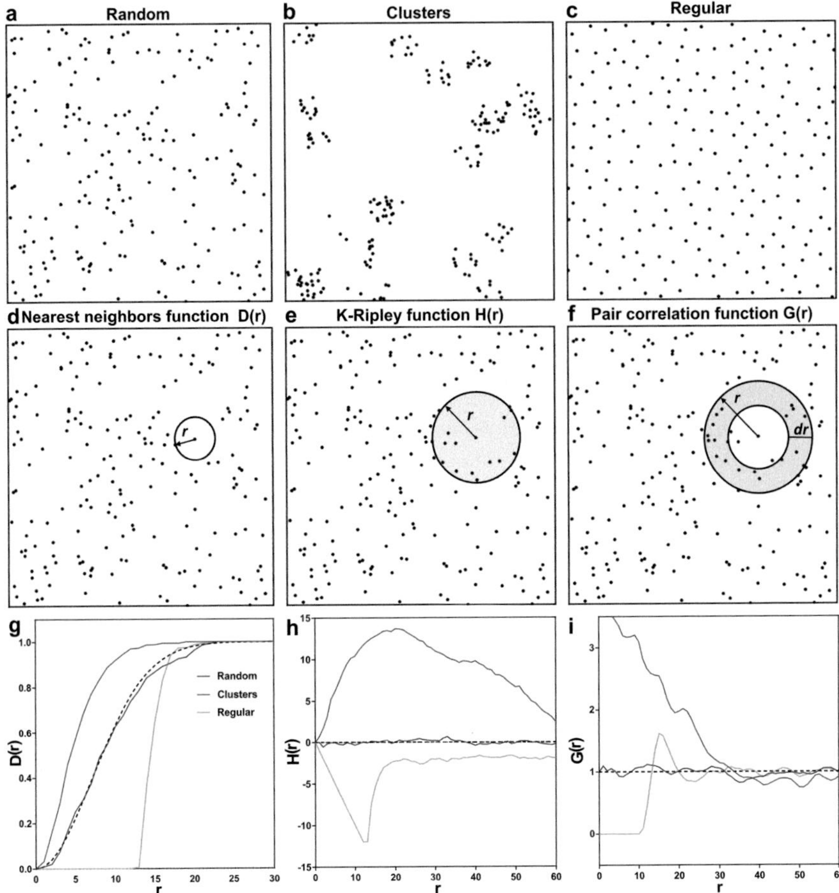

Fig. 4.3 Examples of spatial point patterns and statistical analysis. (**a**) Random. (**b**) Clusters. (**c**) Regular. (**d**) Schematic illustration of nearest neighbor approach; *r* represents the distance between nearest-neighboring points. (**e**) Schematic illustration of K-Ripley function approach; *r* represents the radius of the circle. (**f**) Schematic illustration of pair correlation function approach; *r* represents the radius of the ring and *dr* its width. (**g**) Nearest neighbor analysis. *Blue curve*: random pattern in panel (**a**) is analyzed. *Red curve*: clustered pattern in panel (**b**) is analyzed. *Green curve*: regular pattern in panel (**c**) is analyzed. The *dotted black line* represents the function expected for CSR. (**h**) Results of the K-Ripley analysis, display as H-function, applied to the images in panels (**a–c**). (**i**) Results of the pair correlation function applied to the images in panels (**a–c**). The R toolbox *spatstat* (Baddeley and Turner 2005) has been utilized to generate the spatial point processes and to perform their analysis

Box 4.2. Point Processes Spatial Statistical Analysis

A spatial point process $X \in A \subset \mathbb{R}^2$ represents the locations of an object of study (event) in the finite bidimensional space A. Spatial point processes are used as statistical model to determine which pattern the object of study follows over the area in which it is distributed. Here three main techniques for point pattern analysis are described:

Nearest neighbor method

The distance between each point of the set and its closest neighbor is measured. The nearest neighbor distance distribution function $D(r)$ is the cumulative function of the distances between any point of the process and its nearest neighbor. For CSR,

$$D(r) = 1 - e^{\lambda \pi r^2}$$

where λ is the density of the points in the studied region and r is the distance between nearest-neighboring events. Deviations of the function to the theoretical curve for CSR indicate clusterization or regularity (see Fig. 4.3g) (Baddeley 2007; Cressie 1993).

K-Ripley function

K-Ripley function analyzes the characteristics of the pattern at different distance scales. The number of points located within a certain radius r from each point of the space is, in fact, calculated for increasing radii of the circle. The results are then compared to the ones typical of the complete spatial randomness to identify the degree of point clustering. Considering a circle of radius r (ignoring edges), K-Ripley function can be estimated by

$$K(r) = \lambda^{-1} \sum_{i} \sum_{j \neq i} \frac{I\left(d_{ij} \leq r\right)}{N}$$

where λ is the density of the points in the studied region, d_{ij} is the distance between the ith and the jth points, $I(x)$ is the indicator function with value one when x is true and zero if x is false, and N is the number of points in the area of study. Once Ripley's K function is estimated, the complete spatial randomness can be tested considering that for a point process exhibiting CSR, $K(r) = \pi r^2$ for all r. To have a constant value under CSR, K-Ripley function can be expressed as L-function: $L(r) = \sqrt{\frac{K(r)}{\pi}}$ for $r \geq 0$. A common choice to display the L-function is to use the normalized form $H(r) = L(r) - r$. For CSR, $H(r) = 0$, while $H(r) > 0$ indicates clustering and $H(r) < 0$ indicates regularity (see Fig. 4.3h) (Dixon 2002).

(continued)

Box 4.2 (continued)
Pair correlation function
Pair correlation function $G(r)$ can be related to the K-Ripley function $K(r)$ through

$$G(r) = \frac{1}{2\pi r}\frac{dK(r)}{dr} \text{ for } r \geq 0$$

where r is the ring radius. In a complete CSR process, $G(r) = 1$, $G(r) > 1$ implies clustering, while $G(r) < 1$ suggests a regular pattern. Differently from the K-Ripley function, which is based on cumulative statistics, where all points within the circle are counted, pair correlation function is based on ring statistics, where only the points located in the ring of width dr are counted. The advantage of ring statistics is given by the better determination of a pattern across different scales as the distribution pattern at long distances from the center of the ring is not affected by the distribution patterns at shorter scales (see Fig. 4.3i) (Wiegand and Moloney 2013).

4.4.2 Analysis of Plasma Membrane Protein Distributions by Spatial Statistics

The possibility to apply spatial statistics methods to define the distribution pattern of plasma membrane components depends on the resolution of the microscope; typically high resolution, close to the single-molecule scale, is required. Most often, the described spatial statistical methods are applied to single-molecule localization microscopy images (Deschout et al. 2014b). These images are well suited for spatial statistical analysis, given that the set of single-molecule localizations represents a point process in the 2D space. Nevertheless, other super-resolution imaging techniques can be combined with the mentioned methods for cluster analysis if a sufficiently high resolution is achieved for distinguishing the investigated objects. In STED images, for example, the coordinates of plasma membrane proteins have been extracted from the image by measuring their center of mass and were consequently used for spatial statistical analysis (Kellner et al. 2007). Alternatively, NSOM has contributed to the exploration of plasma membrane compartmentalization (van Zanten et al. 2010), and it has been combined with the nearest neighbor approach to analyze the distribution pattern of the domains formed by the dendritic cell-specific intercellular adhesion molecule (ICAM) grabbing nonintegrin (DC-SIGN) involved in pathogen recognition (de Bakker et al. 2007).

In the context of protein compartmentalization and cell signaling, K-Ripley function has been applied on PALM images to identify the nature of clustering of the G-protein-coupled receptor (GPCR) β_2-adrenergic receptor (Scarselli et al. 2012). This study demonstrated that clusters are present in resting cardiomyocytes and that these are not affected by cholesterol removal, while actin polymerization inhibition leads to the reduction of cluster numbers, excluding β_2-adrenergic receptor enrichment in lipid raft domains. Arrestins are a family of proteins involved in signal transduction through GPCRs that also play a role in endocytosis of these and other receptors. Recently, a method to quantify arrestin2 clustering, occurring after GPCR stimulation, has been proposed (Truan et al. 2013). This approach employs dSTORM combined with nanobody labeling (Ries et al. 2012). Arrestin2 distribution analysis and cluster size evaluation were carried out by Ripley's function calculation in combination with image-based cluster analysis, which accounts for the fluorescence intensity. The increase of arrestin2 cluster size (initially larger than 100 nm) upon blocking of actin polymerization indicates a role for actin in the organization of arrestin at the cell surface.

An algorithm exploiting pair correlation function has been developed to identify clusters of proteins in images acquired by PALM. This algorithm is called pair correlation (PC)-PALM (Sengupta and Lippincott-Schwartz 2012; Sengupta et al. 2013) and allows the quantitative description of spatial protein organization. PC-PALM has been used to probe the heterogeneous organization of plasma membrane proteins (Sengupta et al. 2011). This analysis revealed that GPI-anchored proteins are aggregated in nanoclusters smaller than 60 nm, whose properties are sensitive to cholesterol and sphingomyelin levels as well as to the integrity of the actin cytoskeleton. The same strategy has also been used to demonstrate the alteration of GPI-anchored protein organization and actin polymerization upon cell treatment with pharmacologically relevant levels of ethanol (Tobin et al. 2014) and the distinct protein distribution of two splicing isoforms of the glycine receptor (Notelaers et al. 2014).

Concerning protein compartmentalization in tetraspanin domains, dSTORM imaging has been used to investigate whether tetraspanin-directed arrangements of integrins can affect cell adhesion (Termini et al. 2014). Here, pair autocorrelation function analysis of dSTORM images, developed by (Veatch et al. 2012), has been used to determine the distribution of the tetraspanin CD82 and to assess the role of palmitoylation in CD82 clustering. In addition, density-based spatial clustering of applications with noise (DBSCAN) (Ester et al. 1996), an algorithm that allows density-based detection of clusters, proved that integrin aggregation and density in clusters is modulated by CD82, allowing proper cell adhesion to the extracellular matrix.

In the context of membrane trafficking, the role of galactin-3 in the clustering of cargo proteins for clathrin-independent endocytosis has been recently demonstrated (Lakshminarayan et al. 2014). In particular, galactin-3 has been shown to trigger the formation of clathrin-independent carriers (CLICs) upon interaction with glycosylated proteins containing N-acetylglucosamine saccharides, such as CD44. In this case, dSTORM imaging and *K*-Ripley function analysis revealed that the aggregation of galectin-3 in small clusters (75 ± 2 nm) at the cell surface is glycosphingolipid-dependent and that the N-glycosylation on CD44 is necessary for its clustering and uptake.

The relationship between host-cell actin and the glycoprotein influenza viral membrane protein hemagglutinin was studied by FPALM and pair correlation function demonstrating the localization of hemagglutinin in actin-rich membrane domains (Gudheti et al. 2013). Furthermore, blinking microscopy was used to investigate the relationship between these clusters and C-type lectin underlining their role in virus binding (Itano et al. 2012). Additionally, Gaussian fits to individual clusters have been used to characterize human immunodeficiency virus 1 (HIV1) interaction with host-cell proteins, like the interferon-induced transmembrane protein tetherin (Lehmann et al. 2011).

Spatial distribution analysis of proteins has also improved the characterization of events occurring in the specialized plasma membrane structures present in neuronal and immunological synapses where signal transmission takes place through a multitude of events highly coordinated in time and space.

4.4.2.1 The Neuronal Synapse

Neuronal synapses are specialized junctions (10–20 nm) that allow the transmission of chemical signals between neurons or between neurons and nonneuronal cells, for instance, muscle cells. In the central nervous system, interneuronal chemical synapses are characterized by the presence of specialized zones in both presynaptic and postsynaptic neurons (Fig. 4.4a). The specialized presynaptic element is called active zone and comprises a multitude of proteins involved in the exocytosis of synaptic vesicles (Südhof 2012). The postsynaptic element is referred to as postsynaptic density (PSD), which consists of membrane and cytoplasmic proteins, including neurotransmitter receptors, scaffolding proteins, and adhesion molecules involved in synaptic signal transduction and cell adhesion (Okabe 2007).

At the presynaptic active zone, particular interest has been directed to the organization of SNARE (soluble *N*-ethylmaleimide-sensitive factor attachment receptor) proteins implicated in synaptic vesicle fusion and neuronal exocytosis such as syntaxin 1 and SNAP-25 (25 kDa synaptosome-associated protein) (Milovanovic and Jahn 2015). STED microscopy in combination with fluorescence recovery after photobleaching (FRAP) has shown that syntaxin-1 exists both as clusters of 50–60 nm and free-diffusing single molecules (Sieber et al. 2007). The organization of syntaxin within the clusters and the distribution of the single molecules outside clusters were subsequently elucidated by distribution-based clustering of large spatial databases (DBCLASD) algorithm (Xu et al. 1998) and dSTORM imaging (Bar-On et al. 2012). These investigations demonstrated a nonuniform distribution of syntaxin within the clusters, where protein density decreases toward the periphery of the clusters compared to the center and that single molecules are mostly located around the clusters. Cross-species pair correlation function, applied on two-color super-resolution images, gave detailed information on the assembly of a complex constituted by syntaxin-1, SNAP-25, and Secretory 1 (Sec1)/mammalian

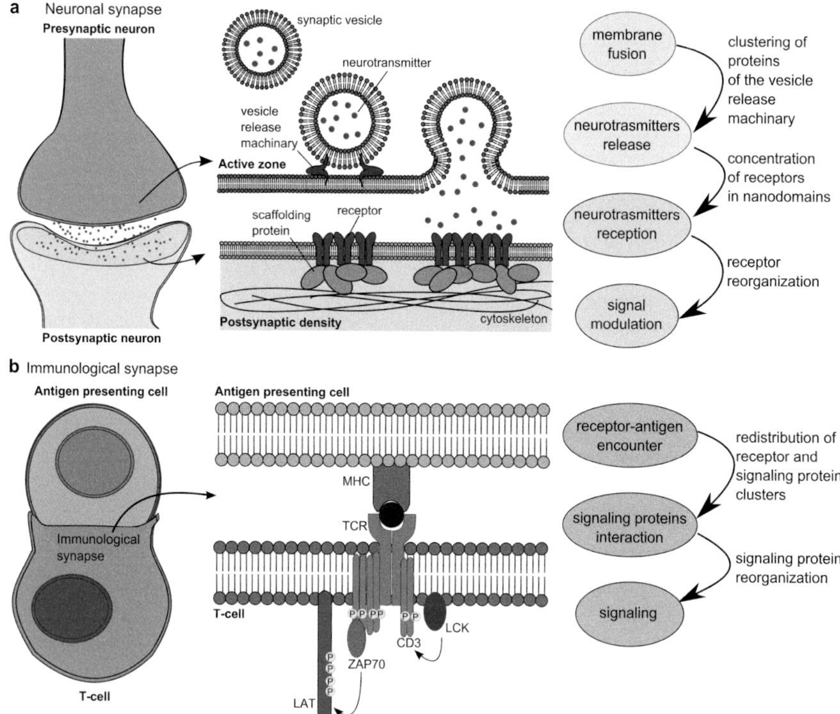

Fig. 4.4 Schematic of specialized plasma membrane structures: the neuronal and the immunological synapse. (**a**) Neuronal synapse. Neurotransmitter release occurs at the active zone of the presynaptic neuron; vesicle release is mediated by vesicle release machinery, which includes the SNARE complex. At the postsynaptic density, neurotransmitter receptors, like AMPA receptor, and scaffolding proteins, like PSD95, organize in domains to facilitate neurotransmission. (**b**) Immunological synapse (image adapted from Friedl et al. (2005)). Upon antigen recognition by TCR, CD3 is phosphorylated by the kinase LCK. This leads to ZAP70 recruitment, which phosphorylates the adaptor LAT

uncoordinated-18 (Munc18)-like protein (Pertsinidis et al. 2013). These complexes appear to be preassembled in $\leq$100 nm microdomains at the plasma membrane of primary neurons. In parallel, the influence of phosphatidylinositol 4, 5-bisphosphate (PI(4,5)P_2) (a phospholipid mainly present in the inner leaflet of the plasma membrane) on SNARE protein clustering has been investigated to further elucidate SNARE-dependent membrane fusion and exocytosis (James et al. 2008). Confocal microscopy in combination with colocalization analysis has revealed that fusion events driven by syntaxin occur at the level of PI(4,5)P_2 clusters (Aoyagi et al. 2005). Here electrostatic protein-lipid interactions steer the organization of these microdomains (van den Bogaart et al. 2011) that have been proposed to act as molecular beacons for vesicle recruitment (Honigmann et al. 2013). In addition, it has been shown through PiMP imaging (Munck et al. 2012) and colocalization analysis that also PI(3,4,5)P_2 facilitates syntaxin 1A clustering at the presynaptic membrane, again through electrostatic interactions (Khuong et al. 2013).

At the postsynaptic density, the arrangement of neurotransmitter receptors, ion channels, and scaffolding proteins is of particular interest given that synapse function can be affected by PSD architecture. One of the first quantitative measurements of receptor organization on super-resolution microscopy images was the arrangement of nicotinic acetylcholine receptors. This receptor, imaged by STED microscopy and analyzed by *K*-Ripley function (Kellner et al. 2007), has been found to be organized in nanoclusters, whose size is cholesterol dependent. Spatial distribution of the sodium potassium pump (Na^+/K^+ ATPase) and dopamine D1 receptor was investigated by nearest neighbor analysis on STED images (Blom et al. 2012). Quantification of their spatial relationship revealed joint and separated confinement of D1 receptors and Na^+/K^+ ATPase pumps, providing novel insights into the modulation of synaptic transmission. Combination of super-resolution microscopy techniques and electron microscopy has shed light on the AMPA receptor arrangement at the PSD (Nair et al. 2013). It was shown that the AMPA receptor has a nano-organization in clusters of ~70 nm (containing ~20 receptors) that is dynamic in time and space and is regulated by the scaffolding protein PSD95 (Nair et al. 2013). In this regard, pair correlation-based analysis of live-cell PALM images has been adopted to map the spatial distribution of glutamate receptors and PSD95 within single PSDs (MacGillavry et al. 2013). The results from these studies indicate that AMPA receptors cluster within the PSD by distributing preferentially at the site of PSD95 domains. This heterogeneous distribution enables the shaping of postsynaptic responses. Additionally, high-density single-molecule tracking has been adopted to understand AMPA receptor molecular interactions (Hoze et al. 2012) and motility in relation to its conformational state (Constals et al. 2015).

4.4.2.2 The Immunological Synapse

The immunological synapse is an ordered and dynamic interface between two cells of the immune system (Xie et al. 2013). For instance, an immunological synapse is established between T cells and antigen-presenting cells (APCs). At this interface, proteins with various roles, like receptors, signaling molecules, and scaffolds, trigger multiple events that finally lead to an immune response (Fig. 4.4b). We refer to appropriate reviews for details on the process of T-cell activation and signal transduction (Smith-Garvin et al. 2009; Malissen et al. 2014). Briefly, the activation of T cells begins with the interaction between the T-cell receptor (TCR) and the peptide exposed by the major histocompatibility complex (MHC) class I or class II, located on the surface of the APC. The TCR forms a multisubunit complex with the cluster of differentiation 3 (CD3) that contains the dimers CD$\varepsilon\gamma$, CD$\varepsilon\delta$, and CD$\xi\xi$, whose cytoplasmic regions present the immunoreceptor tyrosine-based activation motifs (ITAMs). The binding between the TCR and the MHC induces the phosphorylation of ITAMs by Src protein tyrosine kinases including the lymphocyte-specific protein kinase (Lck). Phosphorylation of ITAMs allows the recruitment and activation of the zeta chain-associated protein kinase of 70 kDa (ZAP-70) and the subsequent phosphorylation of the linker for activation of T cells (LAT). LAT functions as a platform for the signaling proteins involved in the next steps of the signaling cascade. The signal propagation induced by LAT activation results in cytokine secretion.

Given the high coordination and organization of the proteins involved in T-cell activation, it is critically important to assess their spatial arrangement for a better comprehension of the mechanisms leading to the immunological response (Rossy et al. 2012). For this purpose, super-resolution microscopy techniques and spatial statistical analysis have been used to bring new insights into the nanoscale spatiotemporal organization of the immunological synapse. Because of the importance of LAT recruitment and activation, different studies focused on the analysis of its localization. Initially, the organization of TCR in relation with the distribution of LAT has been investigated by employing K-Ripley function analysis applied on both high-speed PALM and transmission electron microscopy (TEM) (Lillemeier et al. 2010). From the application of these techniques, in combination with fluorescence cross-correlation spectroscopy, the presence of preclustered TCR and LAT on quiescent T cells was demonstrated. According to this work, the domains in which TCR and LAT are concentrated (here called protein islands having a width of 100–220 nm) are separated, and after T-cell activation, they concatenate with each other, but they do not merge. Two-color PALM and pair correlation analysis were adopted to further study TCR clustering and its interaction with other signaling proteins including LAT and ZAP70 (Sherman et al. 2011). Differently from previous observations, this analysis showed that TCR and LAT nanoclusters are mixed in resting T cells, and stimulation causes a modest growth in LAT cluster size. When investigating LAT clusters employing K-Ripley function on PALM and dSTORM images (Williamson et al. 2011), the preexisting LAT domains present in the resting cells were shown not to be phosphorylated nor recruited by the TCR. Instead, LAT molecules were recruited from subcellular vesicles and phosphorylated. This is however contradictory to another study –alternatively using chimeric CD4-LAT, which demonstrated that surface localized LAT was phosphorylated (Balagopalan et al. 2013). The differences in the results obtained by these different studies could be related to the different experimental conditions. Nevertheless the employment of a variety of microscopy techniques and distribution analysis underlines the effort and the importance of protein organization assessment.

4.5 Conclusions and Outlook

The importance of quantitative measurements in biology together with the high-level development in microscopy technologies has favored the dawn of novel analysis paradigms aimed at protein pattern characterization, which is also explicated in the availability of the described tools for their use on image processing packaging like ImageJ/Fiji (Schindelin et al. 2012). PC-PALM analysis plugins are available in the package “GDSC-SMLM” provided by the University of Sussex (http://sites.imagej.net/GDSC-SMLM/); *K*-Ripley function analysis can be run using “Icy-Spatial Analysis” (http://icy.bioimageanalysis.org/plugin/Spatial_Analysis); other distribution analysis tools can be found in the “BioVoxxel Toolbox” (http://fiji.sc/BioVoxxel_Toolbox) or in the R toolbox *spatstat* (Baddeley and Turner 2005). Despite still being an emerging field, protein distribution analysis has already provided valuable insights into the mechanisms of protein compartmentalization at the plasma membrane. The improved understanding of these mechanisms has in turn enabled a better understanding of initiation and transduction of signaling processes. Nevertheless, the application of spatial statistics to super-resolution images has still to deal with difficulties related to the possible presence of artifacts in the images. In single-molecule localization microscopy images, for instance, distribution analyses can lead to erroneous results if various parameters, such as labeling density, photoactivation, or photoswitching rate of dyes, the image reconstruction process, are not kept under control (Endesfelder and Heilemann 2014). Inefficient sample preparation (Whelan and Bell 2015) or false multiple localizations, due to inappropriate imaging and image processing conditions, (Annibale et al. 2011; Burgert et al. 2015) can result in the formation of artificial clusters. Too low labeling density may lead to an incorrect detection of homogeneous patterns; in fact, the overall protein organization could be corrupted by the fact that just a subset of molecules in the sample is imaged. Concurrently with spatial distribution analysis, other analytical methods have supported investigations toward protein organization, including oligomerization analysis (Godin et al. 2011) and protein interaction analysis (Helmuth et al. 2010; Shivanandan et al. 2013). Together these techniques allow us to gradually build a more detailed picture of protein arrangement in the plasma membrane starting from the oligomerization state of proteins, over the organization in clusters, to the definition of their interacting partners. While the picture that has emerged is complex, it is far from complete. Valuable information could be obtained from a more global analysis of the plasma membrane combined with the local investigation of cluster properties. As in other disciplines, like histopathology (Nawaz et al. 2015) and epidemiology (Haque et al. 2014), hotspot analysis through Getis–Ord G-statistics (Burt et al. 2009; Getis and Ord 1992) could be employed. In the context of analyzing protein distributions on the plasma membrane, one could think of multiplexing several readouts like calcium imaging and spatial analysis. The signaling readout could then be used as an additional weight for the distances of active receptors to determine the localization of hot spots for signaling initiation on the cell surface and thus mapping the cell surface based on its functionality. In the future, the integration of different readouts and analysis paradigms will create a more comprehensive description of the processes and arrangements on the plasma membrane.

Acknowledgments The authors would like to thank Dr. Susana Rocha and Dr. Vinoy Vijayan for their fruitful discussion and Dr. Donna Stolz for the silica particles to isolate plasma membranes. This work is financially supported by VIB, VIB Bio Imaging Core facility, the Hercules Foundation for heavy infrastructure (Hercules AKUL058/HER/08/021, AKUL/09/037, and AKUL13/39 (ISPAMM)), KU Leuven (IDO/12/020), the federal government (IAP P7/16), and SAO-FRA (S#14017). SM was supported by a grant from KU Leuven (CREA/12/22).

References

Abbe E (1873) Beiträge zur Theorie des Mikroskops und der mikroskopischen Wahrnehmung. Arch Für Mikrosk Anat 9:413–418. doi:10.1007/BF02956173

Adam V, Moeyaert B, David CC et al (2011) Rational design of photoconvertible and biphotochromic fluorescent proteins for advanced microscopy applications. Chem Biol 18:1241–1251. doi:10.1016/j.chembiol.2011.08.007

Annibale P, Vanni S, Scarselli M et al (2011) Identification of clustering artifacts in photoactivated localization microscopy. Nat Methods 8:527–528. doi:10.1038/nmeth.1627

Aoyagi K, Sugaya T, Umeda M et al (2005) The activation of exocytotic sites by the formation of phosphatidylinositol 4,5-bisphosphate microdomains at syntaxin clusters. J Biol Chem 280:17346–17352. doi:10.1074/jbc.M413307200

Baddeley A (2007) Spatial point processes and their applications. In: Stochastic geometry. Springer, Berlin Heidelberg, pp 1–75

Baddeley A, Turner R (2005) spatstat: An R package for analyzing spatial point patterns. J Stat Softw 12(6). doi:10.18637/jss.v012.i06

Baddeley D, Cannell MB, Soeller C (2011) Three-dimensional sub-100 nm super-resolution imaging of biological samples using a phase ramp in the objective pupil. Nano Res 4:589–598. doi:10.1007/s12274-011-0115-z

Balagopalan L, Barr VA, Kortum RL et al (2013) Cutting edge: cell surface linker for activation of T cells is recruited to microclusters and is active in signaling. J Immunol Baltim Md 190:3849–3853. doi:10.4049/jimmunol.1202760

Bar-On D, Wolter S, van de Linde S et al (2012) Super-resolution imaging reveals the internal architecture of nano-sized syntaxin clusters. J Biol Chem 287:27158–27167. doi:10.1074/jbc.M112.353250

Barreiro O, Zamai M, Yáñez-Mó M et al (2008) Endothelial adhesion receptors are recruited to adherent leukocytes by inclusion in preformed tetraspanin nanoplatforms. J Cell Biol 183:527–542. doi:10.1083/jcb.200805076

Belardi B, O'Donoghue GP, Smith AW et al (2012) Investigating cell surface galectin-mediated cross-linking on glycoengineered cells. J Am Chem Soc 134:9549–9552. doi:10.1021/ja301694s

Betzig E, Chichester RJ (1993) Single molecules observed by near-field scanning optical microscopy. Science 262:1422–1425. doi:10.1126/science.262.5138.1422

Betzig E, Trautman JK (1992) Near-field optics: microscopy, spectroscopy, and surface modification beyond the diffraction limit. Science 257:189–195. doi:10.1126/science.257.5067.189

Betzig E, Patterson GH, Sougrat R et al (2006) Imaging intracellular fluorescent proteins at nanometer resolution. Science 313:1642–1645. doi:10.1126/science.1127344

Blom H, RöNnlund D, Scott L et al (2012) Nearest neighbor analysis of dopamine D1 receptors and Na+-K+-ATPases in dendritic spines dissected by STED microscopy. Microsc Res Tech 75:220–228. doi:10.1002/jemt.21046

Bolte S, Cordelières FP (2006) A guided tour into subcellular colocalization analysis in light microscopy. J Microsc 224:213–232. doi:10.1111/j.1365-2818.2006.01706.x

Boscher C, Dennis JW, Nabi IR (2011) Glycosylation, galectins and cellular signaling. Curr Opin Cell Biol 23:383–392. doi:10.1016/j.ceb.2011.05.001

Boucheix C, Rubinstein E (2001) Tetraspanins. Cell Mol Life Sci CMLS 58:1189–1205

Brewer CF, Miceli MC, Baum LG (2002) Clusters, bundles, arrays and lattices: novel mechanisms for lectin–saccharide-mediated cellular interactions. Curr Opin Struct Biol 12:616–623. doi:10.1016/S0959-440X(02)00364-0

Brown DA, Rose JK (1992) Sorting of GPI-anchored proteins to glycolipid-enriched membrane subdomains during transport to the apical cell surface. Cell 68:533–544

Burgert A, Letschert S, Doose S, Sauer M (2015) Artifacts in single-molecule localization microscopy. Histochem Cell Biol 144:123–131. doi:10.1007/s00418-015-1340-4

Burt JE, Barber GM, Rigby DL (2009) Elementary statistics for geographers, 3rd edn. Guilford Press, New York

Cardona A, Saalfeld S, Preibisch S et al (2010) An integrated micro- and macroarchitectural analysis of the drosophila brain by computer-assisted serial section electron microscopy. PLoS Biol 8, e1000502. doi:10.1371/journal.pbio.1000502

Chaney LK, Jacobson BS (1983) Coating cells with colloidal silica for high yield isolation of plasma membrane sheets and identification of transmembrane proteins. J Biol Chem 258:10062–10072

Chaudhary N, Gomez GA, Howes MT et al (2014) Endocytic crosstalk: cavins, caveolins, and caveolae regulate clathrin-independent endocytosis. PLoS Biol 12, e1001832. doi:10.1371/journal.pbio.1001832

Constals A, Penn AC, Compans B et al (2015) Glutamate-induced AMPA receptor desensitization increases their mobility and modulates short-term plasticity through unbinding from stargazin. Neuron 85:787–803. doi:10.1016/j.neuron.2015.01.012

Cox S, Rosten E, Monypenny J et al (2012) Bayesian localization microscopy reveals nanoscale podosome dynamics. Nat Methods 9:195–200. doi:10.1038/nmeth.1812

Cressie ACN (1993) Statistics for spatial data. Wiley and Sons, New York, Revised Edition

de Bakker BI, de Lange F, Cambi A et al (2007) Nanoscale organization of the pathogen receptor DC-SIGN mapped by single-molecule high-resolution fluorescence microscopy. Chem Phys Chem 8:1473–1480. doi:10.1002/cphc.200700169

de Lange F, Cambi A, Huijbens R et al (2001) Cell biology beyond the diffraction limit: near-field scanning optical microscopy. J Cell Sci 114:4153–4160

Dertinger T, Colyer R, Iyer G et al (2009) Fast, background-free, 3D super-resolution optical fluctuation imaging (SOFI). Proc Natl Acad Sci U S A 106:22287–22292. doi:10.1073/pnas.0907866106

Deschout H, Zanacchi FC, Mlodzianoski M et al (2014a) Precisely and accurately localizing single emitters in fluorescence microscopy. Nat Methods 11:253–266. doi:10.1038/nmeth.2843

Deschout H, Shivanandan A, Annibale P et al (2014b) Progress in quantitative single-molecule localization microscopy. Histochem Cell Biol 142:5–17. doi:10.1007/s00418-014-1217-y

Diaz-Rohrer B, Levental KR, Levental I (2014) Rafting through traffic: membrane domains in cellular logistics. Biochim Biophys Acta BBA – Biomembr 1838:3003–3013. doi:10.1016/j.bbamem.2014.07.029

Dixon PM (2002) Ripley's K function. In: El-Shaarawi AH, Piegorsch WW (eds) Encyclopedia of environmetrics. John Wiley & Sons, Ltd, Chichester, pp 1796–1803

Donnert G, Keller J, Wurm CA et al (2007) Two-color far-field fluorescence nanoscopy. Biophys J 92:L67–L69. doi:10.1529/biophysj.107.104497

Ehrig J, Petrov EP, Schwille P (2011) Near-critical fluctuations and cytoskeleton-assisted phase separation lead to subdiffusion in cell membranes. Biophys J 100:80–89. doi:10.1016/j.bpj.2010.11.002

Endesfelder U, Heilemann M (2014) Art and artifacts in single-molecule localization microscopy: beyond attractive images. Nat Methods 11:235–238. doi:10.1038/nmeth.2852

Espenel C, Margeat E, Dosset P et al (2008) Single-molecule analysis of CD9 dynamics and partitioning reveals multiple modes of interaction in the tetraspanin web. J Cell Biol 182:765–776. doi:10.1083/jcb.200803010

Ester M, Kriegel H, Sander J, Xu X (1996) A density-based algorithm for discovering clusters in large spatial databases with noise. AAAI Press, Palo Alta, pp 226–231. doi: 10.1023/A:1009745219419

Feigenson GW (2007) Phase boundaries and biological membranes. Annu Rev Biophys Biomol Struct 36:63–77. doi:10.1146/annurev.biophys.36.040306.132721

Flors C, Hotta J, Uji-i H et al (2007) A stroboscopic approach for fast photoactivation−localization microscopy with dronpa mutants. J Am Chem Soc 129:13970–13977. doi:10.1021/ja0747041

Fölling J, Bossi M, Bock H et al (2008) Fluorescence nanoscopy by ground-state depletion and single-molecule return. Nat Methods 5:943–945. doi:10.1038/nmeth.1257

Fornasiero EF, Rizzoli SO (2014) Super-resolution microscopy techniques in the neurosciences. Humana Press, Totowa

Friedl P, den Boer AT, Gunzer M (2005) Tuning immune responses: diversity and adaptation of the immunological synapse. Nat Rev Immunol 5:532–545. doi:10.1038/nri1647

Fujiwara T, Ritchie K, Murakoshi H et al (2002) Phospholipids undergo hop diffusion in compartmentalized cell membrane. J Cell Biol 157:1071–1082. doi:10.1083/jcb.200202050

Gambin Y, Ariotti N, McMahon K-A et al (2014) Single-molecule analysis reveals self assembly and nanoscale segregation of two distinct cavin subcomplexes on caveolae. eLife 3, e01434. doi:10.7554/eLife.01434

Geisler C, Schönle A, von Middendorff C et al (2007) Resolution of λ /10 in fluorescence microscopy using fast single molecule photo-switching. Appl Phys A 88:223–226. doi:10.1007/s00339-007-4144-0

Getis A, Ord JK (1992) The analysis of spatial association by use of distance statistics. Geogr Anal 24:189–206. doi:10.1111/j.1538-4632.1992.tb00261.x

Godin AG, Costantino S, Lorenzo L-E et al (2011) Revealing protein oligomerization and densities in situ using spatial intensity distribution analysis. Proc Natl Acad Sci 108:7010–7015. doi:10.1073/pnas.1018658108

Gong W, Si K, Chen N, Sheppard CJR (2010) Focal modulation microscopy with annular apertures: a numerical study. J Biophotonics 3:476–484. doi:10.1002/jbio.200900110

Gorter E, Grendel F (1925) On bimolecular layers of lipoids on the chromocytes of the blood. J Exp Med 41:439–443

Greig-Smith P (1952) The Use of random and contiguous quadrats in the study of the structure of plant communities. Ann Bot 16:293–316

Gudheti MV, Curthoys NM, Gould TJ et al (2013) Actin mediates the nanoscale membrane organization of the clustered membrane protein influenza hemagglutinin. Biophys J 104:2182–2192. doi:10.1016/j.bpj.2013.03.054

Gustafsson MG (2000) Surpassing the lateral resolution limit by a factor of two using structured illumination microscopy. J Microsc 198:82–87

Gustafsson MGL (2005) Nonlinear structured-illumination microscopy: wide-field fluorescence imaging with theoretically unlimited resolution. Proc Natl Acad Sci U S A 102:13081–13086. doi:10.1073/pnas.0406877102

Gustafsson MGL, Shao L, Carlton PM et al (2008) Three-dimensional resolution doubling in wide-field fluorescence microscopy by structured illumination. Biophys J 94:4957–4970. doi:10.1529/biophysj.107.120345

Hamel V, Guichard P, Fournier M et al (2014) Correlative multicolor 3D SIM and STORM microscopy. Biomed Opt Exp 5:3326–3336. doi:10.1364/BOE.5.003326

Haque U, Overgaard HJ, Clements ACA et al (2014) Malaria burden and control in Bangladesh and prospects for elimination: an epidemiological and economic assessment. Lancet Glob Health 2:e98–e105. doi:10.1016/S2214-109X(13)70176-1

Heilemann M, van de Linde S, Schüttpelz M et al (2008) Subdiffraction-resolution fluorescence imaging with conventional fluorescent probes. Angew Chem Int Ed 47:6172–6176. doi:10.1002/anie.200802376

Heilemann M, van de Linde S, Mukherjee A, Sauer M (2009) Super-resolution imaging with small organic fluorophores. Angew Chem Int Ed Engl 48:6903–6908. doi:10.1002/anie.200902073

Hein B, Willig KI, Hell SW (2008) Stimulated emission depletion (STED) nanoscopy of a fluorescent protein-labeled organelle inside a living cell. Proc Natl Acad Sci 105:14271–14276. doi:10.1073/pnas.0807705105

Heintzmann R (2003) Saturated patterned excitation microscopy with two-dimensional excitation patterns. Micron Oxf Engl 34:283–291

Heintzmann R, Cremer CG (1999) Laterally modulated excitation microscopy: improvement of resolution by using a diffraction grating. Proc SPIE 3568:185–196. doi: 10.1117/12.336833

Heintzmann R, Jovin TM, Cremer C (2002) Saturated patterned excitation microscopy–a concept for optical resolution improvement. J Opt Soc Am A Opt Image Sci Vis 19:1599–1609

Hell SW, Kroug M (1995) Ground-state-depletion fluorescence microscopy: a concept for breaking the diffraction resolution limit. Appl Phys B 60:495–497. doi:10.1007/BF01081333

Hell SW, Wichmann J (1994) Breaking the diffraction resolution limit by stimulated emission: stimulated-emission-depletion fluorescence microscopy. Opt Lett 19:780–782

Helmuth JA, Paul G, Sbalzarini IF (2010) Beyond co-localization: inferring spatial interactions between sub-cellular structures from microscopy images. BMC Bioinformatics 11:1–12. doi:10.1186/1471-2105-11-372

Hemler ME (2003) Tetraspanin proteins mediate cellular penetration, invasion, and fusion events and define a novel type of membrane microdomain. Annu Rev Cell Dev Biol 19:397–422. doi:10.1146/annurev.cellbio.19.111301.153609

Hemler ME (2005) Tetraspanin functions and associated microdomains. Nat Rev Mol Cell Biol 6:801–811. doi:10.1038/nrm1736

Henriques R, Lelek M, Fornasiero EF et al (2010) QuickPALM: 3D real-time photoactivation nanoscopy image processing in ImageJ. Nat Methods 7:339–340. doi:10.1038/nmeth0510-339

Hess ST, Girirajan TPK, Mason MD (2006) Ultra-high resolution imaging by fluorescence photoactivation localization microscopy. Biophys J 91:4258–4272. doi:10.1529/biophysj.106.091116

Honigmann A, van den Bogaart G, Iraheta E et al (2013) Phosphatidylinositol 4,5-bisphosphate clusters act as molecular beacons for vesicle recruitment. Nat Struct Mol Biol 20:679–686. doi:10.1038/nsmb.2570

Horejsi V, Hrdinka M (2014) Membrane microdomains in immunoreceptor signaling. FEBS Lett 588:2392–2397. doi:10.1016/j.febslet.2014.05.047

Hoze N, Nair D, Hosy E et al (2012) Heterogeneity of AMPA receptor trafficking and molecular interactions revealed by superresolution analysis of live cell imaging. Proc Natl Acad Sci 109:17052–17057. doi:10.1073/pnas.1204589109

Huang B, Wang W, Bates M, Zhuang X (2008) Three-dimensional super-resolution imaging by stochastic optical reconstruction microscopy. Science 319:810–813. doi:10.1126/science.1153529

Ipsen JH, Karlström G, Mouritsen OG et al (1987) Phase equilibria in the phosphatidylcholine-cholesterol system. Biochim Biophys Acta 905:162–172

Itano MS, Steinhauer C, Schmied JJ et al (2012) Super-resolution imaging of C-type lectin and influenza hemagglutinin nanodomains on plasma membranes using blink microscopy. Biophys J 102:1534–1542. doi:10.1016/j.bpj.2012.02.022

James DJ, Khodthong C, Kowalchyk JA, Martin TFJ (2008) Phosphatidylinositol 4,5-bisphosphate regulates SNARE-dependent membrane fusion. J Cell Biol 182:355–366. doi:10.1083/jcb.200801056

Juette MF, Gould TJ, Lessard MD et al (2008) Three-dimensional sub–100 nm resolution fluorescence microscopy of thick samples. Nat Methods 5:527–529. doi:10.1038/nmeth.1211

Kellner RR, Baier CJ, Willig KI et al (2007) Nanoscale organization of nicotinic acetylcholine receptors revealed by stimulated emission depletion microscopy. Neuroscience 144:135–143. doi:10.1016/j.neuroscience.2006.08.071

Khuong TM, Habets RLP, Kuenen S et al (2013) Synaptic PI(3,4,5)P3 is required for Syntaxin1A clustering and neurotransmitter release. Neuron 77:1097–1108. doi:10.1016/j.neuron.2013.01.025

Kusumi A, Sako Y (1996) Cell surface organization by the membrane skeleton. Curr Opin Cell Biol 8:566–574. doi:10.1016/S0955-0674(96)80036-6

Kusumi A, Sako Y, Yamamoto M (1993) Confined lateral diffusion of membrane receptors as studied by single particle tracking (nanovid microscopy). Effects of calcium-induced differentiation in cultured epithelial cells. Biophys J 65:2021–2040

Kusumi A, Suzuki KGN, Kasai RS et al (2011) Hierarchical mesoscale domain organization of the plasma membrane. Trends Biochem Sci 36:604–615. doi:10.1016/j.tibs.2011.08.001

Kusumi A, Tsunoyama TA, Hirosawa KM et al (2014) Tracking single molecules at work in living cells. Nat Chem Biol 10:524–532. doi:10.1038/nchembio.1558

Lagache T, Lang G, Sauvonnet N, Olivo-Marin J-C (2013) Analysis of the spatial organization of molecules with robust statistics. PLoS One 8, e80914. doi:10.1371/journal.pone.0080914

Lakshminarayan R, Wunder C, Becken U et al (2014) Galectin-3 drives glycosphingolipid-dependent biogenesis of clathrin-independent carriers. Nat Cell Biol 16:592–603. doi:10.1038/ncb2970

Lang T, Rizzoli SO (2010) Membrane protein clusters at nanoscale resolution: more than pretty pictures. Phys Chem Chem Phys 25:116–124. doi:10.1152/physiol.00044.2009

Lehmann M, Rocha S, Mangeat B et al (2011) Quantitative multicolor super-resolution microscopy reveals tetherin HIV-1 interaction. PLoS Pathog 7, e1002456. doi:10.1371/journal.ppat.1002456

Levental I, Lingwood D, Grzybek M et al (2010) Palmitoylation regulates raft affinity for the majority of integral raft proteins. Proc Natl Acad Sci U S A 107:22050–22054. doi:10.1073/pnas.1016184107

Li R, Gundersen GG (2008) Beyond polymer polarity: how the cytoskeleton builds a polarized cell. Nat Rev Mol Cell Biol 9:860–873. doi:10.1038/nrm2522

Lidke K, Rieger B, Jovin T, Heintzmann R (2005) Superresolution by localization of quantum dots using blinking statistics. Opt Express 13:7052–7062

Lillemeier BF, Mörtelmaier MA, Forstner MB et al (2010) TCR and Lat are expressed on separate protein islands on T cell membranes and concatenate during activation. Nat Immunol 11:90–96. doi:10.1038/ni.1832

London E (2002) Insights into lipid raft structure and formation from experiments in model membranes. Curr Opin Struct Biol 12:480–486. doi:10.1016/S0959-440X(02)00351-2

Ludwig A, Howard G, Mendoza-Topaz C et al (2013) Molecular composition and ultrastructure of the caveolar coat complex. PLoS Biol 11, e1001640. doi:10.1371/journal.pbio.1001640

MacGillavry HD, Song Y, Raghavachari S, Blanpied TA (2013) Nanoscale scaffolding domains within the postsynaptic density concentrate synaptic AMPA receptors. Neuron 78:615–622. doi:10.1016/j.neuron.2013.03.009

Malissen B, Grégoire C, Malissen M, Roncagalli R (2014) Integrative biology of T cell activation. Nat Immunol 15:790–797. doi:10.1038/ni.2959

Mandula O, Šestak IŠ, Heintzmann R, Williams CKI (2014) Localisation microscopy with quantum dots using non-negative matrix factorisation. Opt Express 22:24594–24605

Marin R, Rojo JA, Fabelo N et al (2013) Lipid raft disarrangement as a result of neuropathological progresses: a novel strategy for early diagnosis? Neuroscience 245:26–39. doi:10.1016/j.neuroscience.2013.04.025

Meister M, Tikkanen R (2014) Endocytic trafficking of membrane-bound cargo: a flotillin point of view. Membranes 4:356–371. doi:10.3390/membranes4030356

Milovanovic D, Jahn R (2015) Organization and dynamics of SNARE proteins in the presynaptic membrane. Front Physiol 6:89. doi:10.3389/fphys.2015.00089

Mollinedo F, Gajate C (2015) Lipid rafts as major platforms for signaling regulation in cancer. Adv Biol Regul 57:130–146. doi:10.1016/j.jbior.2014.10.003

Munck S, Miskiewicz K, Sannerud R et al (2012) Sub-diffraction imaging on standard microscopes through photobleaching microscopy with non-linear processing. J Cell Sci 125:2257–2266. doi:10.1242/jcs.098939

Nair D, Hosy E, Petersen JD et al (2013) Super-resolution imaging reveals that AMPA receptors inside synapses are dynamically organized in nanodomains regulated by PSD95. J Neurosci 33:13204–13224. doi:10.1523/JNEUROSCI.2381-12.2013

Nawaz S, Heindl A, Koelble K, Yuan Y (2015) Beyond immune density: critical role of spatial heterogeneity in estrogen receptor-negative breast cancer. Mod Pathol Off J U S Can Acad Pathol Inc. doi:10.1038/modpathol.2015.37

Nicolson GL (2014) The fluid-mosaic model of membrane structure: still relevant to understanding the structure, function and dynamics of biological membranes after more than 40 years. Biochim Biophys Acta 1838:1451–1466. doi:10.1016/j.bbamem.2013.10.019

Nieuwenhuizen RPJ, Lidke KA, Bates M et al (2013) Measuring image resolution in optical nanoscopy. Nat Methods 10:557–562. doi:10.1038/nmeth.2448

Notelaers K, Rocha S, Paesen R et al (2014) Membrane distribution of the glycine receptor α3 studied by optical super-resolution microscopy. Histochem Cell Biol 142:79–90. doi:10.1007/s00418-014-1197-y

Okabe S (2007) Molecular anatomy of the postsynaptic density. Mol Cell Neurosci 34:503–518. doi:10.1016/j.mcn.2007.01.006

Otto GP, Nichols BJ (2011) The roles of flotillin microdomains–endocytosis and beyond. J Cell Sci 124:3933–3940. doi:10.1242/jcs.092015

Palade GE (1953) Fine structure of blood capillaries. J Appl Phys 24:1424

Parton RG, del Pozo MA (2013) Caveolae as plasma membrane sensors, protectors and organizers. Nat Rev Mol Cell Biol 14:98–112. doi:10.1038/nrm3512

Parton RG, Simons K (2007) The multiple faces of caveolae. Nat Rev Mol Cell Biol 8:185–194. doi:10.1038/nrm2122

Pavani SRP, Thompson MA, Biteen JS et al (2009) Three-dimensional, single-molecule fluorescence imaging beyond the diffraction limit by using a double-helix point spread function. Proc Natl Acad Sci U S A 106:2995–2999. doi:10.1073/pnas.0900245106

Pelkmans L, Bürli T, Zerial M, Helenius A (2004) Caveolin-stabilized membrane domains as multifunctional transport and sorting devices in endocytic membrane traffic. Cell 118:767–780. doi:10.1016/j.cell.2004.09.003

Pertsinidis A, Mukherjee K, Sharma M et al (2013) Ultrahigh-resolution imaging reveals formation of neuronal SNARE/Munc18 complexes in situ. Proc Natl Acad Sci 110:E2812–E2820. doi:10.1073/pnas.1310654110

Peters KR, Carley WW, Palade GE (1985) Endothelial plasmalemmal vesicles have a characteristic striped bipolar surface structure. J Cell Biol 101:2233–2238

Pike LJ (2006) Rafts defined: a report on the Keystone symposium on lipid rafts and cell function. J Lipid Res 47:1597–1598. doi:10.1194/jlr.E600002-JLR200

Ramadoss J, Pastore MB, Magness RR (2013) Endothelial caveolar subcellular domain regulation of endothelial nitric oxide synthase. Clin Exp Pharmacol Physiol 40:753–764. doi:10.1111/1440-1681.12136

Rayleigh L (1903) On the theory of optical images, with special reference to the microscope. J R Microsc Soc 23:474–482. doi:10.1111/j.1365-2818.1903.tb04831.x

Ries J, Kaplan C, Platonova E et al (2012) A simple, versatile method for GFP-based super-resolution microscopy via nanobodies. Nat Methods 9:582–584. doi:10.1038/nmeth.1991

Ripley BD (1977) Modelling spatial patterns. J R Stat Soc B 39:172–212

Ristanović Z, Kerssens MM, Kubarev AV et al (2015) High-resolution single-molecule fluorescence imaging of zeolite aggregates within real-life fluid catalytic cracking particles. Angew Chem Int Ed Engl 54:1836–1840. doi:10.1002/anie.201410236

Ritchie K, Iino R, Fujiwara T et al (2003) The fence and picket structure of the plasma membrane of live cells as revealed by single molecule techniques (Review). Mol Membr Biol 20:13–18

Rossy J, Williamson DJ, Benzing C, Gaus K (2012) The integration of signaling and the spatial organization of the T cell synapse. Immunol Mem 3:352. doi:10.3389/fimmu.2012.00352

Rust MJ, Bates M, Zhuang X (2006) Sub-diffraction-limit imaging by stochastic optical reconstruction microscopy (STORM). Nat Methods 3:793–796. doi:10.1038/nmeth929

Saka SK, Honigmann A, Eggeling C et al (2014) Multi-protein assemblies underlie the mesoscale organization of the plasma membrane. Nat Commun. doi:10.1038/ncomms5509

Scarselli M, Annibale P, Radenovic A (2012) Cell type-specific β2-adrenergic receptor clusters identified using photoactivated localization microscopy are not lipid raft related, but depend on actin cytoskeleton integrity. J Biol Chem 287:16768–16780. doi:10.1074/jbc.M111.329912

Schermelleh L, Heintzmann R, Leonhardt H (2010) A guide to super-resolution fluorescence microscopy. J Cell Biol 190:165–175. doi:10.1083/jcb.201002018

Schindelin J, Arganda-Carreras I, Frise E et al (2012) Fiji: an open-source platform for biological-image analysis. Nat Methods 9:676–682. doi:10.1038/nmeth.2019

Schroeder R, London E, Brown D (1994) Interactions between saturated acyl chains confer detergent resistance on lipids and glycosylphosphatidylinositol (GPI)-anchored proteins: GPI-anchored proteins in liposomes and cells show similar behavior. Proc Natl Acad Sci U S A 91:12130–12134

Sengupta P, Lippincott-Schwartz J (2012) Quantitative analysis of photoactivated localization microscopy (PALM) datasets using pair-correlation analysis. BioEssays News Rev Mol Cell Dev Biol 34:396–405. doi:10.1002/bies.201200022

Sengupta P, Jovanovic-Talisman T, Skoko D et al (2011) Probing protein heterogeneity in the plasma membrane using PALM and pair correlation analysis. Nat Methods 8:969–975. doi:10.1038/nmeth.1704

Sengupta P, Jovanovic-Talisman T, Lippincott-Schwartz J (2013) Quantifying spatial organization in point-localization superresolution images using pair correlation analysis. Nat Protoc 8:345–354. doi:10.1038/nprot.2013.005

Sezgin E, Gutmann T, Buhl T et al (2015) Adaptive lipid packing and bioactivity in membrane domains. PLoS One 10, e0123930. doi:10.1371/journal.pone.0123930

Shao L, Kner P, Rego EH, Gustafsson MGL (2011) Super-resolution 3D microscopy of live whole cells using structured illumination. Nat Methods 8:1044–1046. doi:10.1038/nmeth.1734

Sharonov A, Hochstrasser RM (2006) Wide-field subdiffraction imaging by accumulated binding of diffusing probes. Proc Natl Acad Sci 103:18911–18916. doi:10.1073/pnas.0609643104

Sherman E, Barr V, Manley S et al (2011) Functional nanoscale organization of signaling molecules downstream of the T cell antigen receptor. Immunity 35:705–720. doi:10.1016/j.immuni.2011.10.004

Shivanandan A, Radenovic A, Sbalzarini IF (2013) MosaicIA: an ImageJ/Fiji plugin for spatial pattern and interaction analysis. BMC Bioinformatics 14:349. doi:10.1186/1471-2105-14-349

Shivanandan A, Unnikrishnan J, Radenovic A (2015) Accounting for limited detection efficiency and localization precision in cluster analysis in single molecule localization microscopy. PLoS One 10, e0118767. doi:10.1371/journal.pone.0118767

Shvets E, Ludwig A, Nichols BJ (2014) News from the caves: update on the structure and function of caveolae. Curr Opin Cell Biol 29:99–106. doi:10.1016/j.ceb.2014.04.011

Sieber JJ, Willig KI, Kutzner C et al (2007) Anatomy and dynamics of a supramolecular membrane protein cluster. Science 317:1072–1076. doi:10.1126/science.1141727

Simons K, Ikonen E (1997) Functional rafts in cell membranes. Nature 387:569–572. doi:10.1038/42408

Simons K, Sampaio JL (2011) Membrane organization and lipid rafts. Cold Spring Harb Perspect Biol. doi:10.1101/cshperspect.a004697

Simons K, Toomre D (2000) Lipid rafts and signal transduction. Nat Rev Mol Cell Biol 1:31–39. doi:10.1038/35036052

Simons K, Van Meer G (1988) Lipid sorting in epithelial cells. Biochemistry (Mosc) 27:6197–6202. doi:10.1021/bi00417a001

Singer SJ, Nicolson GL (1972) The fluid mosaic model of the structure of cell membranes. Science 175:720–731

Smith-Garvin JE, Koretzky GA, Jordan MS (2009) T cell activation. Annu Rev Immunol 27:591–619. doi:10.1146/annurev.immunol.021908.132706

Solis GP, Hoegg M, Munderloh C et al (2007) Reggie/flotillin proteins are organized into stable tetramers in membrane microdomains. Biochem J 403:313–322. doi:10.1042/BJ20061686

Sparrow CM (1916) On spectroscopic resolving power. Astrophys J 44:76. doi:10.1086/142271

Steinhauer C, Forthmann C, Vogelsang J, Tinnefeld P (2008) Superresolution microscopy on the basis of engineered dark states. J Am Chem Soc 130:16840–16841. doi:10.1021/ja806590m

Stipp CS, Kolesnikova TV, Hemler ME (2003) Functional domains in tetraspanin proteins. Trends Biochem Sci 28:106–112. doi:10.1016/S0968-0004(02)00014-2

Südhof TC (2012) The presynaptic active zone. Neuron 75:11–25. doi:10.1016/j.neuron.2012.06.012

Termini CM, Cotter ML, Marjon KD et al (2014) The membrane scaffold CD82 regulates cell adhesion by altering α4 integrin stability and molecular density. Mol Biol Cell 25:1560–1573. doi:10.1091/mbc.E13-11-0660

Tobin SJ, Cacao EE, Hong DWW et al (2014) Nanoscale effects of ethanol and naltrexone on protein organization in the plasma membrane studied by photoactivated localization microscopy (PALM). PLoS One 9, e87225. doi:10.1371/journal.pone.0087225

Torreno-Pina JA, Castro BM, Manzo C et al (2014) Enhanced receptor–clathrin interactions induced by N-glycan–mediated membrane micropatterning. Proc Natl Acad Sci 111:11037–11042. doi:10.1073/pnas.1402041111

Truan Z, Tarancón Díez L, Bönsch C et al (2013) Quantitative morphological analysis of arrestin2 clustering upon G protein-coupled receptor stimulation by super-resolution microscopy. J Struct Biol 184:329–334. doi:10.1016/j.jsb.2013.09.019

van den Bogaart G, Meyenberg K, Risselada HJ et al (2011) Membrane protein sequestering by ionic protein-lipid interactions. Nature 479:552–555. doi:10.1038/nature10545

van Zanten TS, Cambi A, Garcia-Parajo MF (2010) A nanometer scale optical view on the compartmentalization of cell membranes. Biochim Biophys Acta BBA – Biomembr 1798:777–787. doi:10.1016/j.bbamem.2009.09.012

Veatch SL, Keller SL (2003) Separation of liquid phases in giant vesicles of ternary mixtures of phospholipids and cholesterol. Biophys J 85:3074–3083. doi:10.1016/S0006-3495(03)74726-2

Veatch SL, Machta BB, Shelby SA et al (2012) Correlation functions quantify super-resolution images and estimate apparent clustering due to over-counting. PLoS One. doi:10.1371/journal.pone.0031457

Whelan DR, Bell TDM (2015) Image artifacts in single molecule localization microscopy: why optimization of sample preparation protocols matters. Sci Rep. doi:10.1038/srep07924

Wiegand T, Moloney KA (2013) Handbook of spatial point-pattern analysis in ecology. CRC Press, Boca Raton

Williamson DJ, Owen DM, Rossy J et al (2011) Pre-existing clusters of the adaptor Lat do not participate in early T cell signaling events. Nat Immunol 12:655–662. doi:10.1038/ni.2049

Willig KI, Harke B, Medda R, Hell SW (2007) STED microscopy with continuous wave beams. Nat Methods 4:915–918. doi:10.1038/nmeth1108

Wright MD, Tomlinson MG (1994) The ins and outs of the transmembrane 4 superfamily. Immunol Today 15:588–594. doi:10.1016/0167-5699(94)90222-4

Xie J, Tato CM, Davis MM (2013) How the immune system talks to itself: the varied role of synapses. Immunol Rev 251:65–79. doi:10.1111/imr.12017

Xu X, Ester M, Kriegel H-P, Sander J (1998) A distribution-based clustering algorithm for mining in large spatial databases. In: 14th international conference on data engineering, 1998. Proceedings of 14th International Conference on Data Engineering(ICDE'98), pp 324–331

Yamada E (1955) The fine structure of the gall bladder epithelium of the mouse. J Biophys Biochem Cytol 1:445–458

Yáñez-Mó M, Barreiro O, Gordon-Alonso M et al (2009) Tetraspanin-enriched microdomains: a functional unit in cell plasma membranes. Trends Cell Biol 19:434–446. doi:10.1016/j.tcb.2009.06.004

Chapter 5
Image Informatics Strategies for Deciphering Neuronal Network Connectivity

Jan R. Detrez, Peter Verstraelen, Titia Gebuis, Marlies Verschuuren, Jacobine Kuijlaars, Xavier Langlois, Rony Nuydens, Jean-Pierre Timmermans, and Winnok H. De Vos

Abstract Brain function relies on an intricate network of highly dynamic neuronal connections that rewires dramatically under the impulse of various external cues and pathological conditions. Amongst the neuronal structures that show morphological plasticity are neurites, synapses, dendritic spines and even nuclei. This structural remodelling is directly connected with functional changes such as intercellular

---

J.R. Detrez • P. Verstraelen • M. Verschuuren • J.-P. Timmermans
Laboratory of Cell Biology and Histology, Department of Veterinary Sciences, University of Antwerp, Groenenborgerlaan 171, 2020 Antwerp, Belgium
e-mail: jan.detrez@uantwerpen.be; peter.verstraelen@uantwerpen.be; marlies.verschuuren@uantwerpen.be; jean-pierre.timmermans@uantwerpen.be

T. Gebuis
Department of Molecular and Cellular Neurobiology, Center for Neurogenomics and Cognitive Research, VU University Amsterdam, De Boelelaan 1085, 1081 HV Amsterdam, The Netherlands
e-mail: t.gebuis@vu.nl

J. Kuijlaars
Neuroscience Department, Janssen Research and Development, Turnhoutseweg 30, 2340 Beerse, Belgium

Laboratory for Cell Physiology, Biomedical Research Institute (BIOMED), Hasselt University, Agoralaan, 3590 Diepenbeek, Belgium
e-mail: jkuijla@its.jnj.com

X. Langlois • R. Nuydens
Neuroscience Department, Janssen Research and Development, Turnhoutseweg 30, 2340 Beerse, Belgium
e-mail: xlangloi@its.jnj.com; rnuydens@its.jnj.com

W.H. De Vos (✉)
Laboratory of Cell Biology and Histology, Department of Veterinary Sciences, University of Antwerp, Groenenborgerlaan 171, 2020 Antwerp, Belgium

Cell Systems and Cellular Imaging, Department Molecular Biotechnology, Ghent University, Coupure Links 653, 9000 Ghent, Belgium
e-mail: winnok.devos@uantwerpen.be

W.H. De Vos et al. (eds.), *Focus on Bio-Image Informatics*, Advances in Anatomy, Embryology and Cell Biology 219, DOI 10.1007/978-3-319-28549-8_5

communication and the associated calcium bursting behaviour. In vitro cultured neuronal networks are valuable models for studying these morpho-functional changes. Owing to the automation and standardization of both image acquisition and image analysis, it has become possible to extract statistically relevant readouts from such networks. Here, we focus on the current state-of-the-art in image informatics that enables quantitative microscopic interrogation of neuronal networks. We describe the major correlates of neuronal connectivity and present workflows for analysing them. Finally, we provide an outlook on the challenges that remain to be addressed, and discuss how imaging algorithms can be extended beyond in vitro imaging studies.

5.1 Introduction

Development of the central nervous system entails formation and maintenance of intricate neuronal networks. Synaptic activity and the associated opening of gated ion channels initiate precisely calibrated calcium transients in neuronal cells, which drive short-term and long-term morphological changes, such as dendritic growth and arborization (Bading 2013). This dynamic, cytoskeleton-based remodelling of neuronal appendages, also known as neuronal plasticity, is a key process for virtually all long-lasting adaptations of the brain, such as learning, addiction or chronic pain sensation (Alvarez and Sabatini 2007). While resulting from very different molecular triggers (e.g. the production of toxic protein oligomers, cytoskeletal dysregulation, etc.), disrupted neuronal plasticity represents a pathological hallmark that is shared by numerous psychiatric and neurodegenerative diseases, including schizophrenia, autism spectrum disorder and Alzheimer's disease (Lin and Koleske 2010; Penzes et al. 2011). Thus, understanding the intricacies of neuronal connectivity may not only be instrumental in gaining insights into its physiological importance, but also in resolving stages of disease development.

5.1.1 Models for Studying Neuronal Connectivity

Because of the complexity and long-distance wiring of neurons in the brain, neuronal connectivity is ideally studied within the entire organ. Boosted by the differential power of stochastic multispectral labelling technologies like Brainbow and derivatives (Cai et al. 2013), multiple imaging approaches have been developed that enable connectivity studies in whole fixed and even living brain. Microscopic imaging in awake animals has been achieved with implanted cranial windows that can be accessed after restraining the animal, or using miniature head-mounted microscopes in freely moving animals (Chen et al. 2013; Dombeck et al. 2007). However, the imaging depth of such studies is limited to the optical penetration power of multi-photon microscopes ($\sim$ 1 mm) (Nemoto 2014). Recent advances in tissue clearing and re-invention of light-sheet illumination microscopy have enabled 3D microscopic imaging of intact fixed brains at unprecedented speed (Kim et al. 2013). One of the aims of these efforts is to build a digital atlas from the vast datasets to enable mapping the connectivity between and within brain regions (The Allen Institute 2015; Harvard 2015). However, the methods for acquiring and analysing such datasets are far from standard, the size of the datasets is massive and interpretation, let alone quantification, is non-trivial (Peng et al. 2013).

For live cell imaging studies, acute or organotypic brain slices circumvent the need for extended animal suffering and monitoring of multiple physiological parameters typically accompanying in vivo manipulation (Cho et al. 2007). While maintaining a reasonable level of tissue architecture, this approach improves the experimental access and allows precise control of the extracellular environment. Nevertheless, afferent signals from distant brain regions are inevitably lost and physiological processes cannot be associated with behavioural information. A major disadvantage that is shared by both intact brain and slice model approaches is that it is difficult to standardize the quantitative readout when it comes down to studying connectivity. The inter-individual variability between model organisms creates a tremendous bias and impedes easy extraction of morphological and functional cues. This, together with the need for large amounts of biological material, precludes their use from routine screening in preclinical drug screening campaigns, which is why in vitro models have been established. The advantage of using neuronal cells is that multiple cell cultures can be grown in parallel, allowing multiplex experiments with internal controls. Although existing 3D anatomical connections are lost during the preparation of primary neurons (e.g. extracted from mouse embryos), the cells preserve numerous morphological and functional properties of in vivo neuronal networks (Cohen et al. 2008; Dotti et al. 1988; Fletcher et al. 1994; Papa et al. 1995). For example, it has been shown that primary cultures recapitulate synchronous calcium bursting behaviour, when cultured in a 96-well plate format, making this platform highly attractive for high-throughput pharmacological and genetic manipulation (Verstraelen et al. 2014; Cornelissen et al. 2013). To overcome species differences, recent efforts have also led to the use of human induced pluripotent stem cells (iPSC)-derived neuronal cultures (Takahashi and Yamanaka 2006; Imamura and Inoue 2012). iPSC technology circumvents ethical obstructions regarding human embryonic stem cells and allows cultivating patient-derived neurons, thereby eliminating the need for artificial disease models.

5.1.2 *Correlates of Neuronal Connectivity*

Cultivated neuronal networks display both morphological and functional features that can be used to quantitatively describe the degree of connectivity (Fig. 5.1). The outgrowth of axons and dendrites, collectively called neurites, is a morphological feature that provides information about the general health of the neurons and the connectivity within the neuronal network. Consequently, this feature has been used

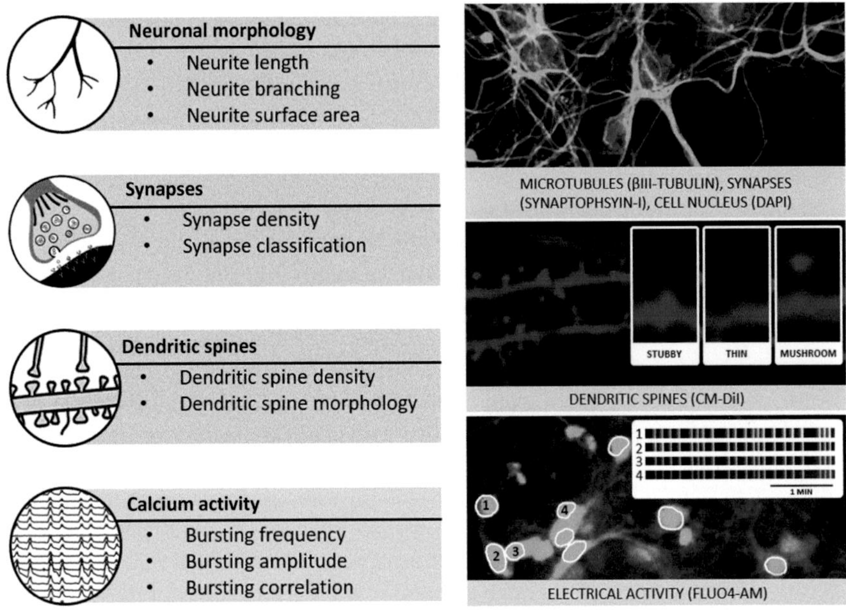

Fig. 5.1 Correlates of neuronal network connectivity. The main morphological (neuronal morphology, synapses and dendritic spines) and functional (calcium activity) correlates of in vitro neuronal network connectivity are depicted. Immunocytochemical labelling of cytoskeletal proteins, such as β-III-tubulin, allows quantifying the neuronal morphology, while labelling of synaptic proteins provides information about the synapse density or the type of neurotransmitter they process. Dendritic spines are specialized compartments that contain excitatory synapses and can be highlighted with lipophilic dyes (e.g. CM-DiI). Both density and morphology of spines correlate with synaptic strength and hence network connectivity. Calcium imaging (e.g. using the calcium-sensitive dye Fluo-4 AM) allows studying the spontaneous electrical activity of neurons

in high-throughput compound toxicity screening and safety evaluation of drugs and environmental chemicals (Harrill et al. 2013; Popova and Jacobsson 2014; Sirenko et al. 2014). Different approaches to quantify neuronal morphology (e.g. neurite outgrowth, neurite bifurcations and Sholl analysis) are discussed in Sect. 5.2.2.

Neuronal communication is established through the formation of synapses. A synapse consists of three major compartments: a presynaptic compartment, a postsynaptic compartment and the synaptic cleft. Pre- and postsynaptic compartments are highly specialized morphological structures containing specific proteins that can be used as markers for assessing neuronal connectivity. As such, fluorescent labelling and quantification of synaptic proteins may provide valuable information about the number of synapses, and therefore serve as an indicator of the connectivity in the network. This is discussed in Sect. 5.2.3.

While inhibitory synapses are made directly on the dendritic shaft, the postsynaptic compartment of excitatory synapses is predominantly located on highly specialized structures, called dendritic spines. These spines are small (0.5–3 μm) protrusions from the dendritic shaft that were first described by Ramon y Cajal in 1891 (Cajal 1891). The exact functions of spines are still debated, but the general view is that they compartmentalize the local electrical and biochemical processes of a single synapse (Sala and Segal 2014). They are highly dynamic structures that change in shape, volume and density in response to cues that influence synaptic strength. Throughout the continuum of spine shapes, different morphological stages (thin, stubby or mushroom shape) can be discriminated, which can change within a matter of minutes via rearrangements of the actin cytoskeleton [Fig. 5.1; (Dent et al. 2011; Lai and Ip 2013; Maiti et al. 2015)]. The synaptic receptors on spines are connected to a local cytoskeletal network via the assembly of scaffold proteins, called the postsynaptic density (PSD). Thin spines contain relatively small PSDs and emerge and disappear over a few days, whereas mushroom spines with larger PSDs may persist for months. Spine density and morphology are becoming increasingly popular as readouts for neuronal network connectivity and alterations in both features have been described in numerous neurological disorders, including Alzheimer's disease, schizophrenia, intellectual disabilities and autism spectrum disorders (Penzes et al. 2011).

While morphological correlates provide a static impression of connectivity, they do not inform on the actual synaptic communication taking place within a network. It is only by direct assessment of this electrical activity that one can grasp the true degree of functional connectivity (discussed in Sect. 5.3). Cultivated neurons are known to exhibit spontaneous electrical activity, which tends to evolve from stochastic activity of individual neurons into robust, synchronized network activity (Cohen et al. 2008; Verstraelen et al. 2014). Neuronal electrical activity can be visualized by means of voltage or calcium sensors, both of which are available as synthetic dyes or genetically encoded fluorescent proteins (Broussard et al. 2014; Fluhler et al. 1985; Jin et al. 2012; Paredes et al. 2008). Such a functional approach not only allows assessing the effect of chronic treatments on neuronal connectivity, but can also provide information about acute responses to pharmacological perturbations.

5.1.3 *From Snapshots to Numbers: Towards High-Content Neuro-Imaging*

Both primary and iPSC-derived neuronal networks can be cultivated in multi-well plates, starting from a limited amount of biological material. In combination with automated fluorescence microscopy, these networks make an attractive model for upscaling to a high-content screening (HCS) platform (Cornelissen et al. 2013; Schmitz et al. 2011). Of vital importance for such a platform is robust measurement of the endpoint of interest. Manual quantification is not only labour-intensive, but also prone to observer bias, which hampers reproducibility of the data. To eliminate this bias and boost throughput, automation of image analysis is inevitable. However, the design and implementation of generic automated image analyses are non-trivial since the experimental conditions, such as microscope settings, type of stains, cell type and cell densities that are used, introduce a strong variability in image quality (Meijering 2010). Nevertheless, with sufficient standardization of the sample preparation and image acquisition protocols, and adequate pre-processing of the raw image datasets, the major correlates of neuronal connectivity can be quantified in an unbiased way. In the following paragraphs, we discuss the main image analysis strategies for quantification of morphological and functional endpoints.

5.2 Measuring Morphological Correlates: From Networks to Spines

As mentioned above, neurons exhibit strong morphological plasticity. Relevant dynamic changes that can be quantified are neuronal morphology, synapse development and the emergence and remodelling of dendritic spines. The analysis of each of these features differs, but they all rely on a generic workflow that consists of four major steps, namely pre-processing (image restoration), segmentation (object detection), rectification (visual verification and correction) and analysis (feature extraction). We will first briefly introduce some of the generic methods in image pre-processing that apply to all analysis pipelines, after which we will focus on the more dedicated algorithms for extracting morphological data.

5.2.1 Basic Image Pre-processing

The principal task of image pre-processing is to correct for systematic errors and imperfections that have been introduced by the image acquisition system. These errors include image blur (imposed by the point-spread function), noise (photon and detector noise) and intensity gradients (due to spatiotemporal illumination inhomogeneity). Various algorithms have been introduced to tackle these issues. One of the first pre-processing steps that is often used is deconvolution (Heck et al. 2012). It is also known as image restoration since it aims at reversing the image formation process, thereby improving the signal-to-noise ratio (SNR) and image resolution (Sarder and Nehorai 2006). Image noise predominantly results from the stochastic nature of the photon-counting process at the detectors (i.e. Poisson noise), and the intrinsic thermal and electronic fluctuations of the acquisition devices (i.e. Gaussian noise). Gaussian noise can be easily removed by conventional spatial filtering techniques (e.g. mean filtering or Gaussian smoothing). This works fast, but generally tends to reduce noise at the expense of sharpness. More advanced [e.g. wavelet-based (Zhang et al. 2008)] methods that correct for Poisson noise have been described as well. Heterogeneous illumination and nonlinearities in the acquisition path are usually corrected for by subtracting an image of an empty region (flat-field correction) or by local background subtraction (pseudo-flat field correction).

5.2.2 *Neuronal Morphology*

The necessity for analysing neuronal morphology has led to the development of a variety of image analysis strategies that mainly differ in their level of accuracy and throughput [for an overview of tools see Parekh et al. (Parekh and Ascoli 2013); Fig. 5.2]. Tracing methods tend to delineate individual neuronal extensions, with high accuracy, but typically demand well-contrasted individual neurons. Thus, either isolated neurons or sparsely labelled neuronal networks are warranted. The latter is typically achieved by means of stochastic labelling methods (e.g. Golgi-staining or DiI) or transgene mouse models [e.g. Thy1-YFP (Feng et al. 2000) or Brainbow mice (Livet et al. 2007)]. Tracing is done either manually or semi-automatically, assisted by global image processing operations and/or local path finding algorithms. An alternative group of methods to define neuronal morphology rely on global, intensity-based thresholding. The advantage of such methods is that they can be applied easily to sparsely labelled networks but also to completely stained, dense networks (using pan-neuronal markers, such as β-III-tubulin or MAP2). Once the neuron is segmented, different metrics can be derived depending on the density of labelled cells. For sparse labelling methods, a fairly simple technique to gauge the complexity of individual neuronal morphology is based on Sholl analysis. In addition, more detailed metrics of single neurons can be obtained such as neurite length and dendritic branching. For pan-labelled neuronal networks, an estimate of these neuron-specific parameters can be given, provided a neuron-specific nuclear counterstaining is available.

5.2.2.1 Sparsely Labelled Neurons

Starting from the camera lucida, an optical superposition system that was used to draw the outline of nerve cells by hand, several efforts have been made to generate digital reconstructions of neuronal morphology. The first tools that became available [e.g. Neuron_Morpho (Brown et al. 2005) and Neurolucida (MBF Bioscience 2015b)] enabled the manual delineation of neurites in a single plane. Although more recent methods allow the segmentation of neuronal processes in 3D by delineating 2D projected images (Peng et al. 2014), manual annotation is slow and labour-intensive, and therefore not amenable to upscaling.

Although the nomenclature and classification of automated neuron tracing algorithms are not consistent in literature, from an image informatics perspective, we discern global image processing methods, local tracing methods, and more modern algorithms that use a combination of both.

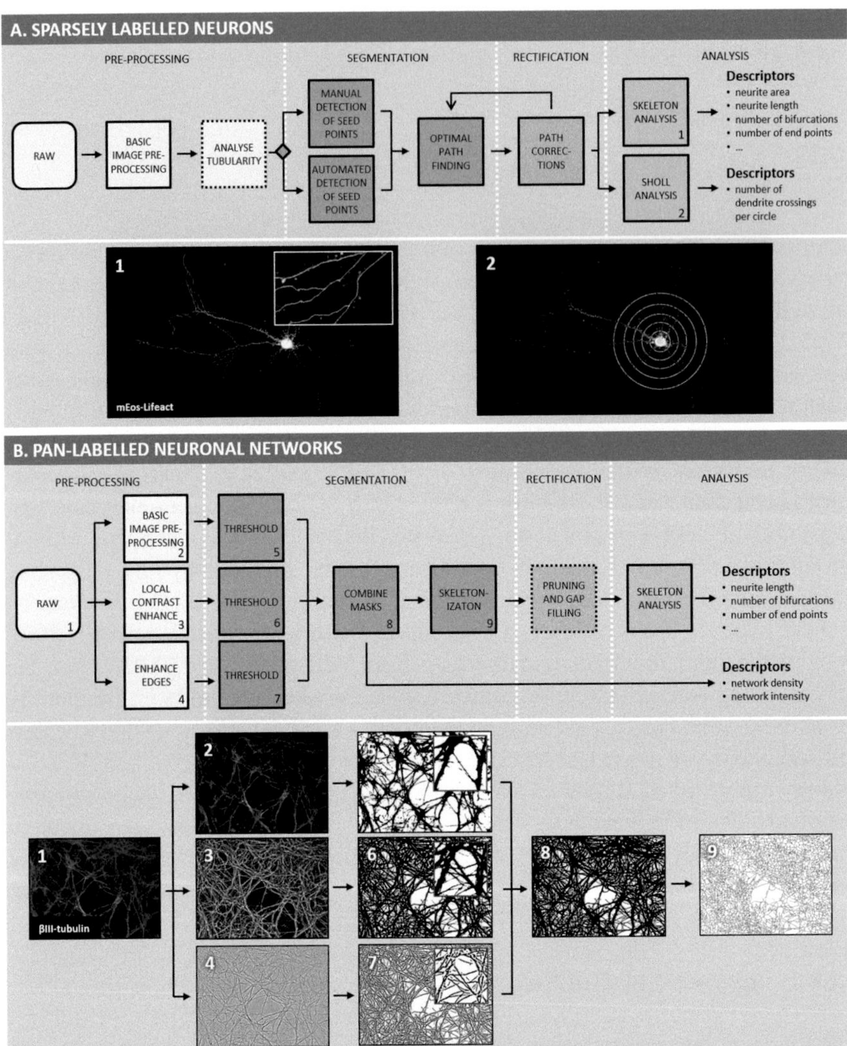

Fig. 5.2 Morphological analysis of sparsely labelled and pan-labelled neuronal networks. **A.** To acquire a detailed view of individual neurite length, sparsely labelled neurons can be traced using semi-automated and automated algorithms. The traced neuron can then be subjected to skeleton analysis to derive detailed information about the neuron's morphology, or to Sholl analysis. The latter method describes the complexity of the neuronal morphology by the number of intersections of the neurites with a group of concentric circles drawn around the cell soma. **B.** This panel shows a multi-tier global segmentation method for analysing pan-labelled neuronal cultures, as implemented in MorphoNeuroNet (Pani et al. 2014). A combination of intensity-based (*2*, *3*) and edge-based (*4*) pre-processing algorithms enables the detection of neurites with variable thickness and fluorescence intensity

Early attempts to automate the neurite reconstruction process are based on a global intensity threshold, followed by voxel thinning or a medial axis transform to obtain the neurite skeleton (Koh et al. 2002; Wearne et al. 2005). As a result of the global threshold, these methods experience difficulties in the presence of signal inhomogeneities, and the iterative nature of the voxel thinning process is computationally intensive.

More recent methods are based on a semi-automatic *modus operandi*, which relies on local computer-aided identification of putative neurites, *in tandem* with manual interaction and/or correction. These local exploratory algorithms, also referred to as neuron tracing, better accommodate for gradual changes in neuron morphology and image quality. Various methods have been developed for the local detection of neurite structures. Amongst these, ridge detectors such as a Hessian filter, which compute a square matrix of second order partial derivatives for every pixel of the image, are used to measure the local tubularity. The directionality of the neurite is obtained by calculating the eigenvectors from the obtained Hessian matrix. The eigenvector with the smallest absolute eigenvalue points in the direction of the vessel (i.e. the direction with the smallest intensity variations). NeuronJ (Meijering et al. 2004) relies on this algorithm to determine the optimal path (that with the lowest cost) between manually defined start- and endpoints (seeds). This approach is also known as live-wire segmentation. Although NeuronJ was conceived for 2D images, the cost function can readily be extended to 3D by using voxel cubes instead of 2D kernels for the Hessian [as implemented in NeuroMantic (Myatt et al. 2012) and AutoNeuron for Neurolucida (MBF Bioscience 2015b)]. Other implementations to locally reconstruct neuronal morphology rely on the modelling of deformable templates and the iterative addition of structural components (e.g. cylinders) (Schmitt et al. 2004; Zhao et al. 2011; Al-Kofahi et al. 2002). Since these local tracing methods produce one branch at the time, a separate branch point detection method is required to complete the reconstruction (Al-Kofahi et al. 2008). Alternatively, model-free local tracing strategies, such as Rayburst sampling (Rodriguez et al. 2006) and voxel scooping (Rodriguez et al. 2009), are able to trace multiple branches from a single seed (typically the cell soma). Although these methods enable fully automated segmentation of homogeneously stained neurons, spurious gaps or branches can still occur when the implemented pre-processing steps fail to accurately separate foreground and background. To address this issue, algorithms have been developed to retrospectively attach disconnected branches based on parameters such as orientation, distance, curvature and intensity (Chothani et al. 2011). An alternative approach is to directly combine local tracing algorithms with global processing methods to find multiple seed points at critical points (such as terminations, bifurcations and inflections) and to guide the finer-scale tracing process (Peng et al. 2011; Xie et al. 2010). While automation of the neurite tracing process continues to improve, human intervention is often still required to steer the tracing process.

Once the neurites are segmented, morphological information can be extracted from the segmented neuron. An old, but still widely used method to study segmented neurons is Sholl analysis (Binley et al. 2014). This method counts how many times the neurites intersect a series of concentric shells that are drawn around the cell soma. Consequently, highly bifurcated neurite networks will return high Sholl values. This tool, while still widely used, has been criticized for its limited sensitivity and inability to correct for branches that cross the same circle multiple times, and those that extend tangentially and do not cross a circle at all. This is why current methods tend to focus more on extracting metrics that can be derived from the backbone, such as neurite length and bifurcation points.

5.2.2.2 Pan-Labelled Neuronal Networks

Because neurite tracing relies on the precise delineation of individual neurons, the throughput of this analysis method is generally low. Detailed neuronal models of neurons, however, are very useful to investigate shape/function relations, or in theoretical neurobiology, in which neuronal morphology is used to describe its electrotonic compartmentalization (Costa et al. 2000). When a higher throughput is required, global methods can be used to segment multiple neurons in the field of view. Although these methods might lack the precision of neuron tracing in case of signal inhomogeneities in the branches, they are well able to detect general changes in neuronal morphology (e.g. neurite length) in response to compounds that affect neurite outgrowth (Pool et al. 2008).

All global segmentation methods rely on binarization (i.e. thresholding) and skeletonization of a pre-processed image [Fig. 5.2B; (Ho et al. 2011)]. The complexity of the pre-processing steps (apart from those mentioned in Sect. 5.2.1) is what truly discriminates different methods, and this is usually based on the image quality and density of the cell culture. Especially in dense networks, the key is to detect both low and high intensity structures of different sizes. To this end, multi-scale or multi-tier object enhancement approaches have been implemented. MorphoNeuroNet (Pani et al. 2014), for example, uses a combination of local contrast enhancement and edge detection algorithms (unsharp masking and Laplace filtering) to highlight less intense parts of the neuronal network. A combination of these images after thresholding generates a more complete mask of the neuronal network than any individual image would. Although this binary mask offers a basic measure of the network density, it is often skeletonized to retrieve more detailed parameters, including neurite length and diameter, the number of bifurcations and endpoints. As the resulting skeleton often contains errors (such as spurious gaps or branches), filling and pruning strategies are often used to rectify these retrospectively (Narro et al. 2007).

In many neuronal network analyses, a measure of cellular density is calculated as well. Cell or soma segmentation is facilitated in the presence of a nuclear counterstain. Indeed, nuclei are preferred as seeds, because of their well-separated distribution and relatively regular shape (this regularity has recently been challenged; cf. Box 5.1). Starting from the nuclear boundaries, regions of interest (ROIs) are then grown to detect the soma.

Box 5.1—Nuclear morphology as a novel correlate of neuronal connectivity Neuronal nuclei have been shown to be extremely mouldable. They can adopt shapes that range from near spherical to complex and highly folded, and this is correlated with neuronal activity (Wittmann et al. 2009). Nuclear folding has been suggested to be necessary for relaying calcium signals to the nucleus, which is fundamental for proper gene expression (Bading 2013). The activity-driven morphological changes of the nucleus are referred to as morphology modulation. Quantification of the internal structure or folding of the nucleus may thus serve as a readout for neuronal connectivity.

Nucleus segmentation is often included in neuronal image analysis pipelines as a starting point for segmenting cell bodies and/or neurites (Meijering 2010). From segmented nuclei in 2D images, nuclear shape descriptors, such as surface and circularity, can easily be derived using general object enhancement and thresholding procedures. As far as the internal nuclear structure is concerned, phenomena, such as folding, have been addressed far less. Nuclear folds are generally visualized using stains for the nuclear lamina and analysed using procedures that often include manual assessment (Wittmann et al. 2009; Lammerding et al. 2006). To describe the internal structure of nuclei in more objective terms, an automatic image analysis procedure has been developed (Righolt et al. 2011) that quantifies the 3D internal structure of nuclei on the basis of a nuclear lamina stain using three descriptors: mean intensity, skewness and mean curvature. To track nuclear morphological changes over time, Gerlich et al. (Gerlich et al. 2001) developed a technique for fully automated quantification and visualization of surfaces from dynamic 3D fluorescent structures in live cells. 3D surface models were constructed for the nuclear membrane and interpolated over time using a process called morphing. These 4D reconstructions, which allow the quantification of volume changes in the nucleus of live cells, could also serve as an indirect measure of nuclear folding. However, both methods require a complex 4D analysis to achieve a level of accuracy that is not necessary for measuring nuclear folding. To make quantification of nuclear folding amenable to upscaling (high-throughput), we implemented a 2D analysis. In our workflow (Fig. 5.3), 3D widefield image stacks of lamin-stained neuronal nuclei are Z-projected and nuclei are detected by means of image thresholding followed by a watershed to dissociate neighbouring nuclei. Second, cross-referencing the nuclei with a marker dedicated to neuronal nuclei (e.g. NeuN) allows the selection of neuronal nuclei only, a process that is necessary in cell cultures, which typically consist of neuronal as well as non-neuronal nuclei such as those of astrocytes. Third, the lamin staining is used for segmentation of nuclear folds. A Laplace filter specifically enhances the edges of nuclear folds as well as the edge of the nucleus. To exclude the latter, the ROIs from the initial nuclear segmentation are eroded and only particles lying within the eroded ROIs are identified as folds. For each segmented ROI, the degree of folding is calculated.

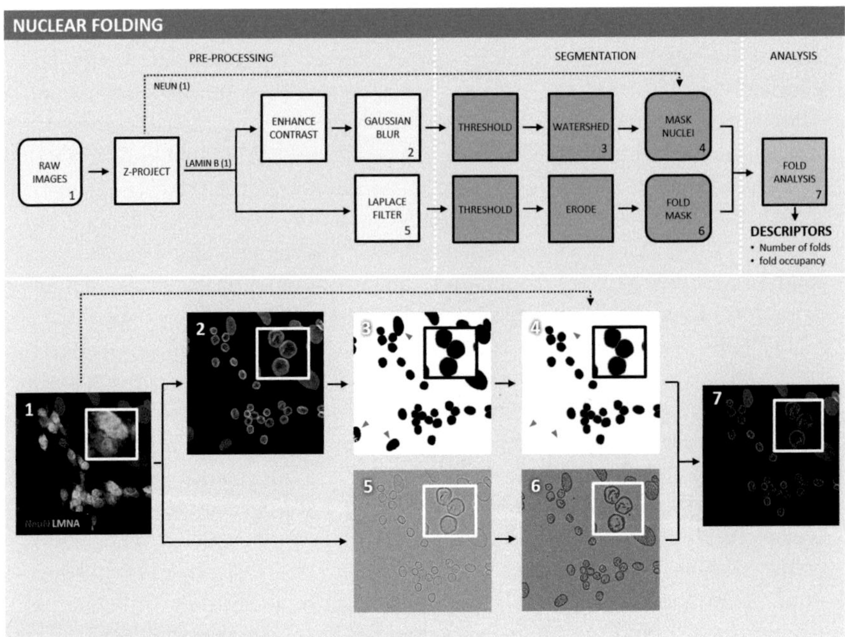

Fig. 5.3 Quantification of nuclear folding. The percentage of nuclear folding can be determined from images of lamin-stained (*red*) neuronal nuclei. First, neuronal nuclei are extracted based on a neuronal marker (*cyan*; *3*, *4*). In parallel, a Laplace filter enhances (*5*) the detection (*6*) of nuclear folds and edges of the nuclei on lamin-stained images. To identify only the ROIs that represent nuclear folds, the nuclear masks (*4*) are eroded and only the ROIs that are confined within these regions are detected

5.2.3 Sampling Synapses

Synapses are small structures that are close to or below the diffraction limit ($< 0.1\,\mu m^2$), which is why their detection is often limited to the quantification of diffraction-limited spots or *puncta* (synapse density). Pan-synaptic labelling is typically achieved by targeting hallmark proteins of the pre- or postsynaptic compartments (e.g. synaptophysin-I, synapsin and PSD95), although synapses that process specific neurotransmitters can be discerned as well using vesicle- or receptor-specific antibodies (e.g. VAChT, VGAT and GluR). Dendritic spines are more pronounced neuronal substructures that only harbour excitatory synapses

(McKinney 2010), but exhibit different shapes that can be quantified and have been suggested to relate to synaptic health. To visualize spines, the same pan-cellular labelling methods are used as those discussed for analysing the neuronal morphology of sparsely labelled neurons.

5.2.3.1 Counting Synaptic *Puncta*

Although numerous spot segmentation approaches have been developed (Meijering 2012), the small size of synapses makes the segmentation process very sensitive to image noise and local variations in contrast (e.g. synaptic structures with a weak signal intensity or those in the presence of intense background signals originating from the soma or thick dendritic branches). Therefore, instead of more conventional noise filtering methods (cf. Sect. 5.2.1), advanced denoising strategies [e.g. the wavelet-based algorithm Multi-Scale Variance Stabilizing Transform (MSVST)] have been proposed to enhance threshold-based segmentation of synaptic structures (Fan et al. 2012).

To further accommodate for local variations in contrast, local adaptive threshold algorithms, whether or not preceded by blob detectors, such as a Mexican hat or Laplace filters, can be used. In essence, the latter algorithms rely on the assumption that synaptic *puncta* can be modelled as 2D Gaussian functions. A potential disadvantage of these operators is that the approximate size of the Gaussian should be specified up front. A solution to this is the use of machine-learning algorithms that estimate the size of the kernel (Schmitz et al. 2011; Feng et al. 2012). As implemented in SynD (Schmitz et al. 2011), particles with a unique local intensity maximum can be used to generate a data-driven single synapse kernel. Alternative solutions are multi-scale spot segmentation (Bretzner and Lindeberg 1998; De Vos et al. 2010) or granulometric analysis to “sieve” image objects with structure elements based on their geometry and size (Prodanov et al. 2006).

In a final step, several criteria can be implemented for filtering false positive results. Particle size filtering and intensity cut-offs can be used to separate true synaptic *puncta* from noise. Other methods also implement distance criteria to exclude particles that are not connected to the neuronal skeleton (Schmitz et al. 2011).

Although there is a limited availability of tools that implement synapse detection, SynD was successfully used in knockout studies aimed at identifying proteins that are involved in synaptic transmission pathways, such as neurotransmitter vesicle fusion (Meijer et al. 2012), and neurotransmitter receptor trafficking (Nair et al. 2013). This tool was later used to evaluate the efficacy of synapto-protective drugs in a micro-fluidics screening platform (Deleglise et al. 2013).

5.2.3.2 Detection of Dendritic Spines

Since dendritic spines are membranous protrusions that form an integral part of the neurite network, their segmentation is usually part of neuronal network segmentation approaches. Therefore, most tools that have been developed for the detection of dendritic spines rely on, or have built-in neurite tracing tools [e.g. NeuronStudio (Rodriguez et al. 2006) and AutoSpine (MBF Bioscience 2015a)].

As for segmentation of the previously discussed morphological parameters, a simple global intensity threshold is inadequate to segment spines, since this approach fails to accurately detect faint or thin spines without distorting the shape of more intense spines. To address this issue, edge-enhancers [e.g. Laplace filtering or unsharp masking (Bai et al. 2007)] and local adaptive threshold algorithms (Cheng et al. 2007; Rodriguez et al. 2008) are used. In contrast to threshold-based methods, another category of spine segmentation algorithms uses a curvilinear structure detector (Zhang et al. 2007). This filter, used in many medical image processing algorithms (e.g. for detecting blood vessels, airways or bones), delineates the dendritic backbones directly on the original image by treating them as 2D line objects. A similar method is then used to detect the centrelines of dendritic spines. After segmentation and skeletonization, most dendritic spines are usually identified as protrusions [Fig. 5.4B; (Bai et al. 2007; Cheng et al. 2007; Koh et al. 2002)]. Some spines, however, become detached in the segmentation process and should be reassigned, e.g. based on the distance from the backbone and on size criteria (Bai et al. 2007). More advanced methods rely on a classifier, built from a library of isolated spines (Zhang et al. 2007).

Although centreline extraction-based approaches offer a reasonable quantification of lateral spines, the limited axial resolution of microscopes makes them unreliable for quantifying spines that are oriented orthogonal to the imaging plane. Therefore, most centreline-based algorithms estimate the spine density from maximum intensity projected images which leads to a substantial underestimation of spine densities (Bai et al. 2007; Cheng et al. 2007; Zhang et al. 2007). While variations in the skeletonization algorithm have led to increased accuracy of spine detection in 3D (Koh et al. 2002; Janoos et al. 2009), these algorithms are computationally expensive. Model-based algorithms such as voxel clustering (Rodriguez et al. 2008) and the marching cubes algorithm (Li et al. 2009) are faster alternatives that identify spines based on a trained classifier. In addition, 3D Gabor wavelets have recently been proposed as a fast method for detecting dendritic spines by clustering candidate voxels according to the response to the wavelet transform (Shi et al. 2014).

None of the existing algorithms are error-free. One common problem is that neighbouring spines are merged on the segmented images as a result of low image resolution or incorrect thresholding. To solve this, one can rely on the fact that voxel intensities are naturally brighter at the centre of spines and dimmer at the edges. Clumped spines can then be delimited based on their 3D intensity vector gradients (Rodriguez et al. 2008). Other methods rely on 3D shape analysis to automatically categorize spines into single spines or touching spines (Li and Deng 2012).

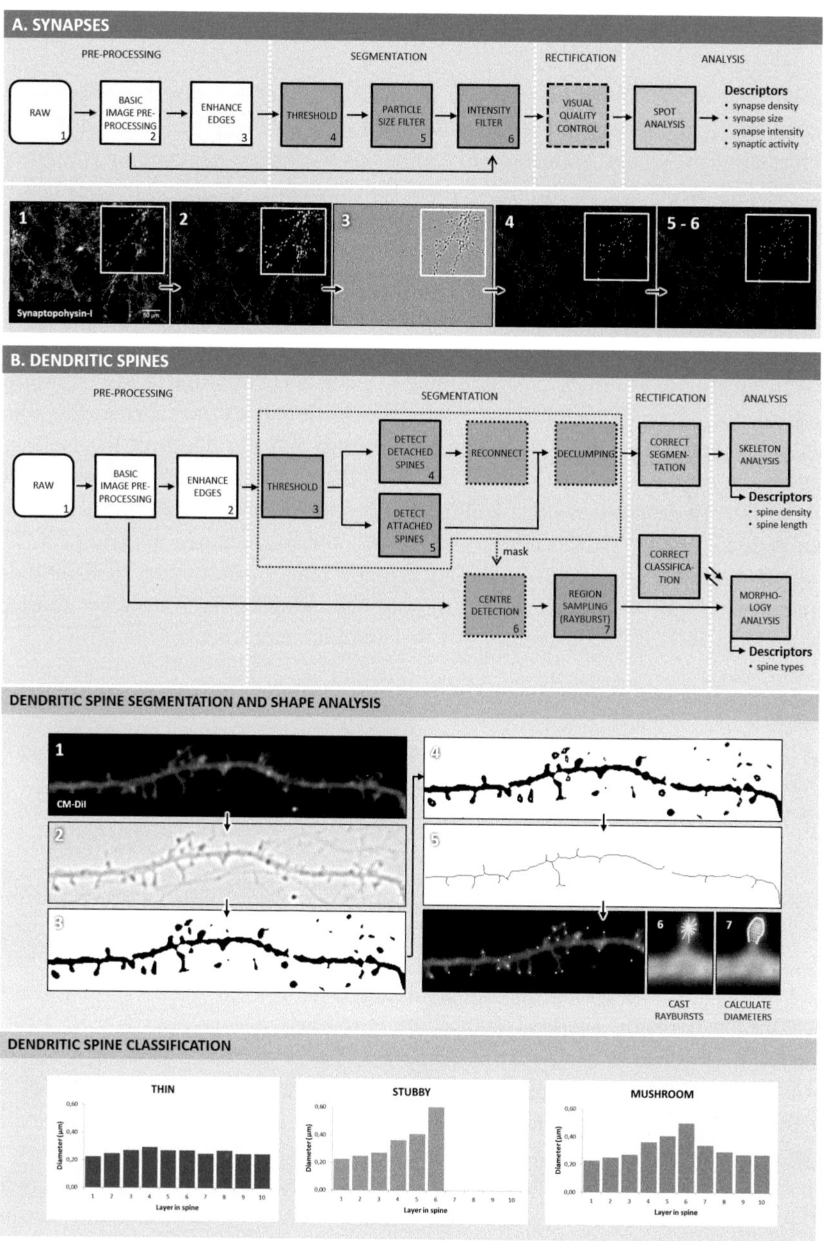

Fig. 5.4 Image analysis of synapses and dendritic spines. **A.** Synapse *puncta* are extracted by means of spot segmentation (Laplace filter). In a next step, false positives can be eliminated from the resulting image, using intensity- and size-based filters. **B.** In the *upper panel*, workflows for extracting dendritic spine density and morphology are shown. In the lower two panels, the process is shown of a centreline-based segmentation method, followed by Rayburst sampling to estimate the diameter in different layers of the spine. The ratio between the width of the spine head and neck can then be used to classify the spine type (stubby: no neck defined; thin: low ratio and mushroom: high ratio)

5.2.4 *Identifying Spine Morphology*

In centreline extraction-based methods, morphology determination is mainly limited to quantifying the length of the segmented dendritic spines. Since small structures, such as dendritic spines, comprise only a few voxels at maximal imaging resolution, quantization errors due to the finite voxel representation in digital images can be significant. Rayburst sampling was introduced to allow more reliable morphometric studies of dendritic spines. This is done by casting a multidirectional core of rays from an interior point (i.e. the centre of mass of the spine) to the spine surface, allowing precise sampling of its anisotropic and irregularly shaped structure. As the ray pattern is casted with sub-voxel accuracy using interpolated pixel intensity values, quantization errors are minimized. Once the contours of the spine are sampled, the spine diameter is calculated for different layers between the spine head and spine neck (Fig. 5.4B). The aspect ratio and the width of the head are then used to resolve the final spine types. Rayburst sampling has been successfully used to detect a decrease in spine volume and dendrite diameter in mouse models for Huntington's disease [R6/2 (Heck et al. 2012)] and Alzheimer's disease [TG2576 (Luebke et al. 2010)]. In addition to its original implementation in NeuronStudio (Rodriguez et al. 2006), the algorithm was also adopted by AutoSpine [part of Neurolucida 360 (MBF Bioscience 2015b)] and FilamentTracer (Andor 2015).

5.3 Sizing the Waves of Activity: Quantifying Calcium Fluxes

5.3.1 *Visualizing Electrical Activity*

Electrical activity exhibited by neurons can be visualized under the microscope using membrane voltage sensors. Classical voltage sensors such as potential sensitive aminonaphthylethenylpyridinium (ANEP) dyes display a spectral shift upon a change in voltage across the membrane (Fluhler et al. 1985); more recently developed genetically encoded sensors such as FlaSh (Siegel and Isacoff 1997), ElectricPk (Barnett et al. 2012) or ArcLight (Jin et al. 2012; Piao et al. 2015) change intensity with voltage. Despite rapid developments in the field (Jin et al. 2010), voltage sensors still do not cover a very high dynamic range and typically have to be measured very fast (up to 60 kHz). This is why electrical activity is still most often measured indirectly, by gauging calcium fluctuations (Herzog et al. 2011; Smetters et al. 1999). The high dynamic range of most calcium sensors allows visualizing electrical activity on a conventional fluorescence microscope at the single-neuron scale, albeit at lower temporal resolution (typically 2–4 Hz) than voltage imaging. Non-ratiometric calcium probes such as Fluo-4 AM display an increase in fluorescence intensity upon calcium binding, while ratiometric probes like Fura-2 exhibit a shift in excitation or emission spectra, allowing precise measurements of intracellular calcium concentration, not biased by uneven dye loading. In addition to synthetic calcium probes, genetically encoded sensors like chameleons or GCaMPs have emerged over the last years (Broussard et al. 2014). These sensors allow long-term follow-up of neuronal activity and their expression can be limited to neurons, e.g. when driven by a synapsin promoter. Also, their spatial localization can be confined to, e.g. synaptic compartments, when fused to synaptic proteins.

5.3.2 *Measuring Calcium Fluxes*

Reliable quantification of dynamic calcium recordings requires integrated image and signal analysis. The workflow of such an analysis is depicted in Fig. 5.5 (upper panel), together with the output from a Fluo-4 AM recording of spontaneous activity in a primary hippocampal culture of 7 days in vitro (DIV, lower panel).

To allow proper assessment of intercellular synchronicity of calcium oscillations, it is essential that individual neurons are properly segmented. This issue is resolved by including a nuclear label since the somas are the most abundant calcium domains. If neuron-specific nuclear tags are available (e.g. nuclear-localized fluorescent proteins expressed under a synapsin promoter), the analysis can immediately proceed to the signal analysis stage. However, synthetic nuclear indicators load all cells and require discrimination between the segmented neurons and astrocytes in the field of view. This can be achieved by exposing the cultures to a high concentration of glutamate, since neurons are known to respond with a very fast

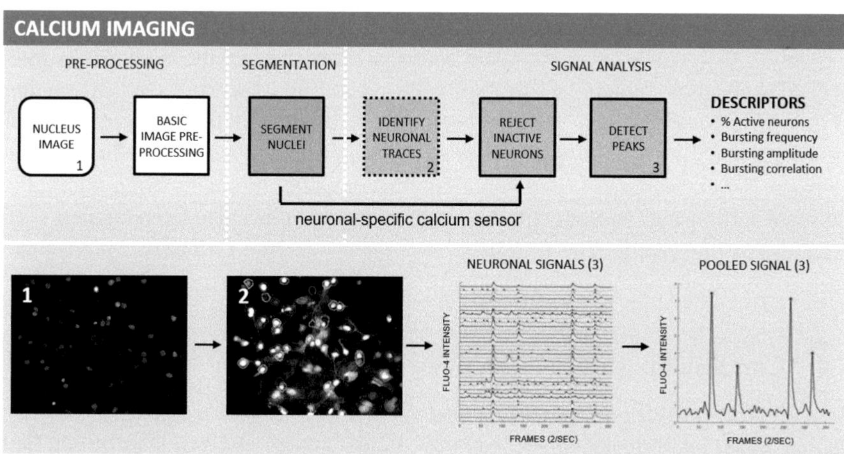

Fig. 5.5 Workflow for analysing calcium recordings from neuronal cultures. The *upper panel* shows image and signal analysis steps to extract numerical data from calcium recordings, while the *lower panel* contains output from a primary hippocampal culture showing both synchronized (corresponding to peaks in the pooled signal) and asynchronous calcium bursts

and prolonged increase in intracellular calcium, while astrocytes exhibit a delayed and transient calcium wave (Pickering et al. 2008). The first step following the extraction of calcium traces from the segmented cells is to define the glutamate addition point (typically the maximum signal). Then, two measures can be used to classify the cellular responses. First, the rise time can be used to detect delayed and slow responses of non-neuronal cells. Second, non-neuronal cells can be discarded based on their relative faster loss in mean fluorescence intensity after glutamate addition.

Similar pre-processing operations to those explained for 2D images (e.g. background subtraction and smoothing) are then performed on the 1D neuronal signals. Inactive neurons are identified based on a signal cut-off and rejected from the downstream analysis. Noise-tolerant peak detection on active neurons returns the location (burst frequency) and amplitude of each peak, as well as the average 50% decay time. Peaks displaying a decay time above a user-supplied maximum are discarded from the analysis and are reported as the number of long decays. Readouts originate from the rejection of inactive neurons (% active neurons) or from peak detection on individual (frequency, amplitude and decay time) or pooled (frequency of synchronized bursts) signals. Another powerful readout is to calculate a Pearson correlation coefficient between all individual neuronal traces, from which an average bursting correlation score can be derived. (Cornelissen et al. 2013).

The proposed image and signal analysis pipeline allows quantifying the effects of chronic pharmacological or genetic treatments on neuronal connectivity with great sensitivity (Verstraelen et al. 2014). For instance, it was shown that deprivation of nerve growth factor (NGF) impaired the synchronization of neuronal activity while increased trophic support by a feeder layer of astrocytes enhanced network formation. Additionally, division of a recording into 2 or 3 stretches allows the evaluation of the acute responses to pharmacological treatments. In this context, it was shown that synchronized network activity is mediated by the NMDA receptor, as NMDA receptor antagonists decreased the synchronicity score. Calcium imaging of in vitro network activity has also been used to study epilepsy by application of the convulsive drug 4-aminopyridine and low magnesium (Pacico and Mingorance-Le Meur 2014). Using an experimental in vitro model of traumatic brain injury, the neuronal response to subsequent glutamate stimulation has also been studied with calcium imaging (Gurkoff et al. 2012).

5.4 Conclusion and Perspectives

In this work we have given an overview of the image analysis algorithms that are used to investigate neuronal connectivity in cell cultures. We discussed the extraction of morphological features, such as the analysis of neuronal morphology and synapses, as well as the measurement of functional parameters used in calcium activity-related imaging studies.

When addressing neuronal morphology, a clear trade-off should be made between accuracy and throughput, and this has to be aligned with the labelling procedure. Whereas neuronal tracing provides an accurate representation of sparsely labelled neurons, it currently still demands manual intervention to rectify segmentation errors. A machine-learning approach that is trained using a manually delineated dataset has recently been proposed to reduce the proofreading time by only highlighting the reconstructions with the lowest confidence (Gala et al. 2014). Further elaboration on this approach may lead to a user-independent self-learning algorithm such as SmartTracing (Chen et al. 2015), in which there is no need for a sample dataset. On the other hand, global segmentation algorithms can be used to delineate neurons and pan-labelled, dense networks in a fully automated mode, albeit with lower accuracy. Recent developments are aimed at combining both global and local segmentation methods to develop fully automated tracing methods that are robust to staining imperfections and noise (Peng et al. 2011). Although early neuronal tracing algorithms were limited to 2D, 3D tracing algorithms are currently fine-tuned in such a way that they can be used to analyse stained neurons in neuronal slices, or even in the intact cleared brain (Chung et al. 2013). To this end, similar stochastic labelling procedures can be used for the sparse labelling of single neurons. Alternatively, more refined labelling strategies (e.g. based on GFP-expressing neurotropic viruses (Wickersham et al. 2007)) that allow trans-synaptic tracing of neurons open doors for more detailed connectome studies. This work further shows that numerous, sometimes redundant, approaches are currently employed to analyse neuronal morphology, making it difficult to select the best method for a given dataset (Peng et al. 2015). In order to compare the accuracy and the computational efficiency of these different methods, the BigNeuron project was launched in March 2015 (Peng et al. 2015). The major goal of this project is to enhance neuron reconstruction by bench-testing multiple algorithms against a large neuron dataset based on the experience of different research groups around the world.

Synapses are analysed by direct labelling of proteins involved in synaptic processing, or by assessing the density and morphology of dendritic spines. Although synaptic *puncta* are easily extracted using blob detectors, pre- and post-processing are often necessary to discriminate the true synaptic *puncta* from noise. Whereas a count of synapses offers an estimate of the number of synaptic proteins, a colocalization analysis of pre- and postsynaptic labels (e.g. VGluT and PSD95) can be performed to define synaptic partners (Kay et al. 2013; Roqué 2011). In addition, FM dyes can be used to selectively stain the presynaptic membrane of living cells to monitor neurotransmitter release and reuptake over time (Fan et al. 2012). The extension of synapse segmentation to 3D is limited by the spatial resolution of confocal microscopes in the axial direction. A solution to this issue is to computationally reconstruct serial ultrathin sections, known as array tomography (Micheva and Smith 2007). Alternatively, 3D superresolution imaging [e.g. 3D STORM (Dani et al. 2010)] can be used for fast volumetric imaging of synapses without the requirement of sectioning.

From an image informatics perspective, dendritic spines are more difficult to detect compared to synapses. This is because the segmentation process has to accommodate for the irregular and variable shape of spines, compared to the more consistent spot pattern that is found for synapse markers. Despite the development of numerous workflows that incorporate parallel analysis lines to increase the detection accuracy of spines, fully automated detection of spines is still a challenge. Similarly, classifying spine morphology requires the input of a human operator for reasons of quality control. Although most image processing algorithms are used to analyse small stacks of in vitro recordings, 3D dendritic spine analysis has also been carried out in tissue slice cultures (Luebke et al. 2010) and in vivo recordings (Fan et al. 2009). Tracking the changes in dendritic spine density and morphology in living animals would not only allow real-time monitoring of the acute effects of drug treatments, but also enable direct correlation of neuronal connectivity parameters with cognitive and behavioural characteristics. Calcium imaging is a valuable tool in the emerging field of iPSC technology to characterize iPSC-derived neurons and to detect phenotypes in patient-derived cultures (Belinsky et al. 2014; Hartfield et al. 2014; Liu et al. 2012; Naujock et al. 2014). Although calcium imaging studies are mostly performed on monocultures, a direct extension of such experiments would be to shift to the co-cultivation of differentially labelled neuronal cultures. This enables the study of cell–cell interactions on calcium bursting behaviour, which might be of interest to investigate the effect of trans-synaptically transmitted toxic proteins (Nussbaum et al. 2013). In addition, calcium imaging can be combined with optogenetics (Deisseroth et al. 2006) or photostimulation (Godwin et al. 1997), so as to perturb specific cells (or even subcellular compartments) and monitor response within a multicellular context. Closing the loop between optical readouts and the generation of these stimuli (i.e. by real-time generation of stimuli based on live image analysis) will provide a powerful strategy to study cause-and-effect relationships in neural circuitry (Grosenick et al. 2015). Although this discussion was limited to calcium imaging of in vitro neuronal networks, obviously such measurements can be expanded to live animals. However, this brings about an additional layer of complexity and imposes challenges, such as correction for motion artefacts and discrimination of calcium signals that originate from different layers in the tissue (Wilt et al. 2009). Tackling these issues, however, will lead to the emergence of further advanced experimental setups, such as those in which mice are subjected to virtual reality systems to study their spatial navigation (Dombeck et al. 2010).

In conclusion, a lot of work has been done to automate the quantification of morphological and functional features of neuronal networks. The ultimate goal of these image analysis algorithms is to provide an accurate, fully automated assessment of neuronal network status. Although there are still challenges to be met in this respect, new methods for tissue preparation and labelling, continuing advances in microscopic imaging systems and further development of image analysis tools will be essential to extract meaningful data from microscopic images.

Acknowledgements This work was supported by the Agency for Innovation by Science and Technology in Flanders (IWT Baekeland fellowship IWT_140775, O&O IWT_150003), the Flemish Institute for Scientific Research (FWO PhD Fellowship 11ZF116N) and the University of Antwerp (UA_29267, UA_29256). We further would like to acknowledge Dr. Steffen Jaensch and Dr. Andreas Ebneth for their valuable comments and discussions.

References

Al-Kofahi KA, Lasek S, Szarowski DH, Pace CJ, Nagy G, Turner JN, Roysam B (2002) Rapid automated three-dimensional tracing of neurons from confocal image stacks. IEEE Trans Inf Technol Biomed 6(2):171–187

Al-Kofahi Y, Dowell-Mesfin N, Pace C, Shain W, Turner JN, Roysam B (2008) Improved detection of branching points in algorithms for automated neuron tracing from 3D confocal images. Cytometry A 73A(1):36–43

Alvarez VA, Sabatini BL (2007) Anatomical and physiological plasticity of dendritic spines. Annu Rev Neurosci 30(1):79–97

Andor (2015) Filament Tracer. http://www.andor.com/scientific-software/imaris-from-bitplane/filamenttracer

Bading H (2013) Nuclear calcium signalling in the regulation of brain function. Nat Publ Group 14(9):593–608

Bai W, Zhou X, Ji L, Cheng J, Wong STC (2007) Automatic dendritic spine analysis in two-photon laser scanning microscopy images. Cytometry A: J Int Soc Anal Cytol 71(10):818–826

Barnett L, Platisa J, Popovic M, Pieribone VA, Hughes T (2012) A fluorescent, genetically-encoded voltage probe capable of resolving action potentials. Plos One 7(9)

Belinsky GS, Rich MT, Sirois CL, Short SM, Pedrosa E, Lachman HM, Antic SD (2014) Patch-clamp recordings and calcium imaging followed by single-cell PCR reveal the developmental profile of 13 genes in iPSC-derived human neurons. Stem Cell Res 12(1):101–118

Binley KE, Ng WS, Tribble JR, Song B, Morgan JE (2014) Sholl analysis: a quantitative comparison of semi-automated methods. J Neurosci Methods 225:65–70

Bretzner L, Lindeberg T (1998) Feature tracking with automatic selection of spatial scales. Comput Vis Image Underst 71(3):385–392

Broussard GJ, Liang R, Tian L (2014) Monitoring activity in neural circuits with genetically encoded indicators. Front Mol Neurosci 7

Brown KM, Donohue DE, D'Alessandro G, Ascoli GA (2005) A cross-platform freeware tool for digital reconstruction of neuronal arborizations from image stacks. Neuroinformatics 3(4):343–359

Cai D, Cohen KB, Luo T, Lichtman JW, Sanes JR (2013) Improved tools for the Brainbow toolbox. Nat Methods 10(6):540–547

Cajal S (1891) Sur la structure de l'écorce cérébrale de quelques mammifères. Typ. de Joseph van In & Cie.; Aug. Peeters, lib. https://books.google.be/books?id=7nD_GwAACAAJ

Chen JL, Andermann ML, Keck T, Xu NL, Ziv Y (2013) Imaging neuronal populations in behaving rodents: paradigms for studying neural circuits underlying behavior in the mammalian cortex. J Neurosci 33(45):17631–17640

Chen H, Xiao H, Liu T, Peng H (2015) SmartTracing: self-learning-based neuron reconstruction. Brain Inform 2:1–10

Cheng J, Zhou X, Miller E, Witt RM, Zhu J, Sabatini BL, Wong STC (2007) A novel computational approach for automatic dendrite spines detection in two-photon laser scan microscopy. J Neurosci Methods 165(1):122–134

Cho S, Wood A, Bowby MR (2007) Brain slices as models for neurodegenerative disease and screening platforms to identify novel therapeutics. Curr Neuropharmacol 5(1):19–33

Chothani P, Mehta V, Stepanyants A (2011) Automated tracing of neurites from light microscopy stacks of images. Neuroinformatics 9(2–3):263–278

Chung K, Wallace J, Kim SY, Kalyanasundaram S, Andalman AS, Davidson TJ, Mirzabekov JJ, Zalocusky KA, Mattis J, Denisin AK, Pak S, Bernstein H, Ramakrishnan C, Grosenick L, Gradinaru V, Deisseroth K (2013) Structural and molecular interrogation of intact biological systems. Nature 497(7449):332–337

Cohen E, Ivenshitz M, Amor-Baroukh V, Greenberger V, Segal M (2008) Determinants of spontaneous activity in networks of cultured hippocampus. Brain Res 1235:21–30

Cornelissen F, Verstraelen P, Verbeke T, Pintelon I, Timmermans JP, Nuydens R, Meert T (2013) Quantitation of chronic and acute treatment effects on neuronal network activity using image and signal analysis: toward a high-content assay. J Biomol Screen 18(7):807–819

Costa LD, Campos AG, Estrozi LF, Rios LG, Bosco A (2000) A biologically-motivated approach to image representation and its application to neuromorphology. In: Proceedings of the biologically motivated computer vision, vol 1811. Springer, Berlin/Heidelberg/Seoul, pp 407–416

Dani A, Huang B, Bergan J, Dulac C, Zhuang X (2010) Superresolution imaging of chemical synapses in the brain. Neuron 68(5):843–856

De Vos WH, Van Neste L, Dieriks B, Joss GH, Van Oostveldt P (2010) High content image cytometry in the context of subnuclear organization. Cytometry Part A: J Int Soc Analytical Cytol 77(1):64–75

Deisseroth K, Feng G, Majewska AK, Miesenbock G, Ting A, Schnitzer MJ (2006) Next-generation optical technologies for illuminating genetically targeted brain circuits. J Neurosci 26(41):10380–10386

Deleglise B, Lassus B, Soubeyre V, Alleaume-Butaux A, Hjorth JJ, Vignes M, Schneider B, Brugg B, Viovy JL, Peyrin JM (2013) Synapto-protective drugs evaluation in reconstructed neuronal network. PloS one 8(8):e71103

Dent EW, Merriam EB, Hu X (2011) The dynamic cytoskeleton: backbone of dendritic spine plasticity. Current Opinion in Neurobiology 21(1):175–181

Dombeck DA, Khabbaz AN, Collman F, Adelman TL, Tank DW (2007) Imaging large-scale neural activity with cellular resolution in awake, mobile mice. Neuron 56(1):43–57

Dombeck DA, Harvey CD, Tian L, Looger LL, Tank DW (2010) Functional imaging of hippocampal place cells at cellular resolution during virtual navigation. Nature Neurosci 13(11):1433–1440

Dotti CG, Sullivan CA, Banker GA (1988) The establishment of polarity by hippocampal-neurons in culture. J Neurosci 8(4):1454–1468

Fan J, Zhou X, Dy JG, Zhang Y, Wong STC (2009) An automated pipeline for dendrite spine detection and tracking of 3D optical microscopy neuron images of in vivo mouse models. Neuroinformatics 7(2):113–130

Fan J, Xia X, Li Y, Dy JG, Wong STC (2012) A quantitative analytic pipeline for evaluating neuronal activities by high-throughput synaptic vesicle imaging. NeuroImage 62(3):2040–2054

Feng GP, Mellor RH, Bernstein M, Keller-Peck C, Nguyen QT, Wallace M, Nerbonne JM, Lichtman JW, Sanes JR (2000) Imaging neuronal subsets in transgenic mice expressing multiple spectral variants of GFP. Neuron 28(1):41–51

Feng L, Zhao T, Kim J (2012) Improved synapse detection for mGRASP-assisted brain connectivity mapping. Bioinformatics (Oxford, England) 28(12):i25–31

Fletcher TL, Decamilli P, Banker G (1994) Synaptogenesis in hippocampal cultures - evidence indicating that axons and dendrites become competent to form synapses at different stages of neuronal development. J Neurosci 14(11):6695–6706

Fluhler E, Burnham VG, Loew LM (1985) Spectra, membrane-binding, and potentiometric responses of new charge shift probes. Biochemistry 24(21):5749–5755

Gala R, Chapeton J, Jitesh J, Bhavsar C, Stepanyants A (2014) Active learning of neuron morphology for accurate automated tracing of neurites. Front Neuroanat 8

Gerlich D, Beaudouin J, Gebhard M, Ellenberg J, Eils R (2001) Four-dimensional imaging and quantitative reconstruction to analyse complex spatiotemporal processes in live cells. Nat Cell Biol 3(9):852–855

Godwin DW, Che DP, OMalley DM, Zhou Q (1997) Photostimulation with caged neurotransmitters using fiber optic lightguides. J Neurosci Methods 73(1):91–106

Grosenick L, Marshel JH, Deisseroth K (2015) Closed-loop and activity-guided optogenetic control. Neuron 86(1):106–139

Gurkoff GG, Shahlaie K, Lyeth BG (2012) In vitro mechanical strain trauma alters neuronal calcium responses: implications for posttraumatic epilepsy. Epilepsia 53(Suppl 1):53–60

Harrill JA, Robinette BL, Freudenrich T, Mundy WR (2013) Use of high content image analyses to detect chemical-mediated effects on neurite sub-populations in primary rat cortical neurons. Neurotoxicology 34:61–73

Hartfield EM, Yamasaki-Mann M, Fernandes HJR, Vowles J, James WS, Cowley SA, Wade-Martins R (2014) Physiological Characterisation of human iPS-derived dopaminergic neurons. Plos One 9(2)

Harvard (2015) The Connectome Project. http://cbs.fas.harvard.edu/science/connectome-project

Heck N, Betuing S, Vanhoutte P, Caboche J (2012) A deconvolution method to improve automated 3D-analysis of dendritic spines: application to a mouse model of Huntington's disease. Brain Struct Funct 217(2):421–434

Herzog N, Shein-Idelson M, Hanein Y (2011) Optical validation of in vitro extra-cellular neuronal recordings. J Neural Eng 8(5)

Ho SY, Chao CY, Huang HL, Chiu TW, Charoenkwan P, Hwang E (2011) NeurphologyJ: an automatic neuronal morphology quantification method and its application in pharmacological discovery. BMC Bioinf 12:230

Imamura K, Inoue H (2012) Research on neurodegenerative diseases using induced pluripotent stem cells. Psychogeriatrics 12(2):115–119

Janoos F, Mosaliganti K, Xu X, Machiraju R, Huang K, Wong STC (2009) Robust 3D reconstruction and identification of dendritic spines from optical microscopy imaging. Med Image Anal 13(1):167–179

Jin L et al (2010) Imaging the brain with optical methods. Springer, Berlin

Jin L, Han Z, Platisa J, Wooltorton JRA, Cohen LB, Pieribone VA (2012) Single action potentials and subthreshold electrical events imaged in neurons with a fluorescent protein voltage probe. Neuron 75(5):779–785

Kay KR, Smith C, Wright AK, Serrano-Pozo A, Pooler AM, Koffie R, Bastin ME, Bak TH, Abrahams S, Kopeikina KJ, McGuone D, Frosch MP, Gillingwater TH, Hyman BT, Spires-Jones TL (2013) Studying synapses in human brain with array tomography and electron microscopy. Nat Protoc 8(7):1366–1380

Kim SY, Chung K, Deisseroth K (2013) Light microscopy mapping of connections in the intact brain. Trends Cogn Sci 17(12):596–599

Koh IY, Lindquist WB, Zito K, Nimchinsky EA, Svoboda K (2002) An image analysis algorithm for dendritic spines. Neural Comput 14(6):1283–1310

Lai KO, Ip NY (2013) Structural plasticity of dendritic spines: the underlying mechanisms and its dysregulation in brain disorders. Biochim Biophys Acta Mol Basis Dis 1832(12):2257–2263

Lammerding J, Fong LG, Ji JY, Reue K, Stewart CL, Young SG, Lee RT (2006) Lamins A and C but not lamin B1 regulate nuclear mechanics. J Biol Chem 281(35):25768–25780

Li Q, Deng Z (2012) A surface-based 3-d dendritic spine detection approach from confocal microscopy images. IEEE Trans Image Process 21(3):1223–1230

Li Q, Zhou X, Deng Z, Baron M, Teylan MA, Kim Y, Wong STC (2009) A novel surface-based geometric approach for 3D dendritic spine detection from multi-photon excitation microscopy images. Proceedings/IEEE International Symposium on Biomedical Imaging: from nano to macro IEEE International Symposium on Biomedical Imaging 10814263:1255–1258

Lin YC, Koleske AJ (2010) Mechanisms of synapse and dendrite maintenance and their disruption in psychiatric and neurodegenerative disorders. Annu Rev Neurosci 33(1):349–378

Liu J, Koscielska KA, Cao Z, Hulsizer S, Grace N, Mitchell G, Nacey C, Githinji J, McGee J, Garcia-Arocena D, Hagerman RJ, Nolta J, Pessah IN, Hagerman PJ (2012) Signaling defects in iPSC-derived fragile X premutation neurons. Hum Mol Genet 21(17):3795–3805

Livet J, Weissman TA, Kang H, Draft RW, Lu J, Bennis RA, Sanes JR, Lichtman JW (2007) Transgenic strategies for combinatorial expression of fluorescent proteins in the nervous system. Nature 450(7166):56–62

Luebke JI, Weaver CM, Rocher AB, Rodriguez A, Crimins JL, Dickstein DL, Wearne SL, Hof PR (2010) Dendritic vulnerability in neurodegenerative disease: insights from analyses of cortical pyramidal neurons in transgenic mouse models. Brain Struct Funct 214(2–3):181–199

Maiti P, Manna J, McDonald MP (2015) Merging advanced technologies with classical methods to uncover dendritic spine dynamics: a hot spot of synaptic plasticity. Neurosci Res 96:1–13

MBF Bioscience (2015a) Autospine. http://www.mbfbioscience.com/autospine

MBF Bioscience (2015b) Neurolucida. http://www.mbfbioscience.com/neurolucida

McKinney RA (2010) Excitatory amino acid involvement in dendritic spine formation, maintenance and remodelling. J Physiol 588(Pt 1):107–116

Meijer M, Burkhardt P, de Wit H, Toonen RF, Fasshauer D, Verhage M (2012) Munc18-1 mutations that strongly impair SNARE-complex binding support normal synaptic transmission. EMBO J 31(9):2156–2168

Meijering E (2010) Neuron tracing in perspective. Cytometry Part A 77A(7):693–704

Meijering E (2012) Cell segmentation: 50 years down the road [life sciences]. IEEE Signal Process Mag 29(5):140–145

Meijering E, Jacob M, Sarria JCF, Steiner P, Hirling H, Unser M (2004) Design and validation of a tool for neurite tracing and analysis in fluorescence microscopy images. Cytometry Part A: J Int Soc Anal Cytol 58(2):167–176

Micheva KD, Smith SJ (2007) Array tomography: a new tool for imaging the molecular architecture and ultrastructure of neural circuits (vol 55, p 25, 2007). Neuron 55(5):824–824

Myatt D, Hadlington T, Ascoli G, Nasuto S (2012) Neuromantic - from semi manual to semi automatic reconstruction of neuron morphology. Front Neuroinf 6(4). doi:10.3389/fninf.2012.00004. http://www.frontiersin.org/neuroinformatics/10.3389/fninf.2012.00004/abstract

Nair R, Lauks J, Jung S, Cooke NE, de Wit H, Brose N, Kilimann MW, Verhage M, Rhee J (2013) Neurobeachin regulates neurotransmitter receptor trafficking to synapses. J Cell Biol 200(1):61–80

Narro ML, Yang F, Kraft R, Wenk C, Efrat A, Restifo LL (2007) NeuronMetrics: software for semi-automated processing of cultured neuron images. Brain Res 1138:57–75

Naujock M, Stanslowsky N, Reinhardt P, Sterneckert J, Haase A, Martin U, Kim KS, Dengler R, Wegner F, Petri S (2014) Molecular and functional analyses of motor neurons generated from human cord-blood-derived induced pluripotent stem cells. Stem Cells Dev 23(24):3011–3020

Nemoto T (2014) Development of novel two-photon microscopy for living brain and neuron. Microscopy (Oxford, England) 63(Suppl 1):i7–i8

Nussbaum JM, Seward ME, Bloom GS (2013) Alzheimer disease: a tale of two prions. Prion 7(1):14–19

Pacico N, Mingorance-Le Meur A (2014) New in vitro phenotypic assay for epilepsy: fluorescent measurement of synchronized neuronal calcium oscillations. Plos One 9(1)

Pani G, De Vos WH, Samari N, de Saint-Georges L, Baatout S, Van Oostveldt P, Benotmane MA (2014) MorphoNeuroNet: an automated method for dense neurite network analysis. Cytometry Part A: J Int Soc Anal Cytol 85(2):188–199

Papa M, Bundman MC, Greenberger V, Segal M (1995) Morphological analysis of dendritic spine development in primary cultures of hippocampal-neurons. J Neurosci 15(1):1–11

Paredes RM, Etzler JC, Watts LT, Zheng W, Lechleiter JD (2008) Chemical calcium indicators. Methods 46(3):143–151

Parekh R, Ascoli GA (2013) Neuronal Morphology goes digital: a research hub for cellular and system neuroscience. Neuron 77(6):1017–1038

Peng H, Long F, Myers G (2011) Automatic 3D neuron tracing using all-path pruning. Bioinformatics 27(13):i239–i247

Peng H, Roysam B, Ascoli GA (2013) Automated image computing reshapes computational neuroscience. BMC Bioinform 14

Peng H, Tang J, Xiao H, Bria A, Zhou J, Butler V, Zhou Z, Gonzalez-Bellido PT, Oh SW, Chen J, Mitra A, Tsien RW, Zeng H, Ascoli GA, Iannello G, Hawrylycz M, Myers E, Long F (2014) Virtual finger boosts three-dimensional imaging and microsurgery as well as terabyte volume image visualization and analysis. Nat Commun 5

Peng H, Hawrylycz M, Roskams J, Hill S, Spruston N, Meijering E, Ascoli GA (2015) BigNeuron: large-scale 3D neuron reconstruction from optical microscopy images. Neuron 87(2):252–256

Penzes P, Cahill ME, Jones KA, VanLeeuwen JE, Woolfrey KM (2011) Dendritic spine pathology in neuropsychiatric disorders. Nat Publ Group 14(3):285–293

Piao HH, Rajakumar D, Kang BE, Kim EH, Baker BJ (2015) Combinatorial mutagenesis of the voltage-sensing domain enables the optical resolution of action potentials firing at 60 hz by a genetically encoded fluorescent sensor of membrane potential. J Neurosci 35(1):372–385

Pickering M, Pickering BW, Murphy KJ, O'Connor JJ (2008) Discrimination of cell types in mixed cortical culture using calcium imaging: a comparison to immunocytochemical labeling. J Neurosci Methods 173(1):27–33

Pool M, Thiemann J, Bar-Or A, Fournier AE (2008) NeuriteTracer: a novel ImageJ plugin for automated quantification of neurite outgrowth. J Neurosci Methods 168(1):134–139

Popova D, Jacobsson SOP (2014) A fluorescence microplate screen assay for the detection of neurite outgrowth and neurotoxicity using an antibody against βIII-tubulin. Toxicology in vitro: an international journal published in association with BIBRA 28(3):411–418

Prodanov D, Heeroma J, Marani E (2006) Automatic morphometry of synaptic boutons of cultured cells using granulometric analysis of digital images. J Neurosci Methods 151(2):168–177

Righolt CH, van 't Hoff MLR, Vermolen BJ, Young IT, Raz V (2011) Robust nuclear lamina-based cell classification of aging and senescent cells. Aging 3(12):1192–1201

Rodriguez A, Ehlenberger DB, Hof PR, Wearne SL (2006) Rayburst sampling, an algorithm for automated three-dimensional shape analysis from laser scanning microscopy images. Nat Protoc 1(4):2152–2161

Rodriguez A, Ehlenberger DB, Dickstein DL, Hof PR, Wearne SL (2008) Automated three-dimensional detection and shape classification of dendritic spines from fluorescence microscopy images. PloS one 3(4):e1997

Rodriguez A, Ehlenberger DB, Hof PR, Wearne SL (2009) Three-dimensional neuron tracing by voxel scooping. J Neurosci Methods 184(1):169–175

Roqué P (2011) In Vitro Neurotoxicology, vol 758. Springer, Berlin, pp 361–390

Sala C, Segal M (2014) Dendritic spines: the locus of structural and functional plasticity. Physiol Rev 94(1):141–188

Sarder P, Nehorai A (2006) Deconvolution methods for 3-D fluorescence microscopy images. IEEE Signal Process Mag 23(3):32–45

Schmitt S, Evers JF, Duch C, Scholz M, Obermayer K (2004) New methods for the computer-assisted 3-D reconstruction of neurons from confocal image stacks. Neuroimage 23(4):1283–1298

Schmitz SK, Hjorth JJJ, Joemai RMS, Wijntjes R, Eijgenraam S, de Bruijn P, Georgiou C, de Jong APH, van Ooyen A, Verhage M, Cornelisse LN, Toonen RF, Veldkamp WJH (2011) Automated analysis of neuronal morphology, synapse number and synaptic recruitment (vol 195, pg 185, 2011). J Neurosci Methods 197(1):190–190

Shi P, Huang Y, Hong J (2014) Automated three-dimensional reconstruction and morphological analysis of dendritic spines based on semi-supervised learning. Biomed Opt Express 5(5):1541–1553

Siegel MS, Isacoff EY (1997) A genetically encoded optical probe of membrane voltage. Neuron 19(4):735–741

Sirenko O, Hesley J, Rusyn I, Cromwell EF (2014) High-content high-throughput assays for characterizing the viability and morphology of human iPSC-derived neuronal cultures. Assay Drug Dev Technol 12(9–10):536–547

Smetters D, Majewska A, Yuste R (1999) Detecting action potentials in neuronal populations with calcium imaging. Methods-Companion Methods Enzymol 18(2):215–221

Takahashi K, Yamanaka S (2006) Induction of pluripotent stem cells from mouse embryonic and adult fibroblast cultures by defined factors. Cell 126(4):663–676

The Allen Institute (2015) Online Public Resources. http://alleninstitute.org/our-research/open-science-resources/

Verstraelen P, Pintelon I, Nuydens R, Cornelissen F, Meert T, Timmermans JP (2014) Pharmacological characterization of cultivated neuronal networks: relevance to synaptogenesis and synaptic connectivity. Cell Mol Neurobiol 34(5):757–776

Wearne SL, Rodriguez A, Ehlenberger DB, Rocher AB, Henderson SC, Hof PR (2005) New techniques for imaging, digitization and analysis of three-dimensional neural morphology on multiple scales. Neuroscience 136(3):661–680

Wickersham IR, Lyon DC, Barnard RJO, Mori T, Finke S, Conzelmann KK, Young JAT, Callaway EM (2007) Monosynaptic restriction of transsynaptic tracing from single, genetically targeted neurons. Neuron 53(5):639–647

Wilt BA, Burns LD, Ho ETW, Ghosh KK, Mukamel EA, Schnitzer MJ (2009) Advances in Light Microscopy for Neuroscience. Annu Rev Neurosci 32:435–506

Wittmann M, Queisser G, Eder A, Wiegert JS, Bengtson CP, Hellwig A, Wittum G, Bading H (2009) Synaptic activity induces dramatic changes in the geometry of the cell nucleus: interplay between nuclear structure, Histone h3 phosphorylation, and nuclear calcium signaling. J Neurosci 29(47):14687–14700

Xie J, Zhao T, Lee T, Myers E, Peng H (2010) Automatic neuron tracing in volumetric microscopy images with anisotropic path searching. In: Medical image computing and computer-assisted intervention: MICCAI international conference on medical image computing and computer-assisted intervention 13(Pt 2):472–479

Zhang Y, Zhou X, Witt RM, Sabatini BL, Adjeroh D, Wong STC (2007) Dendritic spine detection using curvilinear structure detector and LDA classifier. NeuroImage 36(2):346–360

Zhang B, Fadili JM, Starck JL (2008) Wavelets, ridgelets, and curvelets for Poisson noise removal. IEEE Trans Image Process 17(7):1093–1108

Zhao T, Xie J, Amat F, Clack N, Ahammad P, Peng H, Long F, Myers E (2011) Automated reconstruction of neuronal morphology based on local geometrical and global structural models. Neuroinformatics 9(2–3):247–261

Chapter 6
Integrated High-Content Quantification of Intracellular ROS Levels and Mitochondrial Morphofunction

Tom Sieprath, Tobias D.J. Corne, Peter H.G.M. Willems, Werner J.H. Koopman, and Winnok H. De Vos

Abstract Oxidative stress arises from an imbalance between the production of reactive oxygen species (ROS) and their removal by cellular antioxidant systems. Especially under pathological conditions, mitochondria constitute a relevant source of cellular ROS. These organelles harbor the electron transport chain, bringing electrons in close vicinity to molecular oxygen. Although a full understanding is still lacking, intracellular ROS generation and mitochondrial function are also linked to changes in mitochondrial morphology. To study the intricate relationships between the different factors that govern cellular redox balance in living cells, we have developed a high-content microscopy-based strategy for simultaneous quantification of intracellular ROS levels and mitochondrial morphofunction. Here, we summarize the principles of intracellular ROS generation and removal, and we explain the major considerations for performing quantitative microscopy analyses of ROS and mitochondrial morphofunction in living cells. Next, we describe our workflow, and finally, we illustrate that a multiparametric readout enables the unambiguous classification of chemically perturbed cells as well as laminopathy patient cells.

T. Sieprath • T.D.J. Corne • W.H. De Vos (✉)
Cell Systems and Imaging Research Group (CSI), Department of Molecular Biotechnology, Ghent University, Ghent, Belgium

Laboratory of Cell Biology and Histology, Department of Veterinary Sciences, University of Antwerp, Antwerp, Belgium
e-mail: winnok.devos@uantwerpen.be

P.H.G.M. Willems • W.J.H. Koopman
Department of Biochemistry (286), Radboud University Medical Centre (RUMC), Radboud Institute for Molecular Life Sciences (RIMLS), Nijmegen, The Netherlands

W.H. De Vos et al. (eds.), *Focus on Bio-Image Informatics*, Advances in Anatomy, Embryology and Cell Biology 219, DOI 10.1007/978-3-319-28549-8_6

6.1 Principles of Intracellular ROS Generation and Removal

Reactive oxygen species (ROS) are small, short-lived derivatives of molecular oxygen (O_2) of radical and non-radical nature (Halliwell and Gutteridge 2007). Radical ROS variants include superoxide ($O_2^{\bullet -}$), hydroperoxyl ($HO_2^{\bullet}$), hydroxyl ($^{\bullet}OH$), peroxyl ($RO_2^{\bullet}$), alkoxyl ($RO^{\bullet}$), carbonate ($CO_3^{\bullet -}$), carbon dioxide ($CO_2^{\bullet -}$), and singlet oxygen ($O_2{}^1\sum g^+$). Non-radical variants include hydrogen peroxide (H_2O_2), hypobromous acid (HOBr), hypochlorous acid (HOCl), ozone (O_3), singlet oxygen ($O_2{}^1\Delta g$), organic peroxides (ROOH), peroxynitrite ($ONOO^-$), peroxynitrate (O_2NOO^-), nitrosoperoxycarbonate ($ONOOCO_2{}^-$), and peroximonocarbonate ($HOOCO_2{}^-$) (Halliwell and Gutteridge 2007). Of these ROS, $ONOO^-$ and O_2NOO^- are also reactive nitrogen species (RNS). RNS further include nitric oxide ($NO^{\bullet}$), nitrogen dioxide ($NO_2^{\bullet}$), nitrate radical ($NO_3^{\bullet}$) and many other nitrogen derivatives. ROS were originally described as molecular constituents of the defense system of phagocytic cells, but it has become clear that besides their damaging properties, they also function as signaling molecules and mediate a variety of other cellular responses including cell proliferation, differentiation, gene expression, and migration (Lambeth 2004; Bartz and Piantadosi 2010).

6.1.1 *Intracellular ROS Metabolism*

ROS can be generated at various sites in the cell (Fig. 6.1a). This can be either deliberately, e.g., by NADPH oxidases (NOX), or as a byproduct, e.g., during normal cellular respiration in mitochondria (Babior 1999; Turrens 2003; Murphy 2009). The NOX family of NADPH oxidases (NOX1, NOX2, NOX3, NOX4, NOX5, DUOX1, and DUOX2) are proteins that transport electrons (e^-) from NADPH across biological membranes (plasma or endomembranes) (Bedard and Krause 2007; Dupre-Crochet et al. 2013). The activation mechanisms and tissue distribution of the isoforms differ, but they all use O_2 as e^--acceptor, producing $O_2^{\bullet -}$. Through ROS generation, they play a role in many cellular processes including host defense, regulation of gene expression, and cell differentiation (Bedard and Krause 2007). Despite their sometimes significant contribution to the global ROS pools, NOX are not the predominant source of intracellular ROS. Mitochondria are considered the major culprit, in particular under pathological conditions. Mitochondrial ROS are generated as a byproduct of the oxidative phosphorylation (OXPHOS, cf. below).

Irrespective of its source, ROS production generally starts with the reduction of O_2 to $O_2^{\bullet -}$, which is the precursor of most other ROS (Fig. 6.2a). Either spontaneously or, more likely, catalyzed by a superoxide dismutase (SOD), $O_2^{\bullet -}$ is converted into H_2O_2 at a rate close to the diffusion limit ($k = 2 \cdot 10^9\ M^{-1}s^{-1}$ at pH 7.4) (Weisiger and Fridovich 1973; Boveris and Chance 1973; Loschen et al. 1974; Auchère and Rusnak 2002). In turn, H_2O_2 can be converted into water (H_2O) by several enzymes including peroxiredoxins, catalase (CAT), and glutathione

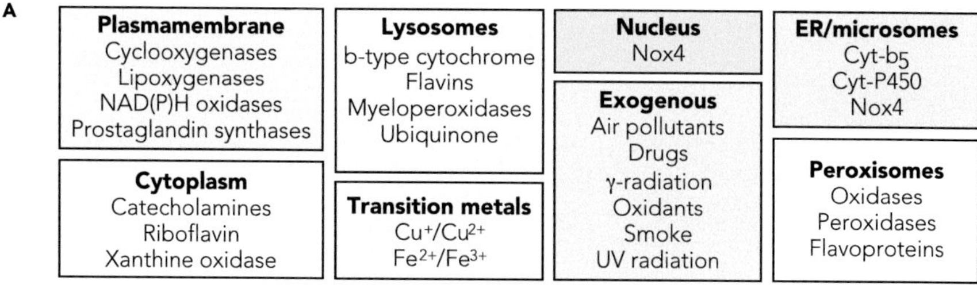

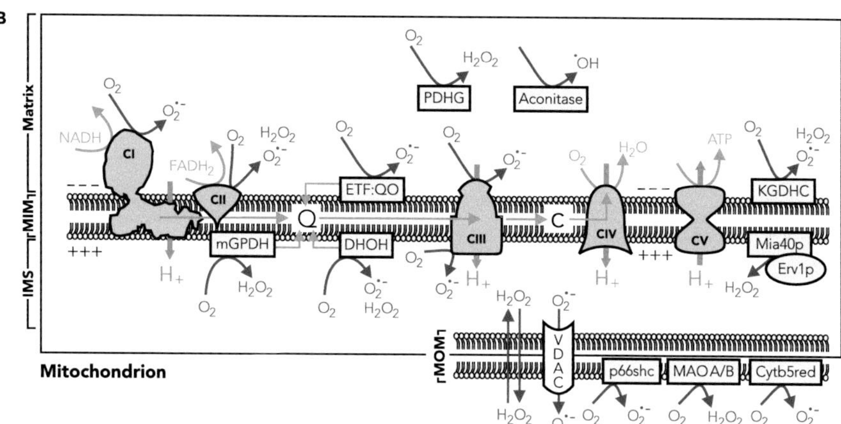

Fig. 6.1 Cellular sources of reactive oxygen species. (**a**) Non-mitochondrial sources of reactive oxygen species. (**b**) Mitochondrial sources of $O_2^{\bullet -}$ and H_2O_2. The scheme depicts the five complexes of the mitochondrial oxidative phosphorylation system involved in ATP production (CI-CV; *blue*) and other mitochondrial ROS-generating proteins (*gray boxes*). Once formed, the anionic $O_2^{\bullet -}$ cannot move across membranes, whereas H_2O_2 can more freely diffuse. Abbreviations: *Cytb5red* cytochrome b5 reductase, *DHOH* dihydroorotate dehydrogenase, *Erv1p/Mia40p* redox system that forms disulfide bridges on proteins to be imported by mitochondria, *ETF:QO* electron transfer lavoprotein-ubiquinone oxidoreductase, *KGDHC* α-ketoglutarate dehydrogenase complex, *MAO* monoamine oxidase, *mGPDH* mitochondrial glycerol-3-phosphate dehydrogenase, *p66shc* 66-kDa src collagen homologue (shc) adaptor protein, *PDHG* pyruvate dehydrogenase complex, *VDAC* voltage-dependent anion channel. The data for this figure was compiled from Giustarini et al. 2009; Koopman et al. 2010; Marchi et al. 2012; Brown and Borutaite 2012; Nathan and Cunningham-Bussel 2013; Woolley et al. 2013; Mailloux et al. 2013

peroxidases (GPXs) (Gupta et al. 2012). Proper function of these systems further requires the action of glutathione reductase (GR), thioredoxin (TRX), thioredoxin reductase (TRXR), glutaredoxin (GRX), peroxiredoxin (PRX), sulfiredoxin (SRX), the glutathione (GSH)-synthesizing enzymes glutathione synthase (GS) and glutamate cysteine ligase (GCL), and ceruloplasmin (Gupta et al. 2012). In addition to enzymatic systems, cells and tissues also contain antioxidants of nonenzymatic nature including glutathione (GSH), thioredoxin (TRX), phytochemicals, vitamins (A,C,E), and taurine (Gupta et al. 2012). The cofactor NADPH (the reduced form of nicotinamide adenine dinucleotide phosphate) is central to cellular ROS removal

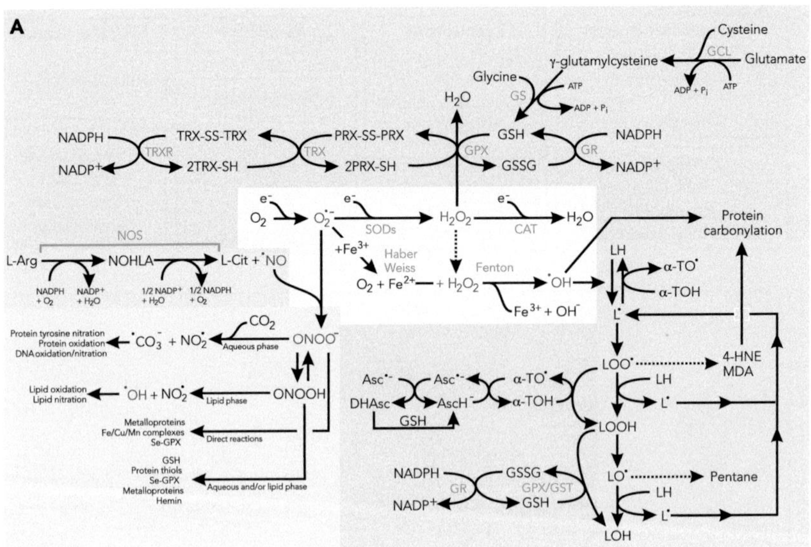

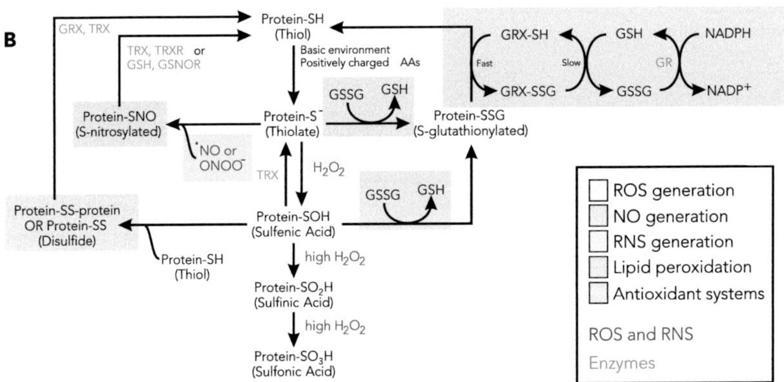

Fig. 6.2 Redox homeostasis in mammalian cells. (**a**) Formation of ROS and NO/RNS, their removal by antioxidant systems and role in lipid peroxidation. (**b**) Reactions of protein thiol (protein-SH) groups leading to reversible S-nitrosilation (protein-SNO), intra- or inter-protein disulfide bond formation (SS) or S-glutationylation (protein-SSG). Abbreviations: *4-HNE* 4-hydroxynonenal; *α-TOH* α-tocopherol, *α-TO*• α-tocopherol radical; *Asc*•− ascorbyl radical, *AscH*− ascorbate, *CAT* catalase, *GCL* glutamate cysteine ligase, *GPX* glutathione peroxidase, *GR* glutathione reductase, *GS* glutathione synthase, *GSH* glutathione, *GSNOR* S-nitrosoglutathione reductase, *GSSG* oxidized glutathione, *MDA* malondialdehyde, *NADPH* reduced nicotinamide adenine dinucleotide phosphate, *NOS* nitric oxide synthase, *NOHLA* Nω-hydroxy-L-arginine, *PRX* peroxiredoxin, *RNS* reactive nitrogen species, *ROS* reactive oxygen species, *SOD* superoxide dismutase, *TRX* thioredoxin, *TRXR* thioredoxin reductase. The data for this figure was compiled from Auchère and Rusnak 2002; Szabó et al. 2007; Sachdev and Davies 2008; Benhar et al. 2009; Rasmussen et al. 2010; Finkel 2011; Traber and Stevens 2011; Pastore and Piemonte 2012; Marí et al. 2013; Nathan and Cunningham-Bussel 2013; Saaranen and Ruddock 2013; Conte Lo and Carroll 2013; Cremers and Jakob 2013; Stangherlin and Reddy 2013; Mailloux et al. 2013; Groitl and Jakob 2014

through the GSH and TRX/PRX systems (Fig. 6.2a). In mitochondria, NADPH is mainly produced via (1) $NADP^+$-dependent isocitrate dehydrogenase and malic enzyme and (2) nicotinamide nucleotide transhydrogenase (Nnt). The latter enzyme utilizes the proton motive force (PMF) to generate NADPH from NADH and $NADP^+$ (Lopert and Patel 2014). Besides the conversion into H_2O_2, $O_2^{\bullet -}$ can also react with nitric oxide ($NO^{\bullet}$), produced in a two-step reaction from L-arginine (L-arg), catalyzed by nitric oxide synthases (NOS). This gives rise to the production of reactive nitrogen species (RNS) peroxynitrite ($ONOO^-$) and peroxynitrous acid (ONOOH). Various other reactions downstream of $ONOO^-$ lead to the formation of $^{\bullet}OH$, $CO_3^{\bullet -}$, and $NO_2^{\bullet}$ (Fig. 6.2a) (Radi et al. 2002; Szabó et al. 2007). In the presence of ferric iron (Fe^{3+}), the $O_2^{\bullet -}$ anion is converted into O_2 and ferrous iron (Fe^{2+}), which can further react with H_2O_2 to reform Fe^{3+}, hydroxide (OH^-), and the highly reactive $^{\bullet}OH$ (Fig. 6.2a) (Thomas et al. 2009). $^{\bullet}OH$ is one of the strongest oxidants in nature and is extremely damaging to biomolecules like DNA, proteins, and lipids (Franco et al. 2008; Marchi et al. 2012). It can initiate formation of lipid ($L^{\bullet}$) and lipid peroxyl ($LOO^{\bullet}$) radicals (lipid peroxidation), which is counterbalanced by the action of various antioxidant systems including vitamin E/α-tocopherol (α-TOH), vitamin C/ascorbate ($AscH^-$), NADPH/$NADP^+$, GSH, GPX/GST, and GR (Fig. 6.2a). Ultimately, sustained stimulation of lipid peroxidation will lead to formation of pentane and the reactive aldehydes malondialdehyde (MDA) and 4-hydroxynonenal (4-HNE). When generated at low levels, 4-HNE can interact with signaling targets, including JNK, P38 MAPK, cell cycle regulators, PKCβ, and PKCδ, leading to numerous cellular responses, ranging from increased expression of the antioxidant enzyme TRXR1 to irreversible cytotoxic injuries and cell death (Chen et al. 2005; Riahi et al. 2010). Mitochondrial aldehyde dehydrogenase 2 (ALDH2) can protect against oxidative stress by detoxification of these cytotoxic aldehydes (Chiu et al. 2015).

ROS can react covalently with certain atomic elements in biological macromolecules (Fig. 6.2b) (Nathan and Cunningham-Bussel 2013). At low ROS levels, these modifications are usually reversible, whereas at high ROS levels, they are not. Reversibility is also confined to specific atoms: reversible modifications occur on selenium (Se; in seleno-Cys) and sulfur (S; in certain Cys and Met), whereas iron-sulfur (Fe-S) clusters and carbon (C) atoms (Arg, Lys, Pro, Thr, and nucleosides) are irreversibly modified. Reactions of primary ROS with proteins include reversible oxidative formation of methionine sulfoxide (by $^{\bullet}OH$) and irreversible formation of 2-oxo-histidine (by H_2O_2/Fe^{2+}), chlorotyrosine (by HOCl), and protein carbonyls (by $^{\bullet}OH$) (Dickinson and Chang 2011). When protein thiol (SH) groups (pKa $\sim$ 8.5) are within a basic environment (such as the mitochondrial matrix) or have their pKa lowered by proximity to positively charged amino acids, they deprotonate and are present in their thiolate (S^-) form (Fig. 6.2b) (Mailloux et al. 2013). Protein thiolate groups reversibly react with ROS (H_2O_2, HOCl) to form protein sulfinic acid (SOH). In the presence of high H_2O_2 levels, the SOH form is subsequently and irreversibly converted into sulfinic acid (SO_2H) and sulfonic acid (SO_3H) forms. The thiolate form can also react with (1) glutathione disulfide (GSSG) to form S-glutathionylated (SSG) proteins and (2) RNS to form S-nitrosated/S-nitrosylated (SNO) proteins

(Benhar et al. 2009; Grek et al. 2013). Starting from the SOH form, the reaction of protein thiols with GSH also leads to formation of S-glutathionylated proteins. By reacting with other SH groups, the SOH form can induce inter- or intramolecular disulfide bond formation (Fig. 6.2b). The SH groups in the SSG, SNO, and disulfide proteins can be reformed via various reactions involving GRX, TRX, TRXR, and NADPH (Fig. 6.2b), allowing redox-dependent cell signaling events (Benhar et al. 2009; Nakamura and Lipton 2011; Murphy 2012; Groitl and Jakob 2014).

There is a subtle balance between the production and removal of the different ROS molecules to maintain their intracellular concentration at a physiological level. Any perturbation to this fragile steady state that increases intracellular ROS provokes oxidative stress, a phenomenon associated with the natural aging process, as well as various multispectral diseases including cancer and laminopathies (Harman 1956; Naderi et al. 2006; Moylan and Reid 2007; Caron et al. 2007; Salmon et al. 2010; Sieprath et al. 2012).

6.1.2 Range of Action of ROS

A surplus of ROS is highly unwanted as it allows them to interact with various cellular constituents. However, to react with biomolecules, ROS need to be able to reach them. Once generated, the range of action of individual ROS differs substantially. For instance, in the presence of GSH (2 mM), values of 50 μm and 1.5 mm were computed for $ONOO^-$ and H_2O_2, respectively (Winterbourn 2008). The same study reported that the range of action for H_2O_2 dropped to < 7 μm, in the presence of 20 μM PRX2 (the main H_2O_2-removing enzyme) and was even lower for $^{\bullet}OH$ (0.35 μm). In aqueous solution, the average 3D diffusion distance or "Kuramoto length" (Δx) was calculated to be < 0.16 μm for $O_2^{\bullet -}$ and between 0.23 and 0.46 μm for H_2O_2 (Koopman et al. 2010). Using the Einstein-Smoluchowski Eq. (6.1), diffusion distances of 50 μm ($O_2^{\bullet -}$, in the absence of SOD), 0.4 μm ($O_2^{\bullet -}$, in the presence of SOD), 3000 μm (H_2O_2), 0.005 μm ($^{\bullet}OH$, in aqueous solution), 0.07 μm ($CO_3^{\bullet -}$), 0.13 μm ($NO_2^{\bullet}$), and 0.07 μm (O_2^1) were predicted (Cardoso et al. 2012).

$$D = \frac{k_B T}{6\pi \eta r} \tag{6.1}$$

where D = diffusion constant, k_B = Boltzmann's constant, T = absolute temperature, η = dynamic viscosity, and r = radius of the spherical particle.

Importantly, several ROS, including $O_2^{\bullet -}$, are charged molecules, which prevents their passive transmembrane permeation. When generated in the mitochondrial matrix, $O_2^{\bullet -}$ is highly unlikely to leave this compartment unless facilitated. Currently, there are no reports of superoxide permeation of the inner membrane. However, it has been proposed that the voltage-dependent anion channel (VDAC) in the mitochondrial outer membrane could mediate $O_2^{\bullet -}$ release from mitochondria

(Han et al. 2003). Taken together, due to their physicochemical properties and the action of (non)enzymatic conversion cascades, various ROS types display different ranges of action within cells and subcellular compartments including mitochondria. This strongly suggests that both ROS-induced damage and signaling are affected by restricted diffusion and compartmentalization (Winterbourn 2008). In this respect, it appears that mitochondria-generated $O_2^{\bullet-}$ acts locally, whereas H_2O_2 and $NO^{\bullet}$, owing to their membrane permeability and relative stability, can function as both a cytosolic and extracellular messenger ($t_{1/2}$ for H_2O_2 is 10^{-2} ms and for $NO^{\bullet}$ between 1 and 30 s, compared to 10^{-3} ms and 10^{-6} ms for $O_2^{\bullet-}$ and $^{\bullet}OH$) (Radi et al. 2002; Boveris et al. 2006; Giorgio et al. 2007; Hamanaka and Chandel 2010). The diffusion properties of H_2O_2 likely depend on its site of generation and (local) conversion, since cytoplasmic microdomains of elevated H_2O_2 levels were demonstrated in cells stimulated with growth factors, suggesting that this type of ROS does not freely diffuse through the cytoplasm (Rhee et al. 2012; Mishina et al. 2012).

6.1.3 Mitochondria Are Prime Sources and Targets of ROS

In total, mitochondria account for 90–95% of the cellular oxygen consumption, and up to 3% of that pool can be converted into $O_2^{\bullet-}$, depending on the mitochondrial functional state or "mitochondrial health" (Marchi et al. 2012). A widely used indicator of mitochondrial health is the magnitude of the membrane potential ($\Delta\psi_m$) across the mitochondrial inner membrane. This potential is central to virtually all major (bioenergetic) functions of the mitochondrion, as it reflects the proton motive force that drives OXPHOS and mitochondrial Ca^{2+} uptake (Turrens 2003). $\Delta\psi_m$ is sustained by the action of the four complexes (complex I–IV) of the electron transport chain (ETC), located on the inner mitochondrial membrane, and the adjoined export of protons into the intermembrane space (Fig. 6.1b). Proton backflow through the F_oF_1-ATPase (complex V) is then used to drive the production of ATP production in the mitochondrial matrix. ROS can be produced at many locations inside the mitochondrion (Fig. 6.1b), but it generally results from electron leakage at complex I of the electron transport chain (ETC) when $\Delta\psi_m$ is highly negative. However, both de- and hyperpolarization have been associated with increased ROS production (Korshunov et al. 1997; Miwa and Brand 2003; Verkaart et al. 2007; Murphy 2009; Lebiedzinska et al. 2010). Various mitochondrial proteins are susceptible to reversible and irreversible redox modifications, allowing local regulation of their function and/or affecting pathological processes. For instance, reversible S-nitrosylation of complex I at Cys39 of the ND3 subunit decreased ROS production, oxidative damage, and tissue necrosis and thereby protected against injury during cardiac ischemia-reperfusion in vivo (Chouchani et al. 2013).

Although a full understanding is still lacking, net mitochondrial morphology, a result of continuous fusion and fission events, appears to be linked to mitochondrial function, ROS generation, and redox state as well (Willems et al. 2015).

An accumulating body of evidence points to direct involvement of ROS (and RNS) in the short-term regulation of mitochondrial morphology and function via non-transcriptional pathways, i.e., through reversible and nonreversible redox modifications (S-nitrosylation, disulfide bond formation) on/in proteins involved in the fission-fusion machinery of mitochondria (Willems et al. 2015). Fragmentation appears correlated with increased ROS production and apoptosis (Koopman et al. 2007; Archer 2013), while a more filamentous phenotype has been linked to nutrient starvation and protection against mitophagy (Rambold et al. 2011).

Given their close relationship, intracellular ROS levels and mitochondrial morphofunction should be studied together in living cells so as to better understand their interconnection during normal and pathological conditions. This is why we have developed a quantitative high-content assay for simultaneous quantification of intracellular ROS, mitochondrial morphology, and $\Delta\psi_m$. Before we explain the workflow in detail, we describe some general considerations required for live cell ROS and mitochondrial imaging.

6.2 Considerations for Quantifying Redox Biology and Mitochondrial Function in Living Cells Using Fluorescence Microscopy

Microplate readers are regularly used to measure fluorescence intensities, but the readout is highly prone to confounding factors, such as variable cell density and autofluorescence. Although flow cytometry measures all cells individually, which greatly increases sensitivity and accuracy, this technique does not provide spatiotemporal information (e.g., no subcellular localization, no time-dependent kinetics) and imposes an operational stress factor (cell detachment) when working with adherent cell cultures. These disadvantages are avoided when using fluorescence microscopy. Microscopy allows gauging redox biology and mitochondrial function in individual adherent cells through time at subcellular resolution and with high sensitivity, both pre- and post-stimulus, i.e., *in fluxo*. However, to enable robust and accurate measurements of intracellular ROS and individual mitochondria, all aspects of the imaging pipeline, from sample preparation to image analysis, have to be thoroughly standardized. In this part, we highlight some of the major considerations.

6.2.1 Cell Culture Conditions

Culture conditions prior to the measurements have to be meticulously controlled in order to obtain robust and reproducible results. For instance, the composition of the culture medium as well as the imaging buffer can greatly affect mitochondrial morphology and function. Nutrient starvation generally leads to a more filamentous mitochondrial phenotype (Rambold et al. 2011; Gomes et al. 2011), while high glucose concentrations have been linked to increased ROS production and mitochondrial fission (Yu et al. 2006; Trudeau et al. 2011). Cells should be seeded at least 24h before actual measurements, at fixed splitting ratios so as to obtain a sub-confluent culture of 70%–80% (substrate occupation) at the time point of measurement. This guarantees optimal performance of downstream image analyses (in particular cell segmentation). Furthermore, imaging buffer/medium should be devoid of potential autofluorescent components, such as phenol red, riboflavin, or tryptophan, in order to reduce nonspecific background intensity (Frigault et al. 2009). Also, to minimize the influence of plate effects, sample distribution should be homogenized or randomized across the plate, and the outer wells should not be used for measurements since they are prone to edge effects (they can however be used to take background images for a downstream flat field correction).

6.2.2 Sensors

A second point of attention pertains to the selection of the appropriate reporter (Table 6.1). As most ROS molecules tend to have a short lifetime (nanoseconds to seconds), fluorescent detection of intracellular redox changes demands sensitive reporter dyes with fast and reversible binding kinetics and high dynamic range (Dikalov and Harrison 2014). Ideally, they also show little or no photobleaching or (photo)toxicity, and loading is quick and easy. Currently available ROS probes can be subdivided into two categories: synthetic small molecule dyes and genetically encoded fluorescent proteins. The most commonly used small molecule ROS probes are dihydroethidium (DHE), mitochondrial-targeted DHE (MitoSOX), and the chemically reduced and acetylated forms of 2′,7′-di-chlorofluorescein (DCF) (Wang et al. 2013). CM-H_2DCFDA (5-(and-6)-chloromethyl-2′,7′-dichlorodihydrofluorescein diacetate, acetyl ester) is a widely used chloromethyl derivative of H_2DCFDA that is used to measure general intracellular ROS levels. It diffuses passively into the cell where its acetate groups are cleaved by intracellular esterases, decreasing its capacity to traverse the cell membrane and thereby trapping it inside the cell. Its thiol-reactive chloromethyl group allows for covalent binding to intracellular components, increasing retention of the dye even further. Following oxidation, highly fluorescent DCF is formed. With an excitation maximum of 502 nm and an emission peak of 523 nm, DCF fluorescence can be readily monitored using standard filter combinations for GFP or FITC (Tarpey et al. 2004; Gomes et al. 2005; Koopman et al. 2006). Other general small molecule ROS probes include Thioltracker® and the CellROX® family of indicators (Life Technologies©). Dyes that are more specific to certain types of ROS exist as well. DHE is generally used as a probe for $O_2^{\bullet-}$ (Zhao et al. 2003). The reaction between DHE and $O_2^{\bullet-}$ generates highly red fluorescent 2-hydroxyethidium (2-OH-E^+; ex. 518 nm, em. 605 nm). Reaction with other oxidants, however, can produce ethidium (E^+),

Table 6.1 Characteristics and usage of common ROS, redox, and mitochondrial probes

Type	Name	Indicator for	Ex/Em (nm)	Remarks	References
Chemical	DHE	$O_2^{\bullet-}$	518/605	Excitation between 350 and 400 nm to differentiate 2-OH-E^+ from E^+	Zielonka and Kalyanaraman (2010), Robinson et al. (2006)
	MitoSOX Red	Mitochondrial $O_2^{\bullet-}$	518/605		Robinson et al. (2008), Forkink et al. (2015)
	CM-H_2DCFDA	General ROS	502/523		Koopman et al. (2006), Sieprath et al. (2015)
	C11-BODIPY	Lipid peroxidation	490/520; 580/590	Oxidized product: 490/520; reduced product: 580/590; also 490/520 and 590 can be used	Drummen et al. (2002)
	MitoPerOx	Mitochondrial lipid peroxidation	490/520; 580/590		Prime et al. (2012)
	TMRM	Mitochondrial morphology and $\Delta\psi_m$	550/576		Nicholls (2012), Koopman et al. (2008)
	MTRs	Mitochondrial morphology	–	Multiple MTRs with different ex/em	Chazotte (2011)
	PF3	H_2O_2	492/515	These probes are all sensitive for H_2O_2, but they exhibit different fluorescence ex/em, making them compatible with other probes	Dickinson et al. (2010a)
	PG1	H_2O_2	460/510		Miller et al. (2007)
	PO1	H_2O_2	540/565		Dickinson et al. (2010a)
	PY1	H_2O_2	519/548		Dickinson et al. (2010a)
	MitoPY1	Mitochondrial H_2O_2	510/528		Dickinson et al. (2013)

(continued)

Table 6.1 (continued)

Type	Name	Indicator for	Ex/Em (nm)	Remarks	References
Protein	HyPer 1	H_2O_2	420 and 500/516	Hyper 2, better dynamic range, slower kinetics; Hyper 3, better dynamic range and better kinetics	Belousov et al. (2006)
	HyPer 2	H_2O_2	420 and 500/516		Markvicheva et al. (2011)
	HyPer 3	H_2O_2	420 and 500/516		Bilan et al. (2013)
	roGFP1	GSH redox potential	400 and 475/509	roGFP2, superior to roGFP1; Grx1-roGFP2, faster kinetics (equilibration time: minutes)	Hanson et al. (2004)
	roGFP2	GSH redox potential	400 and 490/509		Dooley et al. (2004)
	Grx1-roGFP2	GSH redox potential	400 and 490/509		Gutscher et al. (2008)
	Orp1-roGFP2	H_2O_2	400 and 490/509		Gutscher et al. (2009)

which strongly binds DNA, is also red fluorescent (ex. 525 nm, em. 616 nm), and is often present at a much higher concentration (Zielonka and Kalyanaraman 2010). Discrimination between these two can still be possible, however, due to an extra excitation band between 350 and 400 nm for 2-OH-E^+ (Robinson et al. 2006). However, as the ratio E^+/2-OH-E^+ is often 10 or more, contribution of E^+ might still be significant (Zielonka and Kalyanaraman 2010). MitoSOX is a DHE derivative coupled to a positively charged triphenylphosphonium group (TPP^+), enabling efficient targeting to the mitochondria for selective detection of mitochondrial $O_2^{\bullet-}$ (Robinson et al. 2008). The recently described HKSOX1 family of probes is also specific for $O_2^{\bullet-}$ (Hu et al. 2015), and a family of boronate-based sensors (peroxy family, e.g., PF1, PF3, PG1, PO1, PY1, MitoPY1, etc.) targeting to the cytosol or the mitochondria is used for the detection of H_2O_2 (Chang et al. 2004; Miller et al. 2007; Dickinson et al. 2010a; Dickinson et al. 2010b). H_2O_2-mediated removal of a boronate group greatly increases fluorescence of these sensors. They also display a range of fluorescent wavelengths, making them useful for multicolor experiments. For a more extensive overview of small-molecule fluorescent probes for ROS, the reader is referred to Gomes et al. (Gomes et al. 2005). Next to small-molecule fluorescent probes, ROS can also be monitored using genetically encoded fluorescent protein-based probes. While labeling is more complex, usually involving liposome- or virus-based transfection procedures, selectivity of these dyes is generally higher. Moreover, genetic reporters can easily be targeted to a variety of intracellular destinations, and they are maintained for prolonged periods of time allowing long-term and transgeneration follow-up. They are either ROS-sensitive fluorophores or standard fluorophores fused to ROS-sensing domains borrowed from other proteins like SoxR and OxyR, i.e., transcription factors found in *E. coli* that become activated by oxidation with $O_2^{\bullet-}$ or H_2O_2, respectively (Zheng et al. 1998; Fujikawa et al. 2012). In SoxR, the regulatory domain contains a 2Fe-2S cluster, while that of OxyR has several redox active cysteine residues. Both of them undergo a significant conformational change upon activation. In order to translate this into a quantifiable change in fluorescence, these domains are linked to circularly permuted (cp) versions of fluorescent proteins (Topell et al. 1999; Baird et al. 1999). The HyPer family of fluorescent probes was created by inserting a cpYFP in the regulatory domain of OxyR (Belousov et al. 2006; Markvicheva et al. 2011; Bilan et al. 2013). HyPer acts as a ratiometric H_2O_2 probe with 2 excitation maxima (420 and 500 nm) corresponding to the protonated and anionic forms of the protein, respectively, and one emission maximum (516 nm). Upon oxidation, a disulfide bond is formed between Cys 199 and Cys 208, resulting in a decrease of the 420 nm excitation peak and a proportional increase of the 500 nm excitation peak. This ratiometric determination greatly reduces the influence of expression level differences between individual cells. Unfortunately, HyPer is partially sensitive to pH changes. Acidification of the cellular environment leads to protonation, thus mimicking reduction of the probe; alkalization on the other hand mimics oxidation (Belousov et al. 2006). SypHer, a H_2O_2-insensitive HyPer variant (C199S), is a pH sensor and can be used as a control (Poburko et al. 2011). Another family of fluorescent protein-based redox-sensitive probes is the roGFPs, which were created by introducing oxidizable cysteine residues on the outside of the ß-barrel structure of GFP near the location of the chromophore. They can be used to ratiometrically measure intracellular redox balance (GSH/GSSG-ratio). roGFPs, just like HyPer, have 2 excitation peaks, which correspond to their oxidized and reduced states, but in contrast to HyPer, they are considered insensitive to pH (Lukyanov and Belousov 2014). One of the drawbacks of roGFP reporters is their slow kinetics. A fusion between roGFP and glutaredoxin 1 (Grx1-roGFP) resulted in a probe with faster kinetics (Gutscher et al. 2008). However, it still takes minutes or longer to equilibrate with cellular redox potential changes, which is still too slow to detect fast transient events (Meyer and Dick 2010). Another variant is Orp1-roGFP, a fusion between roGFP and the yeast peroxidase Orp1, which functions as an intracellular, ratiometric, pH-stable H_2O_2 probe (Gutscher et al. 2009). Its response to H_2O_2 is similar to that observed with HyPer, although oxidation is slower (Gutscher et al. 2009). Next to direct ROS probes, one could also use probes that assess ROS indirectly by measuring the downstream damage such as lipid peroxidation. C11-BODIPY$^{581/591}$, for example, is a ratiometric sensor for lipid peroxidation (Drummen et al. 2002), while MitoPeroX (a mitochondria-targeted derivative of C11- BODIPY$^{581/591}$) is a ratiometric probe for the specific assessment of mitochondrial phospholipid peroxidation (Prime et al. 2012).

Several fluorescent dyes have also been developed for measuring mitochondrial morphology and $\Delta\psi_m$. They all are cell permeant and become readily sequestered by active mitochondria in a $\Delta\psi_m$-dependent manner (Iannetti et al. 2015). Of these, the red-orange fluorescent tetramethylrhodamine methyl ester (TMRM) is one of the most efficient because it equilibrates fastest across membranes, is least toxic, and demonstrates the lowest aspecific binding (Nicholls 2012). With its excitation and emission maxima being 550nm and 576nm, respectively, it is also compatible with many probes that are fluorescent in the GFP region, like the HyPers and roGFPs, or CM-H_2DCFDA. MitoTracker® dyes (MTRs) should only be used for morphological analysis. Although they are also sequestered to the mitochondria based on $\Delta\psi_m$, they are retained there as a result of reaction with mitochondrial biomolecules, making it impossible to measure dynamic changes of $\Delta\psi_m$ (Dong et al. 2013).

6.2.3 *Microscopy*

When performing live cell imaging, optimal cellular health condition is crucial to ensure that the physiological and biological processes under investigation are not altered in any way. For mammalian cells, the temperature must ideally be kept stable at 37°C, pH should be at a physiological level (~pH = 7.2–7.4), and changes in osmolarity have to be avoided by minimizing evaporation (Frigault et al. 2009). However, to reduce externalization and vacuolization of internalized dyes, measurements are often performed at lower temperatures (Staljanssens et al. 2012). Temperature control can be achieved by means of a large incubator enclosing the whole microscope, or a stage top incubator in combination with objective heaters. To minimize thermally induced focus drift along the z-axis, samples should be allowed to equilibrate on the microscope before imaging. When using bicarbonate-based culture medium, a CO_2 incubation chamber or HEPES-based buffer has to be used to keep pH at a physiological level (Casey 2006). To avoid evaporation of the medium and the resulting changes in osmolarity, relative humidity needs to be kept at nearly 100%. Typically, CO_2 gas is bubbled through a water container to humidify the incubator (Frigault et al. 2009).

As fluorescence excitation induces ROS production (photodamage), it is quintessential to minimize light exposure, especially when aiming at quantification of intracellular ROS and redox-related processes (Dixit and Cyr 2003; Pattison and Davies 2006; Zhao et al. 2008). The most straightforward way to mitigate photodamage is to reduce the illumination load. This can be achieved by limiting the total imaging duration, but this goes at the expense of the signal-to-noise ratio. Hence, the efficiency of light collection should be optimized as well. This can be done using hard-coated filters, high-numerical aperture (NA) lenses, and detectors with high quantum efficiency, such as EM-CCD cameras (Frigault et al. 2009). When scaling a microscopy assay up to a multi-well format, variations in focus levels within and between wells impose another level of complexity. Images need to be perfectly focused to measure morphological and intensity metrics accurately

(Koopman et al. 2008). Stage movements and time-resolved revisiting of regions of interest therefore call for accurate autofocusing methods. Hardware-based autofocusing methods, which rely on laser or LED deflection on the substrate, allow for continuous, real-time correction of the distance drifts between objective and substrate caused by plate imperfections and thermal fluctuations, while software-based methods correct for biological focus variations such as cell and organelle (e.g., mitochondria) positioning by calculating a sharpness or contrast metric in a series of axial recordings (Rabut and Ellenberg 2004; Frigault et al. 2009). Software-based autofocus methods are not recommended for redox biology imaging, because they require multiple exposures, but sometimes they are crucial to fine-tune hardware-driven axial positioning. To minimize phototoxicity, software-based autofocusing should be done using low intensity transmitted light. Another consideration when scaling up to a multi-well format is the time needed for the acquisition of all wells. The measured signal should be stable from the measurement of the first well to the last, but the total acquisition time increases linearly with the number of wells and the number of individual images recorded in each well. The available time window is dependent on several variables, including the dynamics of the process under investigation and the used staining method (for instance, transient staining with a small molecule dye versus stable expression of a genetic marker).

6.3 Method for Simultaneous Quantification of ROS Levels and Mitochondrial Morphofunction

Taking into account the considerations for microscopic assessment of redox biology, we have established a method for the simultaneous quantification of ROS levels and mitochondrial features in living cells using automated wide-field fluorescence microscopy and automated image analyses. The method is optimal for primary human dermal fibroblasts, which, due to their extremely flat morphology and relatively large size, are well suited for analysis by wide-field (non-confocal) fluorescence microscopy (Koopman et al. 2006). Nevertheless, the method should be applicable to a wide variety of adherent cell types. As proof of principle, we have chosen two generic and easily applicable fluorescent indicators, namely, CM-H_2DCFDA for measuring general intracellular ROS levels and TMRM for mitochondrial morphology and $\Delta\psi_m$. An overview of the workflow is given in Fig. 6.3.

6.3.1 Sample Preparation and Image Acquisition

To make the assay amenable to medium- to high-throughput screens, we have optimized a workflow for 96-well plates. To minimize artifacts, cells are seeded

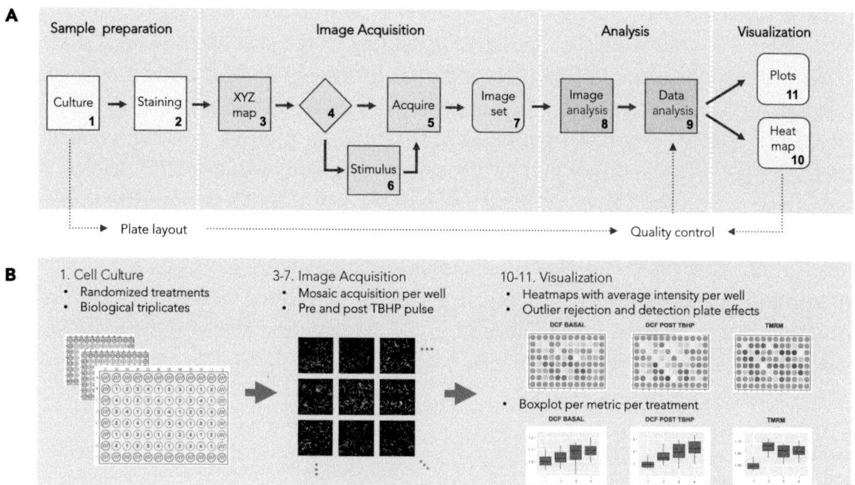

Fig. 6.3 General overview of the high-content method for simultaneous measurement of intracellular basal and induced ROS levels and mitochondrial morphofunction. (**a**) Schematic representation of the major operational blocks and subtasks. (**b**) Illustrated example: cells are seeded in multiple identical 96-well plates. The numbers in the wells represent different treatments, homogeneously distributed along the plate. After staining, 4 images are taken per channel in the center of each well, both pre and post *in fluxo* treatment. The image analysis workflow is described in more detail in Fig. 6.4. Intuitive representation of the intensity results in a 96-well layout permits rapid detection of plate effects (plate-wide gradient patterns – as seen in the DCF basal image) or aberrant wells. After curation of complete experimental data sets, data analysis is performed

at least 24 h in advance, and they are allowed to grow and equilibrate in conditioned medium. When using different conditions (e.g., controls and perturbations), seeding locations are homogeneously distributed so as to minimize plate effects. The outer wells are filled with medium, but are not imaged, as they are highly prone to edge effects. To avoid scattering and cross talk (of excitation and emission) between adjacent wells, we make use of black polystyrene plates with a thin continuous polystyrene film bottom (190 μm $\pm$ 20 μm; Greiner®). A staining protocol was optimized that uses a minimal amount of reporter dyes as overloading may affect cellular health status and cause nonlinear effects due to quenching (Invitrogen 2006). Image acquisition is performed with an automated wide-field microscope using a 20x air plan-corrected objective (NA = 0.75). The first well is sacrificed for determination of the optimal focus plane. As this procedure induces an increase in DCF signal intensity, this well is excluded from further analysis. Next, an acquisition protocol is initialized, using hardware-based autofocus that captures a set of 4 fields per channel in the center of each well. With our setup, the plate acquisition time of this protocol is approximately 10 min. This has proven to be sufficiently short so as to not cause any significant differences between the first and last wells due to the transient nature of the staining or the dynamics of the processes under investigation. Optimization experiments revealed that both the DCFDA and

TMRM signals remain stable from 7 to at least 50 min after loading (coefficient of variation < 2%), allowing for a possible upscaling of the assay to 384-well plates.

After the acquisition protocol is completed, a stimulus/perturbation can be given, using an on- or off-stage automated micropipette or multichannel pipette. We make use of the oxidant *tert*-butyl peroxide (TBHP) as an internal positive control for the CM-H_2DCFDA staining and as a means to measure induced ROS levels. At a fixed time span after TBHP addition (minimum 3 min to allow equilibration), the acquisition protocol is repeated. A maximum of 10 different treatments is used per plate. Then each plate contains 6 technical replicates with 4 images per replicate. In addition, a minimum of 3 identical plates is measured, increasing the number of images per treatment to a minimum of 72. This guarantees sufficient statistical power to detect even small differences.

6.3.2 Image Processing and Data Analysis

After acquisition, raw image data sets are directly backed up to a server, with remote access. A virtual desktop application or command shell interface can be used to organize and analyze image data sets. All image processing is performed in FIJI (http://fiji.sc), a packaged version of ImageJ freeware (W.S. Rasband et al., National Institutes of Health, Bethesda, Maryland, USA, http://rsb.info.nih.gov/ij/, 1997–2015), which runs directly on the server. We have conceived a dedicated script for automated analysis of intracellular and mitochondrial signals and morphological characteristics (RedoxMetrics.ijm), which is available upon request. The image analysis pipeline can be divided into 4 major blocks, which can be adapted to the specific image quality and cell type, namely, (1) preprocessing, (2) object enhancement, (3) segmentation, and (4) analysis (Fig. 6.4):

1. Preprocessing is generic to all channels and involves a flat field correction (FF) to correct for spatiotemporal illumination heterogeneity, which arises from imperfections of the acquisition system. The flat field image is usually acquired separately in a separate (usually outer) well with no cells, but with dyes. Alternatively, a pseudo-flat field image is generated by means of an anisotropic 3D median filter, across all images of a well plate. Obviously, this procedure only works well when there are sufficient images to average across (min. 50).
2. After preprocessing, the objects of interest will be selectively enhanced. Depending on the object, cells (CMDCFDA stains the entire cell), or mitochondria, different enhancement procedures are followed. For cell segmentation, a combination of local background subtraction (rolling ball, RB), noise reduction (in the form of a Gaussian blur operation, GB), and local contrast enhancement (CE) (Zuiderveld 1994) is used. The kernel sizes for these operators are tunable parameters, which are automatically set to optimized values based on the image calibration (pixel size), retrieved from the metadata. In case of mitochondria, a normalized Laplace of Gaussian operator (LG) is applied, for which the optimal

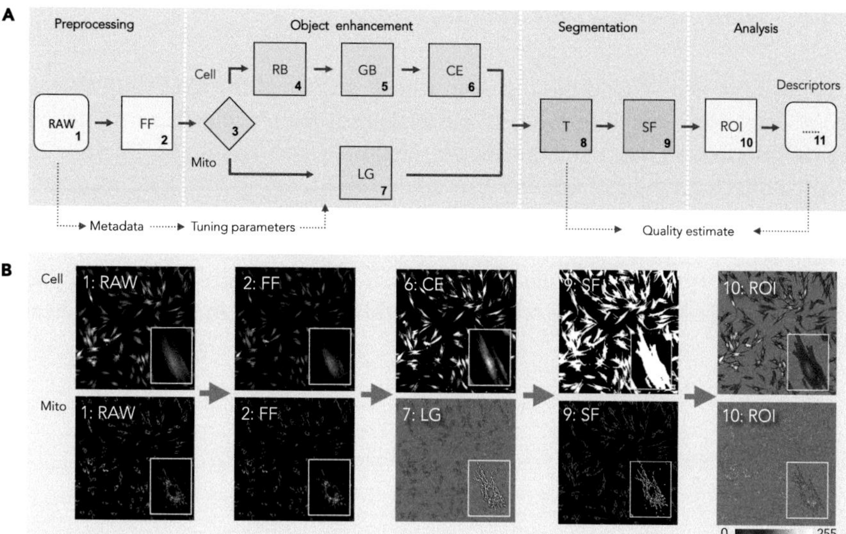

Fig. 6.4 Overview of the image analysis workflow. (**a**) Schematic representation of the entire image analysis process. Abbreviations: *FF* flatfield correction, *RB* rolling ball background subtraction, *GB* Gaussian blur, *CE* local contrast enhancement, *LG* Laplace of Gaussian, *T* thresholding, *SF* size filter, *ROI* region of interest. (**b**) Illustrated example. The *top row* represents the analysis of DCF images, the *bottom row* the analysis of TMRM images

scale is automatically selected based on the most salient features in scale space (De Vos et al. 2010).

3. Automatic segmentation demands implementation of a robust thresholding method. A variety of auto-threshold methods have been conceived (Sahoo et al. 1988; Glasbey 1993), and we have found Huang's algorithm (Huang and Wang 1995), which minimizes image fuzziness (the difference between the original image and its binary version), to work particularly well for both object types. However, an inherent caveat of auto-threshold methods is that they adjust the cut-off values based on the intensity distribution within the image. This introduces an unwanted bias when aiming at comparative quantifications, which is why we calculate threshold values using intensity information from the entire image data set. Alternatively, the threshold value can be set manually. Before proceeding to image analysis, a binary size filter (SF) is applied so as to exclude objects that fall out of the realistic size range.
4. Once generated, regions of interest (ROIs) are used for extracting intensity, texture, and morphological (size and shape) parameters on the flat field corrected images. Both ROI sets are used to analyze signals in both channels, enabling spatial discrimination of intensity fluctuations.

6.3.3 Data Analysis and Visualization

Data analysis and visualization is done with R statistical freeware (http://www.r-project.org). Raw output from the image analysis is read in, together with the plate layout information, and is automatically organized and visualized. Intuitive heat maps, projected onto the original well-plate layout, allow for facile recognition of expected (e.g., dose response) or unwanted (gradient) patterns and outliers (Fig. 6.3). The latter are usually automatically discarded (cf. gray wells in Fig. 6.3), based on quality criteria including minimal cell density or maximum intensity levels. Finally, intensity from the complete experiment is summarized and statistically compared.

6.3.4 Validation

To validate the described workflow, several control experiments have been conducted (Fig. 6.5). To verify the correlation between intracellular ROS levels and DCF fluorescence, and to determine its dynamic range, human fibroblasts were treated for 15 min with increasing concentrations of TBHP before being measured. Within a dose range of 10–160μM TBHP, a linear correlation between ROS level and fluorescent signal was observed (Fig. 6.5c). The same experimental setup was used for TMRM. Fibroblasts were treated for 30 min with increasing concentrations of oligomycin, which induces $\Delta\psi_m$ hyperpolarization, before being measured. This approach equally resulted in a linear increase of the measured signal within the 1–10μg/μl dose range (Fig. 6.5a). Conversely, when fibroblasts were treated with valinomycin *in fluxo*, a gradual, quantifiable decrease of TMRM fluorescence was measured, corresponding to an expected $\Delta\psi_m$ depolarization (Fig. 6.5b). We also compared the microscopy-based method with spectrophotometry and flow cytometry (Fig. 6.5d). Flow cytometry showed a higher dynamic range (as measured after treatment with 20μM TBHP), but also a much larger variability in measurements (also note that this method requires the cells to be in suspension and does not allow for spatiotemporal analysis). Spectrophotometry showed a comparable dynamic range and variability. However, this method, just like flow cytometry, cannot discern morphological details nor is it capable of detecting confounding factors such as abnormal cell density or autofluorescent contaminants in individual wells.

Finally, we validated the generic character of the methodology by replacing the dye combinations CM-H_2DCFDA/TMRM for calcein/MitoSOX. Here, calcein (1μM) was used to generate cell masks and to exclude dead cells from the analysis; MitoSOX (5μM) served to measure mitochondrial $O_2^{\bullet-}$ levels. After staining, primary human fibroblasts were imaged at 1 frame each 6s. Addition of 500 μM TBHP resulted in a clear increase of the MitoSOX signal, when measured per cellular pixel and even more pronounced when expressed per mitochondrial pixel (Fig. 6.6).

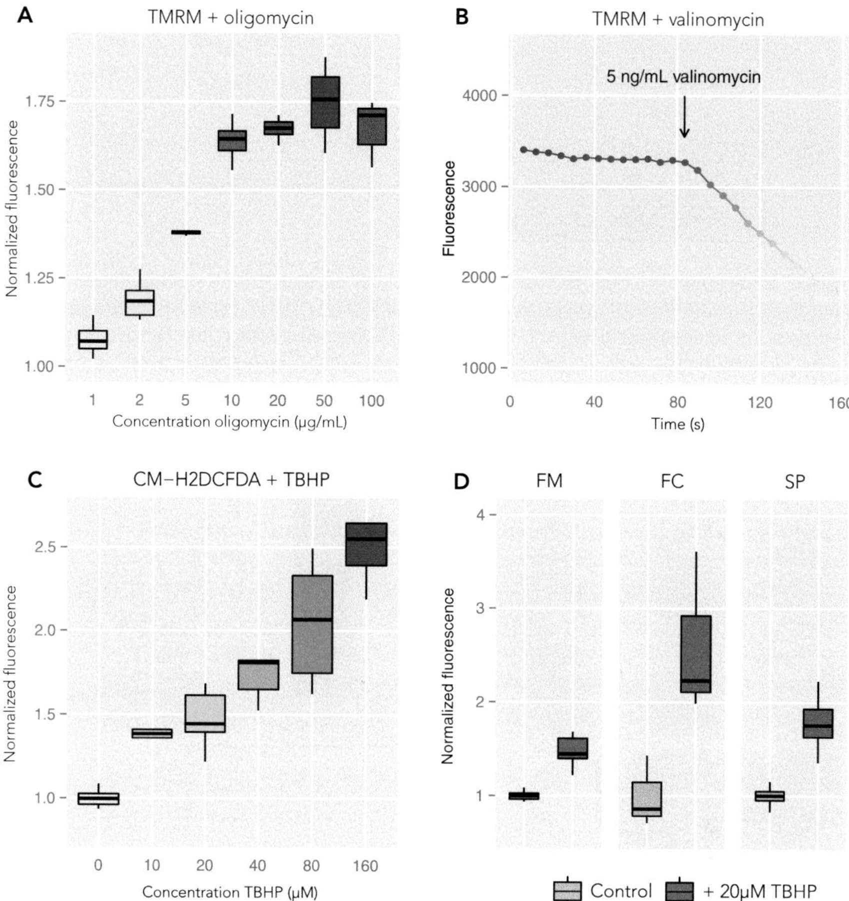

Fig. 6.5 Validation of the CM-H2DCFDA/TMRM-based high content microscopy method. All experiments were conducted in primary human fibroblasts. (**a**) Normalized $\Delta\psi m$ as measured by mitochondrial TMRM signal after treatment with increasing concentrations of oligomycin for 30 min.. (**b**) Live cell imaging of $\Delta\psi m$ in cells stained with TMRM and treated with 5 ng/mL valinomycin after 85 s. (**c**) Normalized levels of intracellular ROS measured as intracellular CM-H2DCFDA signal after treatment with increasing concentrations of TBHP for 15 min. (**d**) Comparison of microscopic, flow cytometric, and spectrophotometric (*plate reader*) measurement of basal intracellular ROS levels using CM-H2DCFDA in control cells (CTR) versus cells treated with 20 μM TBHP

6.3.5 Biological Applications

After optimization and validation, we have performed a number of experiments to illustrate the performance of the high-content microscopy methodology. HIV protease inhibitors (HIV PIs) have been shown to induce increased basal ROS levels (Chandra et al. 2009; Touzet and Philips 2010) and lowered antioxidant

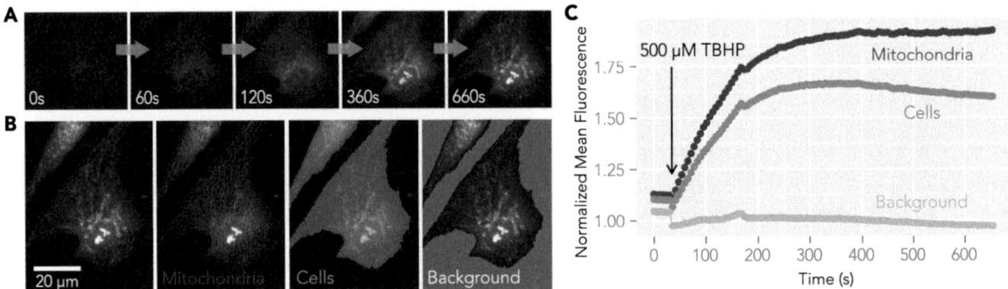

Fig. 6.6 Compatibility of the HC assay with MitoSOX measurements. (**a**) Response of cells stained with 5 μM MitoSOX towards acute addition of 500 μM TBHP. (**b**) Masks used to measure fluorescence intensity in different parts of the image. (**c**) Normalized mean fluorescence in function of time as measured through the different masks shown in (**b**) (The bump observed after approx. 160 s is due to a floating fluorescent particle (probably a detached cell) passing, out of focus, through the field of view inducing a weak increase in signal in the affected images.)

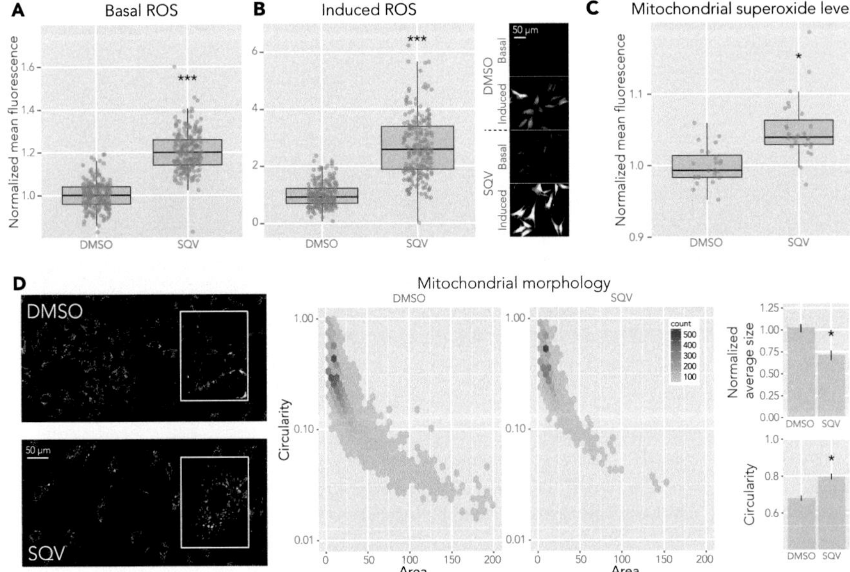

Fig. 6.7 Effect of Saquinavir (SQV; HIV-1 protease inhibitor) on primary human fibroblasts (control is DMSO). (**a**) Normalized basal levels of intracellular ROS as measured by CM-H_2DCFDA and (**b**) response towards induced ROS, measured as relative gain in intensity after 20 mM TBHP addition at different time points after treatment with 20 μM of SQV. (**c**) Normalized $O_2^{\bullet -}$ level as measured by average MitoSOX signal per mitochondrial pixel after treatment with 10 μM SQV. (**d**) SQV treatment (20 μM) causes mitochondrial fragmentation as illustrated by (**e**) a scatterplot of mitochondrial circularity and mitochondrial area (* = p value <0.05; ** = p value < 0.01; *** = p value < 0.001; the range of the Y-axes has been adjusted to optimally display the differences)

defenses (represented by lowered expression of SOD2 (Xiang et al. 2015)). Independently, other reports have linked HIV PI to changed mitochondrial morphofunction (Estaquier et al. 2002; Matarrese et al. 2003; Roumier et al. 2006; Bociąga-Jasik

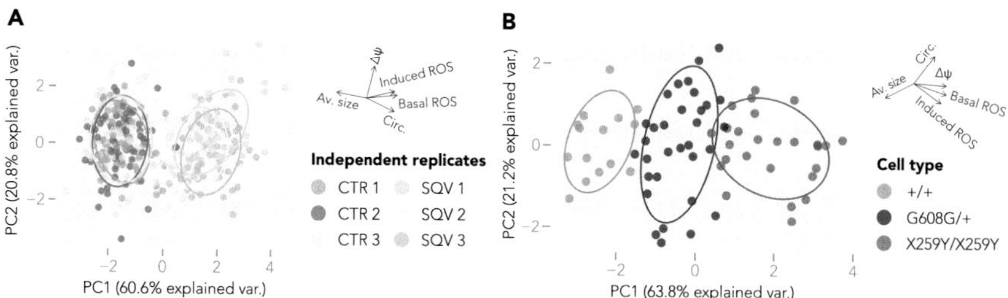

Fig. 6.8 Principal component analysis (PCA) distinguishes chemically perturbed cells and patient cells from healthy controls. (**a**) 2D scatterplot of the first 2 principal components (PCs) from a PCA analysis on 5 variables (basal and induced ROS levels, average mitochondrial size, circularity, and Δψm) from the SQV dataset described in Fig. 6.7. Independent replicates are plotted with a different color. The *brown arrows* represent the directions of the original 5 variables with respect to the principal components. (**b**) 2D scatterplot of the first 2 principal components (PCs) from a PCA analysis on the same 5 variables measured in 2 different laminopathy patient cell lines and healthy primary human fibroblasts

et al. 2013; Xiang et al. 2015). As a case study, we have used the established workflow to assess the effect of HIV PI on both parameters. To this end, primary human fibroblast cells were treated for 72h with 20 μM of the HIV PI saquinavir (SQV – experiment performed in biological triplicate). Subsequently, intracellular basal ROS levels, induced ROS levels, mitochondrial morphology, and $\Delta\psi_m$ were measured. Basal ROS levels were significantly higher compared to control cells treated with DMSO (Fig. 6.7a). Also the induced ROS levels were increased, pointing to lowered antioxidant defenses (Fig. 6.7b). With respect to the mitochondria, SQV treatment induced a highly fragmented phenotype when compared to control cells, illustrated by a higher circularity and lower average size of the individual mitochondria (Fig. 6.7d). $\Delta\psi_m$, measured as average TMRM signal per mitochondrial pixel, was not significantly altered (not shown). Using the dye combination calcein/MitoSOX, we also quantified mitochondrial $O_2^{\bullet-}$ levels. When applied to fibroblasts, treated for 72h with 10 μM of SQV, we measured a significant increase in mitochondrial $O_2^{\bullet-}$ levels compared to control cells treated with DMSO (Fig. 6.7c). A specific increase in this localized ROS variant aligns well with earlier findings showing that SQV induces lower expression of the mitochondrial enzyme SOD2 (Xiang et al. 2015).

Our multiparametric analysis yielded a feature set describing both morphological and intensity characteristics. When performing a principal component analysis (PCA) using a subset of 5 parameters, namely, basal and induced ROS levels, average mitochondrial size, average mitochondrial circularity, and $\Delta\psi_m$, we could unambiguously separate the two conditions (control and SQV-treated) independently in three biological replicates with just the first two principal components, explaining 81.4% of the total variance in the data (Fig. 6.8a). This demonstrates the robustness of our workflow and suggests that the combined readout may serve as a sensitive predictor of cellular health condition. To test this hypothesis, we ran a similar analysis on a group of laminopathy patient cells, for which we have previously shown differential ROS levels and mitochondrial dysfunction (Sieprath et al. 2015). Specifically, we compared fibroblasts from a healthy patient, with those from a patient suffering from Hutchinson-Gilford progeria (G608G/+) (Verstraeten et al. 2006) and fibroblasts from a patient with a lethal laminopathy phenotype due to a nonsense Y259X homozygous mutation in the LMNA gene (Y259X/Y259X) (Muchir et al. 2003). Again, using only 5 variables and 2 PCs, all three cell types could be readily separated (Fig. 6.8b).

6.4 Discussion

In this work, we have described and benchmarked a workflow for simultaneous quantification of intracellular ROS levels and mitochondrial morphofunction. As proof of principle, we showed that SQV induces a significant rise in basal and induced ROS levels in human fibroblasts and that this is accompanied by distinct mitochondrial fragmentation as well as increased mitochondrial $O_2^{\bullet-}$ levels. These findings support earlier evidence in the literature where increased ROS levels or mitochondrial dysfunction were also observed individually upon treatment with type 1 HIV protease inhibitors (Estaquier et al. 2002; Matarrese et al. 2003; Roumier et al. 2006; Chandra et al. 2009; Touzet and Philips 2010; Bociąga-Jasik et al. 2013; Xiang et al. 2015), but it is the first time these parameters were measured simultaneously. The major advantage is that an unambiguous determination of both factors together in space and time allows pinpointing causal relationships. For instance, by including compounds that promote or reduce mitochondrial function (ETC) or dynamics (fusion/fission), one could now directly assess the impact on intracellular ROS levels and vice versa. Another advantage of our method lies in its generic character in a sense that virtually any combination of spectrally compatible fluorescent probes for ROS and mitochondria can be used. It is becoming increasingly clear that, besides the general cell-wide effects of ROS, fast, transient, and highly localized production of low doses of specific intracellular ROS species plays an important role in cellular signal transduction and mitochondrial morphofunction. We have previously used the calcein/MitoSOX approach to study (mitochondrial) $O_2^{\bullet-}$ levels in human fibroblast cells with *LMNA* and *ZMPSTE24* knockdowns (Sieprath et al. 2015). However, any targetable ROS probe that has fast equilibration kinetics could be used. It is in this field that great progress is conceivable. As yet, not many such sensors exist. HyPer and (Orp1)roGFP2 would be good candidates, but their kinetics would still have to be improved to be able to really measure quick transient changes. In extenso, the workflow is easily amenable to the analysis of other organelles (e.g., the ER), and while the method was originally conceived for cell-based assays, it has already been adapted to cater for measurements of redox metabolism and mitochondrial morphology or density in *C. elegans* (Back et al. 2012; Castelein et al. 2014; de Boer et al. 2015; Smith et al. 2015). Despite proven performance in 2D, a challenge resides in the 3D nature of the imaged tissue, demanding confocal instead of wide-field acquisition and more complex 3D segmentation procedures.

While being measured simultaneously, processing and analysis of the redox and mitochondrial parameters are usually done separately to gain unbiased insight into the fundamental underlying processes. However, integration of all the information using data mining techniques allows the calculation of more sensitive fingerprints. In line with this, we have shown that both chemically (SQV) treated cells and laminopathy patient cells could become effectively discriminated using a combination of 5 different metrics. Such a redox fingerprint may become a valuable tool for classification of cells from different pathological conditions or could lead to novel cell-based screening methods for diagnostic purposes. It has been shown that the combination of multiple morphological parameters of the mitochondrial network permits robust classification of different phenotypes using unsupervised and supervised data mining strategies (de Boer et al. 2015; Blanchet et al. 2015). Indeed, hierarchical clustering has allowed for stratifying antiretroviral drug treatments based on mitochondrial morphology fingerprints (de Boer et al. 2015). Likewise, learning methods (logistic regression and support vector machines learning) have been successfully used to discriminate between primary fibroblasts of a healthy individual and a Leigh syndrome patient and to identify potential therapeutic compounds based on their mitochondrial morphofunctional phenotype (Blanchet et al. 2015). These examples demonstrate the potential of integrated image-based redox profiling.

In conclusion, we believe that our method will contribute to a better understanding of the relationship between mitochondrial function and intracellular ROS signaling. This, in turn, will provide invaluable information regarding a wide variety of human pathologies in which mitochondrial function and redox homeostasis are disturbed.

Acknowledgments This research was supported by the University of Antwerp (TTBOF/UA_29267), the Special Research Fund of Ghent University (project BOF/11267/09), the Hercules Foundation (AUGE/ 013), NB-Photonics (Project code 01-MR0110), and the CSBR (Centers for Systems Biology Research) initiative from the Netherlands Organization for Scientific Research (NWO; No: CSBR09/013V).

References

Archer SL (2013) Mitochondrial dynamics–mitochondrial fission and fusion in human diseases. N Engl J Med 369:2236–2251. doi:10.1056/NEJMra1215233

Auchère F, Rusnak F (2002) What is the ultimate fate of superoxide anion in vivo? J Biol Inorg Chem 7:664–667. doi:10.1007/s00775-002-0362-2

Babior B (1999) NADPH oxidase: an update. Blood 93:1464–1476

Back P, De Vos WH, Depuydt GG et al (2012) Free radical biology & medicine. Free Rad Biol Med 52:850–859. doi:10.1016/j.freeradbiomed.2011.11.037

Baird GS, Zacharias DA, Tsien RY (1999) Circular permutation and receptor insertion within green fluorescent proteins. Proc Natl Acad Sci U S A 96:11241–11246

Bartz RR, Piantadosi CA (2010) Clinical review: oxygen as a signaling molecule. Crit Care 14:234. doi:10.1186/cc9185

Bedard K, Krause K-H (2007) The NOX family of ROS-generating NADPH oxidases: Physiology and pathophysiology. Physiol Rev 87:245–313. doi:10.1152/physrev.00044.2005

Belousov VV, Fradkov AF, Lukyanov KA et al (2006) Genetically encoded fluorescent indicator for intracellular hydrogen peroxide. Nat Methods 3:281–286. doi:10.1038/nmeth866

Benhar M, Forrester MT, Stamler JS (2009) Protein denitrosylation: enzymatic mechanisms and cellular functions. Nat Rev Mol Cell Biol 10:721–732. doi:10.1038/nrm2764

Bilan DS, Pase L, Joosen L et al (2013) HyPer-3: a genetically encoded H(2)O(2) probe with improved performance for ratiometric and fluorescence lifetime imaging. ACS Chem Biol 8:535–542. doi:10.1021/cb300625g

Blanchet L, Smeitink JAM, van Emst-de Vries SE et al (2015) Quantifying small molecule phenotypic effects using mitochondrial morpho-functional fingerprinting and machine learning. Sci Rep 5:8035. doi:10.1038/srep08035

Bociąga-Jasik M, Polus A, Góralska J et al (2013) Metabolic effects of the HIV protease inhibitor–saquinavir in differentiating human preadipocytes. Pharmacol Rep 65:937–950

Boveris A, Chance B (1973) The mitochondrial generation of hydrogen peroxide. General properties and effect of hyperbaric oxygen. Biochem J 134:707–716

Boveris A, Valdez LB, Zaobornyj T, Bustamante J (2006) Mitochondrial metabolic states regulate nitric oxide and hydrogen peroxide diffusion to the cytosol. Biochim Biophys Acta 1757:535–542. doi:10.1016/j.bbabio.2006.02.010

Brown GC, Borutaite V (2012) There is no evidence that mitochondria are the main source of reactive oxygen species in mammalian cells. Mitochondrion 12:1–4. doi:10.1016/j.mito.2011.02.001

Cardoso AR, Chausse B, da Cunha FM et al (2012) Mitochondrial compartmentalization of redox processes. Free Rad Biol Med 52:2201–2208. doi:10.1016/j.freeradbiomed.2012.03.008

Caron M, Auclair M, Donadille B et al (2007) Human lipodystrophies linked to mutations in A-type lamins and to HIV protease inhibitor therapy are both associated with prelamin A accumulation, oxidative stress and premature cellular senescence. Cell Death Differ 14:1759–1767. doi:10.1038/sj.cdd.4402197

Casey JR (2006) Why bicarbonate? Biochem Cell Biol 84:930–939. doi:10.1139/o06-184

Castelein N, Muschol M, Dhondt I et al (2014) Experimental gerontology. EXG 56:26–36. doi:10.1016/j.exger.2014.02.009

Chandra S, Mondal D, Agrawal KC (2009) HIV-1 protease inhibitor induced oxidative stress suppresses glucose stimulated insulin release: protection with thymoquinone. Exp Biol Med (Maywood) 234:442–453. doi:10.3181/0811-RM-317

Chang MCY, Pralle A, Isacoff EY, Chang CJ (2004) A selective, cell-permeable optical probe for hydrogen peroxide in living cells. J Am Chem Soc 126:15392–15393. doi:10.1021/ja0441716

Chazotte B (2011) Labeling mitochondria with MitoTracker dyes. Cold Spring Harb Protoc 2011:990–992. doi:10.1101/pdb.prot5648

Chen Z-H, Saito Y, Yoshida Y et al (2005) 4-Hydroxynonenal induces adaptive response and enhances PC12 cell tolerance primarily through induction of thioredoxin reductase 1 via activation of Nrf2. J Biol Chem 280:41921–41927. doi:10.1074/jbc.M508556200

Chiu C-C, Yeh T-H, Lai S-C et al (2015) Neuroprotective effects of aldehyde dehydrogenase 2 activation in rotenone-induced cellular and animal models of parkinsonism. Exp Neurol 263:244–253. doi:10.1016/j.expneurol.2014.09.016

Chouchani ET, Methner C, Nadtochiy SM et al (2013) Cardioprotection by S-nitrosation of a cysteine switch on mitochondrial complex I. Nat Med 19:753–759. doi:10.1038/nm.3212

Conte Lo M, Carroll KS (2013) The redox biochemistry of protein sulfenylation and sulfinylation. J Biol Chem 288:26480–26488. doi:10.1074/jbc.R113.467738

Cremers CM, Jakob U (2013) Oxidant sensing by reversible disulfide bond formation. J Biol Chem 288:26489–26496. doi:10.1074/jbc.R113.462929 de Boer R, Smith RL, De Vos WH et al (2015) Caenorhabditis elegans as a model system for studying drug induced mitochondrial toxicity. PLoS One 10, e0126220. doi:10.1371/journal.pone.0126220

De Vos WH, Van Neste L, Dieriks B et al (2010) High content image cytometry in the context of subnuclear organization. Cytometry A 77:64–75. doi:10.1002/cyto.a.20807

Dickinson BC, Chang CJ (2011) Chemistry and biology of reactive oxygen species in signaling or stress responses. Nat Chem Biol 7:504–511. doi:10.1038/nchembio.607

Dickinson BC, Huynh C, Chang CJ (2010a) A palette of fluorescent probes with varying emission colors for imaging hydrogen peroxide signaling in living cells. J Am Chem Soc 132(16):5906–15

Dickinson BC, Srikun D, Chang CJ (2010b) Mitochondrial-targeted fluorescent probes for reactive oxygen species. Curr Opin Chem Biol 14:50–56. doi:10.1016/j.cbpa.2009.10.014

Dickinson BC, Lin VS, Chang CJ (2013) Preparation and use of MitoPY1 for imaging hydrogen peroxide in mitochondria of live cells. Nat Protoc 8:1249–1259. doi:10.1038/nprot.2013.064

Dikalov SI, Harrison DG (2014) Methods for detection of mitochondrial and cellular reactive oxygen species. Antioxid Redox Signal 20:372–382. doi:10.1089/ars.2012.4886

Dixit R, Cyr R (2003) Cell damage and reactive oxygen species production induced by fluorescence microscopy: effect on mitosis and guidelines for non-invasive fluorescence microscopy. Plant J 36:280–290

Dong H, Cheung SH, Liang Y et al (2013) "Stainomics": identification of mitotracker labeled proteins in mammalian cells. Electrophoresis 34:1957–1964. doi:10.1002/elps.201200557

Dooley CT, Dore TM, Hanson GT et al (2004) Imaging dynamic redox changes in mammalian cells with green fluorescent protein indicators. J Biol Chem 279:22284–22293. doi:10.1074/jbc.M312847200

Drummen GPC, van Liebergen LCM, Op den Kamp JAF, Post JA (2002) C11-BODIPY581/591, an oxidation-sensitive fluorescent lipid peroxidation probe: (micro)spectroscopic characterization and validation of methodology. Free Rad Biol Med 33:473–490. doi:10.1016/S0891-5849(02)00848-1

Dupre-Crochet S, Erard M, Nusse O (2013) ROS production in phagocytes: why, when, and where? J Leukoc Biol 94:657–670. doi:10.1189/jlb.1012544

Estaquier J, Lelièvre J-D, Petit F et al (2002) Effects of antiretroviral drugs on human immunodeficiency virus type 1-induced CD4(+) T-cell death. J Virol 76:5966–5973

Finkel T (2011) Signal transduction by reactive oxygen species. J Cell Biol 194:7–15. doi:10.1083/jcb.201102095

Forkink M, Willems PHGM, Koopman WJH, Grefte S (2015) Live-cell assessment of mitochondrial reactive oxygen species using dihydroethidine. In: Methods in molecular biology. Springer, New York, pp 161–169

Franco R, Schoneveld O, Georgakilas AG, Panayiotidis MI (2008) Oxidative stress, DNA methylation and carcinogenesis. Cancer Lett 266:6–11. doi:10.1016/j.canlet.2008.02.026

Frigault MM, Lacoste J, Swift JL, Brown CM (2009) Live-cell microscopy - tips and tools. J Cell Sci 122:753–767. doi:10.1242/jcs.033837

Fujikawa M, Kobayashi K, Kozawa T (2012) Direct oxidation of the [2Fe-2S] cluster in SoxR protein by superoxide: distinct differential sensitivity to superoxide-mediated signal transduction. J Biol Chem 287:35702–35708. doi:10.1074/jbc.M112.395079

Giorgio M, Trinei M, Migliaccio E, Pelicci PG (2007) Hydrogen peroxide: a metabolic by-product or a common mediator of ageing signals? Nat Rev Mol Cell Biol 8:722A–728

Giustarini D, Dalle-Donne I, Tsikas D, Rossi R (2009) Oxidative stress and human diseases: Origin, link, measurement, mechanisms, and biomarkers. Crit Rev Clin Lab Sci 46:241–281. doi:10.3109/10408360903142326

Glasbey CA (1993) An analysis of histogram-based thresholding algorithms. CVGIP Graph Model Image Proc 55:532–537

Gomes A, Fernandes E, Lima JLFC (2005) Fluorescence probes used for detection of reactive oxygen species. J Biochem Biophys Methods 65:45–80. doi:10.1016/j.jbbm.2005.10.003

Gomes LC, Di Benedetto G, Scorrano L (2011) Essential amino acids and glutamine regulate induction of mitochondrial elongation during autophagy. Cell Cycle 10:2635–2639

Grek CL, Zhang J, Manevich Y et al (2013) Causes and consequences of cysteine S-glutathionylation. J Biol Chem 288:26497–26504. doi:10.1074/jbc.R113.461368

Groitl B, Jakob U (2014) Thiol-based redox switches. Biochimica et Biophysica Acta (BBA) – Proteins and Proteomics 1844:1335–1343. doi:10.1016/j.bbapap.2014.03.007

Gupta SC, Hevia D, Patchva S et al (2012) Upsides and downsides of reactive oxygen species for cancer: the roles of reactive oxygen species in tumorigenesis, prevention, and therapy. Antioxid Redox Signal 16:1295–1322. doi:10.1089/ars.2011.4414

Gutscher M, Pauleau A-L, Marty L et al (2008) Real-time imaging of the intracellular glutathione redox potential. Nat Methods 5:553–559. doi:10.1038/nmeth.1212

Gutscher M, Sobotta MC, Wabnitz GH et al (2009) Proximity-based protein thiol oxidation by H2O2-scavenging peroxidases. J Biol Chem 284:31532–31540. doi:10.1074/jbc.M109.059246

Halliwell B, Gutteridge J (2007) The chemistry of free radicals and related “reactive species.” In: Free Radicals in Biology and Medicine, 4 edn. Oxford University Press, New York, pp 30–78

Hamanaka RB, Chandel NS (2010) Mitochondrial reactive oxygen species regulate cellular signaling and dictate biological outcomes. Trends Biochem Sci 35:505–513. doi:10.1016/j.tibs.2010.04.002

Han D, Antunes F, Canali R et al (2003) Voltage-dependent anion channels control the release of the superoxide anion from mitochondria to cytosol. J Biol Chem 278:5557–5563. doi:10.1074/jbc.M210269200

Hanson GT, Aggeler R, Oglesbee D et al (2004) Investigating mitochondrial redox potential with redox-sensitive green fluorescent protein indicators. J Biol Chem 279:13044–13053. doi:10.1074/jbc.M312846200

Harman D (1956) Aging: a theory based on free radical and radiation chemistry. J Gerontol 11:298–300

Hu JJ, Wong N-K, Ye S et al (2015) Fluorescent probe HKSOX-1 for imaging and detection of endogenous superoxide in live cells and in vivo. J Am Chem Soc. doi:10.1021/jacs.5b01881

Huang LK, Wang M (1995) Image thresholding by minimizing the measures of fuzziness. Pattern Recognit 28:41–51. doi:10.1016/0031-3203(94)E0043-K

Iannetti EF, Willems PHGM, Pellegrini M et al (2015) Toward high-content screening of mitochondrial morphology and membrane potential in living cells. Int J Biochem Cell Biol. doi:10.1016/j.biocel.2015.01.020

Invitrogen (2006) Reactive Oxygen Species (ROS) detection reagents. Invitrogen. 10 Jan 2006. https://tools.thermofisher.com/content/sfs/manuals/mp36103.pdf. Accessed 22 Apr 2015

Koopman WJH, Verkaart S, van Emst-de Vries SE et al (2006) Simultaneous quantification of oxidative stress and cell spreading using 5-(and-6)-chloromethyl-2 “,7 -”dichlorofluorescein. Cytom Part A 69A:1184–1192. doi:10.1002/cyto.a.20348

Koopman WJH, Verkaart S, Visch HJ et al (2007) Human NADH:ubiquinone oxidoreductase deficiency: radical changes in mitochondrial morphology? AJP: Cell Physiol 293:C22–C29. doi:10.1152/ajpcell.00194.2006

Koopman WJH, Distelmaier F, Esseling JJ et al (2008) Computer-assisted live cell analysis of mitochondrial membrane potential, morphology and calcium handling. Methods 46:304–311. doi:10.1016/j.ymeth.2008.09.018

Koopman WJH, Nijtmans LGJ, Dieteren CEJ et al (2010) Mammalian mitochondrial complex I: biogenesis, regulation, and reactive oxygen species generation. Antioxid Redox Signal 12:1431–1470. doi:10.1089/ars.2009.2743

Korshunov SS, Skulachev VP, Starkov AA (1997) High protonic potential actuates a mechanism of production of reactive oxygen species in mitochondria. FEBS Lett 416:15–18

Lambeth JD (2004) NOX enzymes and the biology of reactive oxygen. Nat Rev Immunol 4:181–189. doi:10.1038/nri1312

Lebiedzinska M, Karkucinska-Wieckowska A, Giorgi C et al (2010) Oxidative stress-dependent p66Shc phosphorylation in skin fibroblasts of children with mitochondrial disorders. Bba-Gen Subjects 1797:952–960. doi:10.1016/j.bbabio.2010.03.005

Lopert P, Patel M (2014) Nicotinamide nucleotide transhydrogenase (Nnt) links the substrate requirement in brain mitochondria for hydrogen peroxide removal to the thioredoxin/peroxiredoxin (Trx/Prx) system. J Biol Chem 289:15611–15620. doi:10.1074/jbc.M113.533653

Loschen G, Azzi A, Richter C, Flohé L (1974) Superoxide radicals as precursors of mitochondrial hydrogen peroxide. FEBS Lett 42:68–72

Lukyanov KA, Belousov VV (2014) Biochimica et biophysica acta. BBA – General Subjects 1840:745–756. doi:10.1016/j.bbagen.2013.05.030

Mailloux RJ, McBride SL, Harper M-E (2013) Unearthing the secrets of mitochondrial ROS and glutathione in bioenergetics. Trends Biochem Sci 38:592–602. doi:10.1016/j.tibs.2013.09.001

Marchi S, Giorgi C, Suski JM et al (2012) Mitochondria-ros crosstalk in the control of cell death and aging. J Signal Transduct 2012:329635. doi:10.1155/2012/329635

Marí M, Morales A, Colell A et al (2013) Mitochondrial glutathione: features, regulation and role in disease. Bba-Gen Subjects 1830:3317–3328. doi:10.1016/j.bbagen.2012.10.018

Markvicheva KN, Bilan DS, Mishina NM et al (2011) A genetically encoded sensor for H2O2 with expanded dynamic range. Bioorg Med Chem 19:1079–1084. doi:10.1016/j.bmc.2010.07.014

Matarrese P, Gambardella L, Cassone A et al (2003) Mitochondrial membrane hyperpolarization hijacks activated T lymphocytes toward the apoptotic-prone phenotype: homeostatic mechanisms of HIV protease inhibitors. J Immunol 170:6006–6015

Meyer AJ, Dick TP (2010) Fluorescent protein-based redox probes. Antioxid Redox Sign 13:621–650. doi:10.1089/ars.2009.2948

Miller EW, Tulyathan O, Tulyanthan O et al (2007) Molecular imaging of hydrogen peroxide produced for cell signaling. Nat Chem Biol 3:263–267. doi:10.1038/nchembio871

Mishina NM, Bogeski I, Bolotin DA et al (2012) Can we see PIP3 and hydrogen peroxide with a single probe? Antioxid Redox Signal 17:505–512. doi:10.1089/ars.2012.4574

Miwa S, Brand MD (2003) Mitochondrial matrix reactive oxygen species production is very sensitive to mild uncoupling. Biochem Soc Trans 31:1300–1301

Moylan JS, Reid MB (2007) Oxidative stress, chronic disease, and muscle wasting. Muscle Nerve 35:411–429. doi:10.1002/mus.20743

Muchir A, van Engelen B, Lammens M et al (2003) Nuclear envelope alterations in fibroblasts from LGMD1B patients carrying nonsense Y259X heterozygous or homozygous mutation in lamin A/C gene. Exp Cell Res 291:352–362. doi:10.1016/j.yexcr.2003.07.002

Murphy MP (2009) How mitochondria produce reactive oxygen species. Biochem J 417:1–13. doi:10.1042/BJ20081386

Murphy MP (2012) Mitochondrial thiols in antioxidant protection and redox signaling: distinct roles for glutathionylation and other thiol modifications. Antioxid Redox Signal 16:476–495. doi:10.1089/ars.2011.4289

Naderi W, Lopez C, Pandey S (2006) Chronically increased oxidative stress in fibroblasts from Alzheimer's disease patients causes early senescence and renders resistance to apoptosis by oxidative stress. Mech Ageing Dev 127:25–35. doi:10.1016/j.mad.2005.08.006

Nakamura T, Lipton SA (2011) Redox modulation by S-nitrosylation contributes to protein misfolding, mitochondrial dynamics, and neuronal synaptic damage in neurodegenerative diseases. Cell Death Differ 18:1478–1486. doi:10.1038/cdd.2011.65

Nathan C, Cunningham-Bussel A (2013) Beyond oxidative stress: an immunologist's guide to reactive oxygen species. Nat Rev Immunol 13:349–361. doi:10.1038/nri3423

Nicholls DG (2012) Fluorescence measurement of mitochondrial membrane potential changes in cultured cells. Methods Mol Biol 810:119–133. doi:10.1007/978-1-61779-382-0_8

Pastore A, Piemonte F (2012) S-Glutathionylation signaling in cell biology: progress and prospects. Eur J Pharm Sci 46:279–292. doi:10.1016/j.ejps.2012.03.010

Pattison DI, Davies MJ (2006) Actions of ultraviolet light on cellular structures. EXS 131–157

Poburko D, Santo-Domingo J, Demaurex N (2011) Dynamic regulation of the mitochondrial proton gradient during cytosolic calcium elevations. J Biol Chem 286:11672–11684. doi:10.1074/jbc.M110.159962

Prime TA, Forkink M, Logan A et al (2012) A ratiometric fluorescent probe for assessing mitochondrial phospholipid peroxidation within living cells. Free Rad Biol Med 53:544–553. doi:10.1016/j.freeradbiomed.2012.05.033

Rabut G, Ellenberg J (2004) Automatic real-time three-dimensional cell tracking by fluorescence microscopy. J Microsc 216:131–137. doi:10.1111/j.0022-2720.2004.01404.x

Radi R, Cassina A, Hodara R et al (2002) Peroxynitrite reactions and formation in mitochondria. Free Rad Biol Med 33:1451–1464

Rambold AS, Kostelecky B, Elia N, Lippincott-Schwartz J (2011) Tubular network formation protects mitochondria from autophagosomal degradation during nutrient starvation. Proc Natl Acad Sci U S A 108:10190–10195. doi:10.1073/pnas.1107402108

Rasmussen HH, Hamilton EJ, Liu C-C, Figtree GA (2010) Reversible oxidative modification: implications for cardiovascular physiology and pathophysiology. Trends Cardiovasc Med 20:85–90. doi:10.1016/j.tcm.2010.06.002

Rhee SG, Woo HA, Kil IS, Bae SH (2012) Peroxiredoxin functions as a peroxidase and a regulator and sensor of local peroxides. J Biol Chem 287:4403–4410. doi:10.1074/jbc.R111.283432

Riahi Y, Cohen G, Shamni O, Sasson S (2010) Signaling and cytotoxic functions of 4-hydroxyalkenals. Am J Physiol Endocrinol Metab 299:E879–E886. doi:10.1152/ajpendo.00508.2010

Robinson KM, Janes MS, Pehar M et al (2006) Selective fluorescent imaging of superoxide in vivo using ethidium-based probes. Proc Natl Acad Sci U S A 103:15038–15043. doi:10.1073/pnas.0601945103

Robinson KM, Janes MS, Beckman JS (2008) The selective detection of mitochondrial superoxide by live cell imaging. Nat Protoc 3:941–947. doi:10.1038/nprot.2008.56

Roumier T, Szabadkai G, Simoni A-M et al (2006) HIV-1 protease inhibitors and cytomegalovirus vMIA induce mitochondrial fragmentation without triggering apoptosis. Cell Death Differ 13:348–351. doi:10.1038/sj.cdd.4401750

Saaranen MJ, Ruddock LW (2013) Disulfide bond formation in the cytoplasm. Antioxid Redox Signal 19:46–53. doi:10.1089/ars.2012.4868

Sachdev S, Davies KJA (2008) Production, detection, and adaptive responses to free radicals in exercise. Free Rad Biol Med 44:215–223. doi:10.1016/j.freeradbiomed.2007.07.019

Sahoo PK, Soltani S, Wong A, Chen YC (1988) A survey of thresholding techniques. Comput Vis Graph Image Process 41:233–260

Salmon AB, Richardson A, Perez VI (2010) Update on the oxidative stress theory of aging: does oxidative stress play a role in aging or healthy aging? Free Rad Biol Med 48:642–655. doi:10.1016/j.freeradbiomed.2009.12.015

Sieprath T, Darwiche R, De Vos WH (2012) Lamins as mediators of oxidative stress. Biochem Biophys Res Commun 421:635–639. doi:10.1016/j.bbrc.2012.04.058

Sieprath T, Corne TDJ, Nooteboom M et al (2015) Sustained accumulation of prelamin a and depletion of lamin a/c both cause oxidative stress and mitochondrial dysfunction but induce different cell fates. Nucleus 6:1–11. doi:10.1080/19491034.2015.1050568

Smith RL, De Vos WH, de Boer R et al (2015) In vivo visualization and quantification of mitochondrial morphology in C. Elegans. In: Methods in molecular biology. Springer, New York, pp 367–377

Staljanssens D, De Vos WH, Willems P et al (2012) Time-resolved quantitative analysis of CCK1 receptor-induced intracellular calcium increase. Peptides 34:219–225. doi:10.1016/j.peptides.2011.02.014

Stangherlin A, Reddy AB (2013) Regulation of circadian clocks by redox homeostasis. J Biol Chem 288:26505–26511. doi:10.1074/jbc.R113.457564

Szabó C, Ischiropoulos H, Radi R (2007) Peroxynitrite: biochemistry, pathophysiology and development of therapeutics. Nat Rev Drug Discov 6:662–680. doi:10.1038/nrd2222

Tarpey MM, Wink DA, Grisham MB (2004) Methods for detection of reactive metabolites of oxygen and nitrogen: in vitro and in vivo considerations. Am J Physiol-Reg I 286:R431–44. doi:10.1152/ajpregu.00361.2003

Thomas C, Mackey MM, Diaz AA, Cox DP (2009) Hydroxyl radical is produced via the Fenton reaction in submitochondrial particles under oxidative stress: implications for diseases associated with iron accumulation. Redox Rep 14:102–108. doi:10.1179/135100009X392566

Topell S, Hennecke J, Glockshuber R (1999) Circularly permuted variants of the green fluorescent protein. FEBS Lett 457:283–289

Touzet O, Philips A (2010) Resveratrol protects against protease inhibitor-induced reactive oxygen species production, reticulum stress and lipid raft perturbation. AIDS 24:1437–1447. doi:10.1097/QAD.0b013e32833a6114

Traber MG, Stevens JF (2011) Vitamins C and E: beneficial effects from a mechanistic perspective. Free Rad Biol Med 51:1000–1013. doi:10.1016/j.freeradbiomed.2011.05.017

Trudeau K, Molina AJA, Roy S (2011) High glucose induces mitochondrial morphology and metabolic changes in retinal pericytes. Invest Ophthalmol Vis Sci 52:8657–8664. doi:10.1167/iovs.11-7934

Turrens JF (2003) Mitochondrial formation of reactive oxygen species. J Physiol 552:335–344. doi:10.1113/jphysiol.2003.049478

Verkaart S, Koopman WJH, van Emst-de Vries SE et al (2007) Superoxide production is inversely related to complex I activity in inherited complex I deficiency. Bba-Gen Subjects 1772:373–381. doi:10.1016/j.bbadis.2006.12.009

Verstraeten VLRM, Broers JLV, van Steensel MAM et al (2006) Compound heterozygosity for mutations in LMNA causes a progeria syndrome without prelamin A accumulation. Hum Mol Genet 15:2509–2522. doi:10.1093/hmg/ddl172

Wang X, Fang H, Huang Z et al (2013) Imaging ROS signaling in cells and animals. J Mol Med 91:917–927. doi:10.1007/s00109-013-1067-4

Weisiger RA, Fridovich I (1973) Superoxide dismutase – organelle specificity. J Biol Chem 248:3582–3592

Willems PHGM, Rossignol R, Dieteren CEJ et al (2015) Redox homeostasis and mitochondrial dynamics. Cell Metab 22:207–218. doi:10.1016/j.cmet.2015.06.006

Winterbourn CC (2008) Reconciling the chemistry and biology of reactive oxygen species. Nat Chem Biol 4:278–286. doi:10.1038/nchembio.85

Woolley JF, Stanicka J, Cotter TG (2013) Recent advances in reactive oxygen species measurement in biological systems. Trends Biochem Sci 38:556–565. doi:10.1016/j.tibs.2013.08.009

Xiang T, Du L, Pham P et al (2015) Nelfinavir, an HIV protease inhibitor, induces apoptosis and cell cycle arrest in human cervical cancer cells via the ROS-dependent mitochondrial pathway. Cancer Lett 364:79–88. doi:10.1016/j.canlet.2015.04.027

Yu TZ, Robotham JL, Yoon Y (2006) Increased production of reactive oxygen species in hyperglycemic conditions requires dynamic change of mitochondrial morphology. Proc Natl Acad Sci U S A 103:2653–2658. doi:10.1073/pnas.0511154103

Zhao H, Kalivendi S, Zhang H et al (2003) Superoxide reacts with hydroethidine but forms a fluorescent product that is distinctly different from ethidium: potential implications in intracellular fluorescence detection of superoxide. Free Rad Biol Med 34:1359–1368

Zhao B, Bilski PJ, He Y-Y et al (2008) Photo-induced reactive oxygen species generation by different water-soluble fullerenes (C) and their cytotoxicity in human keratinocytes. Photochem Photobiol 84:1215–1223. doi:10.1111/j.1751-1097.2008.00333.x

Zheng M, Aslund F, Storz G (1998) Activation of the OxyR transcription factor by reversible disulfide bond formation. Science 279:1718–1721

Zielonka J, Kalyanaraman B (2010) Hydroethidine- and MitoSOX-derived red fluorescence is not a reliable indicator of intracellular superoxide formation: Another inconvenient truth. Free Rad Biol Med 48:983–1001. doi:10.1016/j.freeradbiomed.2010.01.028

Zuiderveld K (1994) Contrast limited adaptive histogram equalization. Graphics gems IV. Academic Press Professional Inc, San Diego

Chapter 7
KNIME for Open-Source Bioimage Analysis: A Tutorial

Christian Dietz and Michael R. Berthold

Abstract The open analytics platform KNIME is a modular environment that enables easy visual assembly and interactive execution of workflows. KNIME is already widely used in various areas of research, for instance in cheminformatics or classical data analysis. In this tutorial the KNIME Image Processing Extension is introduced, which adds the capabilities to process and analyse huge amounts of images. In combination with other KNIME extensions, KNIME Image Processing opens up new possibilities for inter-domain analysis of image data in an understandable and reproducible way.

7.1 Introduction

Every day, research involves recording increasing numbers of images as a result of the constantly improving imaging techniques, making them key to life science research. Advanced microscopy allows the acquisition of multidimensional images almost without any user interaction and can therefore generate a plethora of heterogeneous image data. However, to make sense of the generated image data and finally draw conclusions, an exhaustive analysis of the images has to be conducted. In addition to classical image processing techniques, more sophisticated algorithms are increasingly being applied—from the field of machine learning and data mining (Eliceiri et al. 2012). The extracted information is then further analysed with established statistical analysis techniques. For instance, detecting objects within images (i.e. segmentation) and the detailed statistical evaluation of the collected results are essential stages of a typical image analysis process (Saha et al. 2013; Ljosa et al. 2012; Aligeti et al. 2014). For a full exploitation of the outcome, an appropriate visualization of the information or a linkage to other information sources from other domains may be necessary to gain new insights.

C. Dietz (✉) • M.R. Berthold
University of Konstanz, Chair for Bioinformatics and Information Mining, Universitaetsstrasse 10, 78464 Konstanz, Germany
e-mail: christian.dietz@uni-konstanz.de; michael.berthold@uni-konstanz.de

W.H. De Vos et al. (eds.), *Focus on Bio-Image Informatics*, Advances in Anatomy, Embryology and Cell Biology 219, DOI 10.1007/978-3-319-28549-8_7

A large number of monolithic and highly task-oriented software solutions have been proposed to tackle the problems that occur in each step of bioimage analysis tasks (Eliceiri et al. 2012). As a result, researchers are required to choose from a set of stand-alone tools, which have to be orchestrated to solve the given task. Typically, two approaches are used that link these kinds of tools: one approach is to transfer the data manually between the tools while the other approach involves writing a customized program or script to automate a particular process. However, these approaches typically lead to a number of critical problems.

- Transferring the data manually involves a human being and is therefore time-consuming and does not scale with the amount of the acquired images.
- Customized scripts are prone to errors. Furthermore, results calculated with these highly problem-specific scripts are frequently unable to be reproduced or reused by others.

A straightforward, but infeasible solution to the described problems is to build a single monolithic platform that covers the complete range of functionalities required by a bioimage analysis workflow. However, future demands are yet unknown and therefore a closed, proprietary software solution does not scale with the new requirements that evolve with technological advance. Therefore, the open-source community has realized the great need for, and benefit of, closer cooperation by fostering interoperability among individual projects and open, extensible platforms. Following this approach, the open-source analytics platform KNIME (Berthold et al. 2008) provides the ability to seamlessly integrate a diverse and powerful collection of existing software tools and libraries. KNIME is a user-friendly and comprehensive open-source data integration, processing, analysis, and exploration platform designed to handle large amounts of heterogeneous data. It has been developed since 2006 and is used by professionals in industry and academia. As an integration platform, KNIME directly combines the advantages of several different tools and domains. The integrated tools are encapsulated KNIME nodes, the basic processing units in KNIME, which in turn can be combined to form the so-called workflows. KNIME workflows not only inherently document the entire analysis process, but they can also be exported and easily made available to others, who can subsequently reproduce the results or use the workflows as a starting point for their own analysis. To guarantee reproducibility, KNIME makes sure whenever any of the modules change in any way, for example the change of a version of an integrated tool, the previous version of that module is carefully deprecated but remains part of the platform. Hence, workflows published years ago still run with the most recent releases of KNIME. Once a workflow has been created, it can be applied to hundreds of thousands of images and other large data sets—even on small-scale devices thanks to the intelligent caching technology of KNIME. This makes KNIME well suited for high-throughput screenings, in which the analysis results can also be quite large.

The KNIME Image Processing Extension enhances KNIME by providing algorithms and data structures to process and analyse images. To avoid reinventing the wheel, KNIME Image Processing uses and integrates state-of-the-art libraries such as ImageJ1 (Schindelin et al. 2012) and ImageJ2,[1] SCIFIO,[2] OMERO (Allan et al. 2012), ClearVolume (Royer et al. 2015), ImgLib2 (Pietzsch et al. 2012), CellProfiler (Kamentsky et al. 2011), TrackMate[3] and others. These well-known image processing tools can not only exchange data and therefore be used in combination, but it is also possible to link their output to other extensions from completely different domains. For example, once interesting hits have been identified in the image data, the respective molecules can be explored with one of the many KNIME cheminformatics extensions, for instance the KNIME RDKit extension.[4]

An image processing and analysis workflow typically consists of a subset of several consecutive steps: Loading images, (pre)-processing, segmentation, tracking, feature extraction, model learning and the subsequent visualization and statistical analysis of the information gathered in the previous steps. Different problems can be incurred in each of these steps, depending on the image analysis task itself. However, by combining KNIME Image Processing nodes with nodes from other available KNIME extensions, it is easy to orchestrate these comprehensible workflows, which can span multiple domains, to solve the issues in KNIME without needing to program a single line of code. In Sect. 7.2 the main concepts of KNIME and KNIME Image Processing are introduced. Taking this as a basis, Sect. 7.3 goes on to explain an image processing workflow example in a step-by-step process.

7.2 Basic Concepts

This section explains how KNIME and its extensions are downloaded and installed. Next, the KNIME User Interface is described, while the last part of this section covers the most fundamental concepts of KNIME Image Processing, which are important for understanding the image processing workflow explained in Sect. 7.3.

7.2.1 Download and Installation

The open analytics platform KNIME can be downloaded and installed from the KNIME website.[5] KNIME comes packed with an installer for Windows and Mac systems. Linux users simply have to extract KNIME. As KNIME is a plugin-based system, there are several extensions that are not part of the basic KNIME

[1] http://imagej.net.

[2] http://scif.io.

[3] http://fiji.sc/TrackMate.

[4] http://tech.knime.org/community/rdkit.

[5] http://www.knime.org.

installation. These extensions are easily installed via the so-called update-sites. KNIME Image Processing,[6] for example, is installed from the *Trusted Community Contributions* site. For details on how to install additional plugins, please see http://tech.knime.org/community.

7.2.2 KNIME User Interface

Figure 7.1 shows the KNIME User Interface. The KNIME Explorer (A) depicts the various locations where workflows can be stored or uploaded. By default, two locations are available: (1) The KNIME Example server on which several example workflows can be found. (2) The *LOCAL* workspace, which was selected on the first start-up of KNIME. A new workflow can be created with File > New > New KNIME Workflow. This new, empty workflow is accessed via the *LOCAL* workspace. Workflows in KNIME are essentially graphs that connect nodes (atomic processing units in KNIME) and visually model the individual processing steps of a certain task. A Double-Click on the workflow in the *KNIME Explorer (A)* opens it in the *Workflow Editor (C)*. The user is now able to drag and drop nodes from the *Node Repository (B)* onto the canvas of the workflow editor, to compose complex yet clear workflows, for example to process and analyse images. The nodes can then be connected by drawing a line from the output node to the input node, enabling the data to be passed from node to node. Additionally, each KNIME node provides a

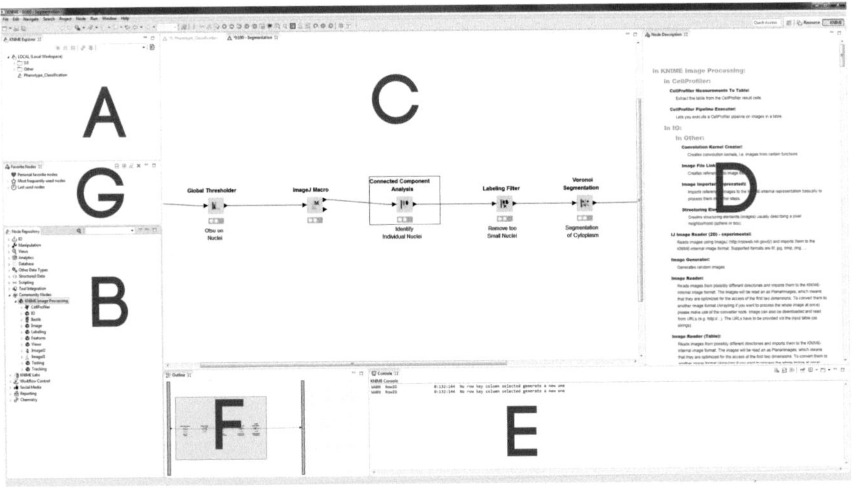

Fig. 7.1 KNIME user interface

[6]http://knime.imagej.net.

Node Description (D) explaining which input data it requires, explanations of the required parameters, what the node does with the incoming data and the output of the node. The *Node Repository (B)* contains all of the KNIME nodes that are part of the currently installed KNIME extensions. The default KNIME Open Analytics Platform installation provides a basic set of nodes for data manipulation, data mining, a selection of data views, node control, time series analytics and basic IO and Database nodes. KNIME nodes for image analysis can be added by installing more KNIME extensions, as described in Sect. 7.2.1. The *KNIME Console (E)* view displays error and warning messages in order to provide feedback to the user. Finally, the *Outline (F)* view provides an overview of the whole workflow even if only a small part is visible in the workflow editor and the *Favorite Nodes (G)* provide quick access to personal favourite, frequently and recently used nodes.

7.2.3 Handling of Images and Labelings in KNIME

A workflow usually starts with a node, which represents a data source, e.g. connecting a database, reading a text file or reading images. The data are transported between the connected nodes, typically organized in data tables, consisting of columns of certain (extensible) data-types and an arbitrary number of rows. A typical data table is depicted in Fig. 7.2, with each column of the table comprising an arbitrary object type, e.g. numbers, text or molecules. KNIME Image Processing adds two new column types to the mix: images and labelings. Labelings represent the segmentation of an image—the partitioning of an image into segments. As opposed to images, labelings store one or more labels for each pixel, instead of numeric values. A label associates each pixel with an object, class value, track number or any other information.

Contrary to what might be assumed initially, images and labelings stored in a single cell of a data table can be of arbitrary dimensionality. For example, a table cell may contain a multi-channel video or Z-stack. To accomplish n-dimensional image processing, KNIME uses ImgLib2 as its underlying programming framework.

7.2.4 Image Processing Specific Dialog Components

7.2.4.1 Dimension Selection

In order to provide the user with the flexibility to choose how images and labelings with more than just two dimensions are to be processed, most of the nodes provided in the KNIME Image Processing Extension offer a so-called *Dimension Selection* dialog (see Fig. 7.3). This dialog enables users to select the dimensions on which an algorithm will operate. For instance in the case of a simple Z-Stack, the *Image Normalization* node can be configured so as to apply normalization to each *X,Y*

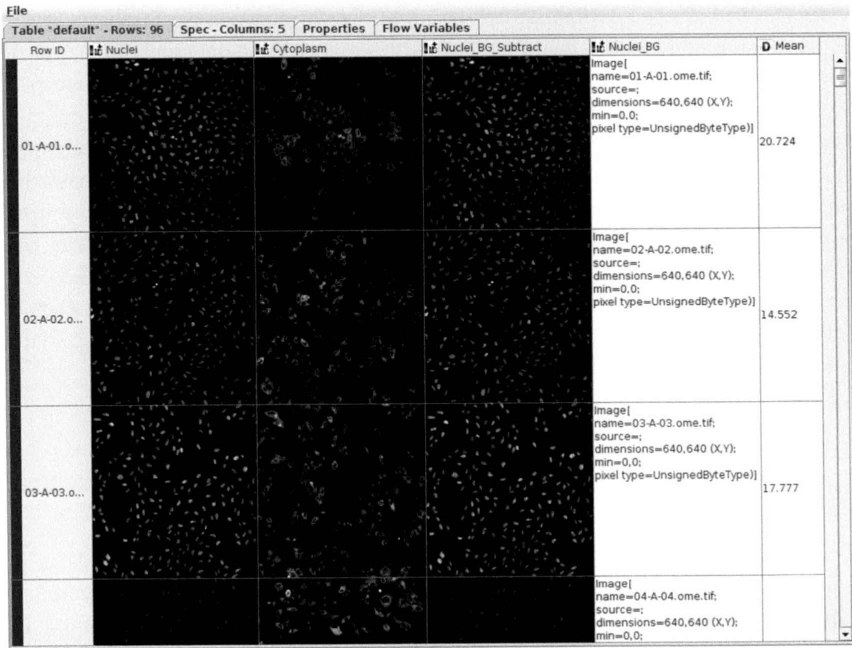

Fig. 7.2 A typical KNIME table with five columns. Each column of the table has a certain data-type, e.g. numbers, text, molecules or images

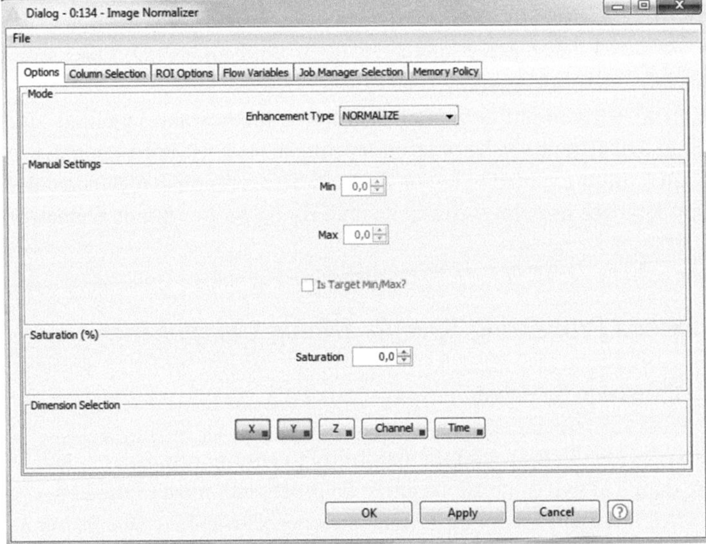

Fig. 7.3 The configuration dialog of the *Image Normalizer* node

plane either independently or for the entire *X,Y,Z* cube by selecting *X,Y* or *X,Y,Z* in the dimension selection.

7.2.4.2 Column Selection

Many KNIME Image Processing nodes, whose input is an image or labeling, operate on a row-to-row basis. This means that—given an input image—another image or labeling is calculated based on the algorithm implemented in the node. The user can determine the layout of the output table of these nodes with a dialog component called *Column Selection*. Generally, a user has three options: the resulting column with images or labelings can either be appended to the incoming table, replace the existing column or an entirely new table can be created.

7.2.5 Visualization of Images and Labelings

KNIME Image Processing enables users to explore images and labelings in more detail, which is especially useful if an image or labeling comprises more than two dimensions (e.g. Z-stacks or videos). The user can access this view by Right Click > Open Image Viewer (see Fig. 7.4) on a KNIME Image Processing node. Another, more specific view is the *Interactive Segmentation View* node. It can be used to validate segmentation, classification or tracking results as it offers an overlay view for images and labelings. Additional visualization plugins can be installed to extend KNIME Image Processing. For instance, the *ClearVolume Integration* offers fast, GPU accelerated 5D volume rendering and can easily be used within KNIME.

7.3 Step by Step to Phenotype Classification

In this section we walk step by step through an image processing workflow. The workflow classifies cells as either positive or negative according to their phenotype (see Fig. 7.5).[7]

The cells in this example stem from images from the publicly available high-content screening image data provided in Ljosa et al. (2012) (human cytoplasm-nucleus translocation assay, available from the Broad Bioimage Benchmark Collection). The images were taken from stably transfected osteosarcoma cells seeded in a 96 well plate and contain the information about the

[7]The entire *Phenotype Classification* workflow is available for download at http://knime.imagej.net/aaec.

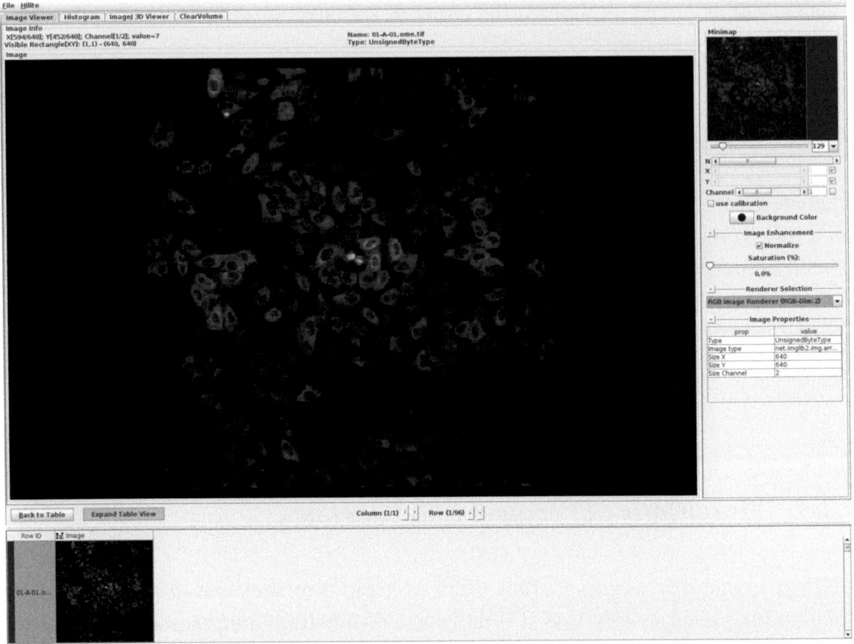

Fig. 7.4 The KNIME Image Processing *Image Viewer* allows users to inspect the images in more detail. Users can browse through the various dimensions of an image, inspect the values of the pixels and obtain information about important meta-data

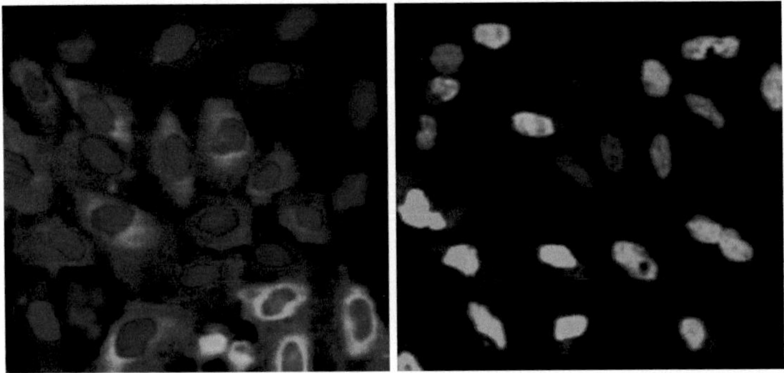

Fig. 7.5 Two images from the publicly available high-content screening image data provided in Ljosa et al. (2012). The *left* image contains positive cells and the *right*, negative cells

translocalization of the Forkhead (FKHR-EGFP) fusion protein from the cytoplasm (Channel 2) to the nucleus (Channel 1).[8]

[8]For detailed information see http://www.broadinstitute.org/bbbc/BBBC013/. Please note: The BMP images available on the website are already split into the individual channels.

Fig. 7.6 The workflow discussed in this tutorial

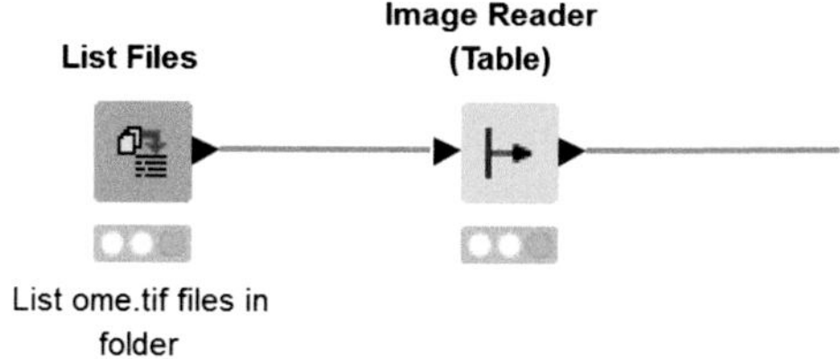

Fig. 7.7 Detailed look inside the *Loading Images* meta node

The example workflow is depicted in Fig. 7.6. The individual parts of the workflow are organized in so-called *meta nodes* to reduce the complexity of the workflow. Meta nodes are nodes that contain subworkflows, i.e. in the workflow they look like a single node and yet they can contain many nodes and even more meta nodes. This provides a series of advantages such as enabling the user to design much larger, more complex workflows and the encapsulation of specific actions.

A Double-Click on the meta node allows the user to have a closer look at what is inside. In the following, the content of each meta node will be explained in more detail. However, it is important to note that the individual parts of the workflow are easily replaced by other nodes possibly more suitable for other image processing tasks. Besides KNIME Image Processing, integrations with, for example, R, Python, Weka and especially KNIME itself offer a wide range of functionality for more complex visualizations and advanced machine learning and data mining techniques.

7.3.1 Loading Images

To date, the proprietary file formats of microscope image analysis software have made it difficult for open-source platforms to load images generically. However, SCIFIO with its integration of the BioFormats (Linkert et al. 2010) library can convert approx. Hundred and twenty five file formats used by various microscope manufacturers, such as Zeiss LSM, Metamorph Stack, Leica LCS or DICOM, into a KNIME compatible format. In KNIME, this functionality can be accessed via the *Image Reader* node, which integrates these libraries. A user can either select the images in the *Image Reader* configuration dialog (Right Click > Configure) or provide URLs to the images as an input table coming from another node, e.g. the *List Files* node. The resulting workflow of the latter approach is illustrated in Fig. 7.7.

The *List Files* node is used to list the URLs of all images of a certain folder. Connecting the node to the *Image Reader* node enables the user to configure the *Image Reader* node (Right-Click on Image Reader > Configure), such that it loads all images into KNIME from these URLs. In this configuration: Tab: Additional Options > File name column in optional column has to be set to the column of the incoming table that contains the URLs to the requested images.

7.3.2 Preprocessing Images

KNIME Image Processing offers a range of general (pre-)processing techniques to enhance image quality: Standard linear and non-linear filters are available as are morphological and binary operations, pixel-wise image arithmetics, edge-detectors, background subtraction algorithms, projections or the nodes for the manipulation of the dimensionality, such as splitting and merging images. Additionally, the *ImageJ-Macro* node, which is part of the KNIME Image Processing—ImageJ Integration,[9] allows the execution of arbitrary ImageJ1 macros on a huge amount of image data.

The image preprocessing used in this tutorial is implemented in the *Preprocessing* meta node (see Fig. 7.6). First of all, the channels of the images are split, which results in a table with two columns, the *Nuclei* and *Cytoplasm* (see Fig. 7.8). Next, each channel is preprocessed individually. The images in the *Nuclei* column suffer from non-uniform illumination, which makes it difficult to apply automated segmentation methods. Therefore in the *Background Subtraction of Images in Nuclei Channel* meta node the quality of the images in the *Nuclei* column is enhanced. Here, a very simple background subtraction technique was chosen, especially to demonstrate how individual KNIME nodes are easily combined to create an existing or completely new processing or analysis technique without any programming. Figure 7.9 depicts the workflow implementing the algorithm: The images in the *Nuclei* column are filtered with a very large kernel (sigma = 100.0) using the *Gaussian Convolution* node and the output is appended as an additional column to the input table. The mean value of the pixel intensities is calculated for each of the filtered images using the *Image Feature* node.

Finally, the resulting background corrected images are obtained by subtracting the sum of the filtered image and the mean of the filtered image at each pixel position from the original image using the *Image Calculator* node.

The configuration of the *Image Calculator* node is shown in Fig. 7.10. The last node in the *Background Subtraction of Images in Nuclei Channel* meta node is the *Image Converter*. This node can be used to normalize and scale the intensity values of the images to a certain range. In this tutorial we normalize and scale the values between 0 and 255 (= *UnsignedByteType*) to reduce the amount of required memory.

[9]For details and installation instructions see https://tech.knime.org/community/imagej.

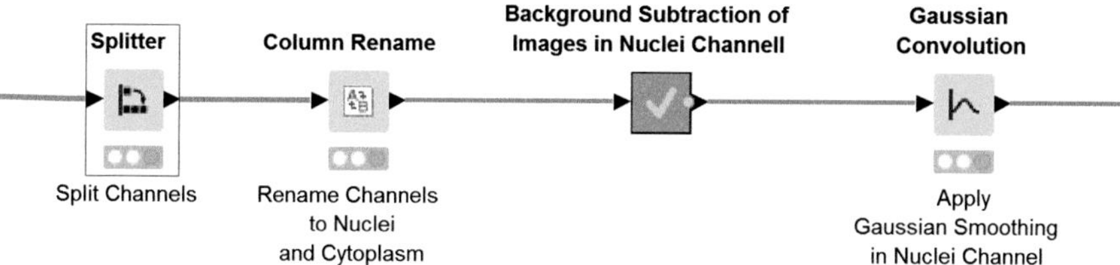

Fig. 7.8 Detailed look into the *Preprocessing* meta node. The images are split into Nuclei and Cytoplasm channels and renamed accordingly

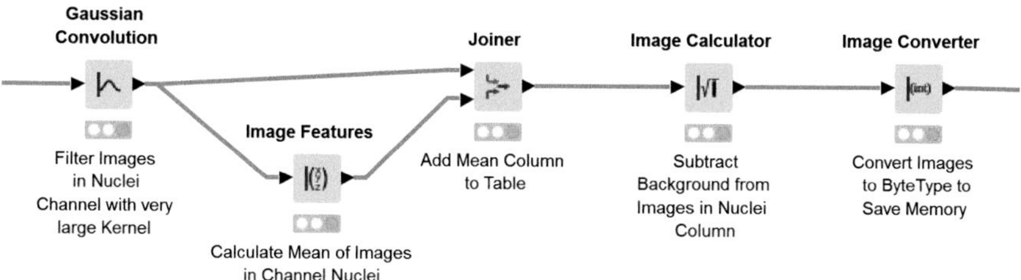

Fig. 7.9 Detailed look into the *Background Subtraction of Images in Nuclei Channel* meta node

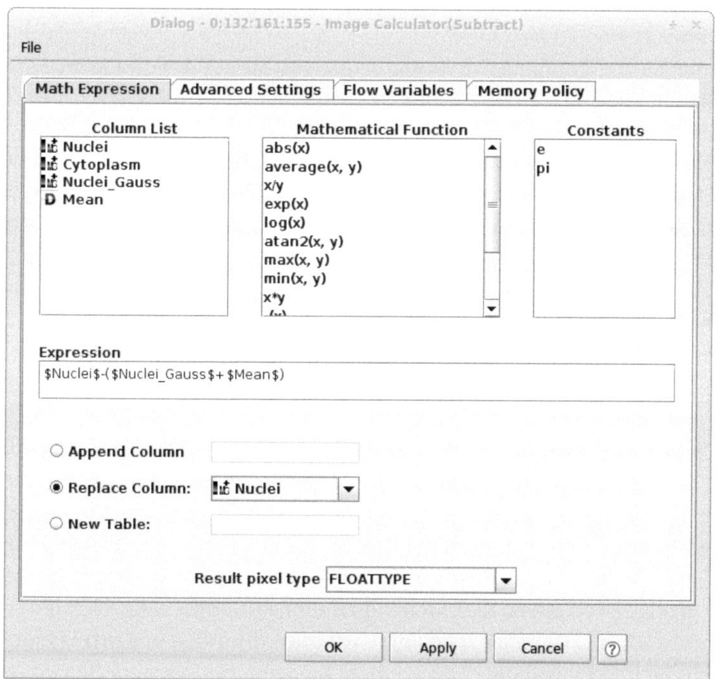

Fig. 7.10 Configuration dialog of the *Image Calculator* node

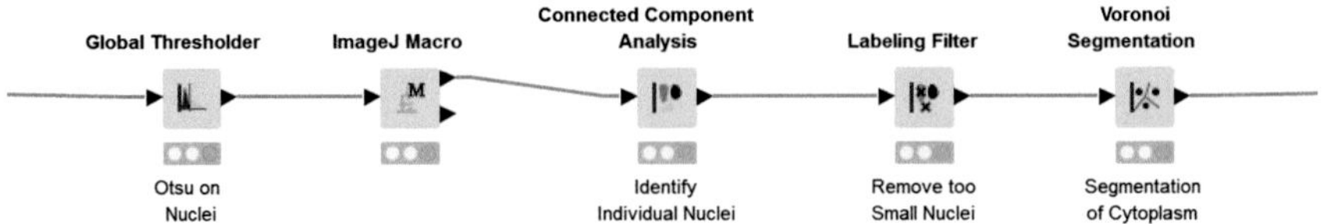

Fig. 7.11 Workflow to segment the nuclei and cytoplasm, respectively

Finally, both the background-corrected images from the *Nuclei* column and the images from the *Cytoplasm* column are filtered with a small Gaussian kernel (sigma $=$ 2.0) in the *Preprocessing* meta node. The results are appended to the table.

7.3.3 Segmentation

In order to classify the cells into *positive* and *negative* ones according to their phenotypes, both the nuclei and their cytoplasm have to be segmented. The subworkflow for this segmentation is encapsulated in the *Segmentation* meta node and is shown in Fig. 7.11.

The images in the column *Nuclei* are segmented using the well-known *Otsu* (Otsu 1975) thresholding algorithm, which is implemented in the *Global Thresholder* node. The output of the node is an image consisting only of black and white pixels. At each position of this binary image indication is given on whether the pixel belongs to a nucleus or to the background of the image. In order to split potential touching objects into the individual nuclei, the ImageJ1 Watershed Macro is executed on the binary images using the *ImageJ1 Macro* node, which is part of the KNIME Image Processing—ImageJ Integration. The subsequent *Connected Component Analysis* node derives a labeling from the binary images, which determines whether each pixel belongs to an individual nuclei as opposed to determining merely whether the pixel belongs to the nuclei or the background. The result is appended to the table. Thanks to the connected *Labeling Filter* node, objects that are either too small or too big can be removed from the labeling by manually defining the expected size of the nuclei. In this workflow we set the minimum size of nuclei to 50. The remaining nuclei now serve as seeding points for the segmentation of the cytoplasm. Starting at each nucleus in parallel, the region growing algorithm implemented in the *Voronoi Segmentation* node extends the seeding segments until no more pixels can be added to the individual segments. This is the case if a pixel has already been added to another segment or the intensity value of a pixel is lower than a manually defined threshold. The *Voronoi Segmentation* was configured to return the segmentation of the cytoplasm without the seeds, obtained with a threshold of 25 and *Fill Holes* activated.

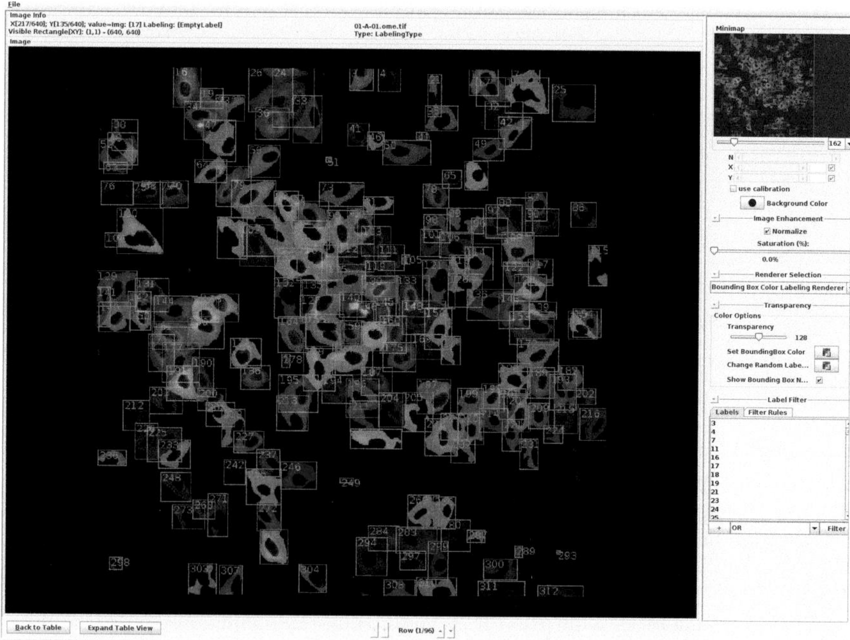

Fig. 7.12 Results of the *Voronoi Segmentation*

Figure 7.12 shows the resulting segmentation of the images in the Cytoplasm channel, using the *Interactive Segmentation View* node.

For other segmentation tasks, KNIME Image Processing offers a wide range of simple and more advanced segmentation techniques. Besides established algorithms such as *Graph Cuts* or *Local Thresholding*, the *KNIME Image Processing—Supervised Image Segmentation (SUISE)* extension comprises nodes for supervised pixel and segment classification.[10]

7.3.4 Feature Extraction

After certain objects have been identified and segmented, features can be calculated from either the derived labelings alone or the combination with their source images. Features numerically describe the individual objects and are instrumental in drawing conclusions from the acquired images. These features are therefore part of most of the image processing and analysis tasks.

[10]For details see the example workflows on http://tech.knime.org/supervised-image-segmentation.

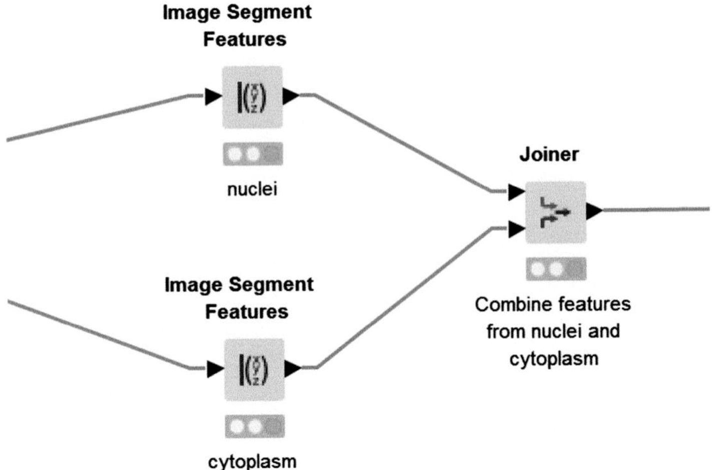

Fig. 7.13 Detailed look into the *Feature Extraction* meta node. First Order Statistics, Geometric and Haralick Features are extracted individually for the Cytoplasm and the Nuclei channel

KNIME Image Processing provides several feature implementations, for example simple first order statistics of the intensity values of a segment (mean intensity, standard deviation, kurtosis, etc.) or geometric properties of a segment (roundness, size, convexity, Zernike Moments (Khotanzad and Hong 1990), Fourier shape descriptors, etc.), as well as more complex texture measurements (Haralick et al. 1973; Tamura et al. 1978, etc.).

Figure 7.14 shows the output table of the *Joiner* node in Fig. 7.13, which combines the results from the preceding *Image Segment Feature* nodes by joining the rows of the individual tables according to their *RowId*. Given the nuclei (Channel 1), the cytoplasm (Channel 2) and their corresponding labelings, which were derived in the previous *Segmentation* meta node, these nodes calculate for each identified object the first order statistics, haralick texture features and several geometric properties. Each row of the output table of the *Feature Extraction* meta node corresponds to the numerical descriptions of a single object (Fig. 7.14).

7.3.5 *Model Learning*

The output of the *Feature Extraction* meta node can now be connected to KNIME nodes, which allow operations to be performed on numerical data.

Typical examples include nodes for statistical testing, machine learning and data mining or visualization. In this example, a supervised classification of the nuclei and cytoplasm is performed based on the calculated features, in order to determine whether it is considered positive or negative (see Fig. 7.15). Therefore, as a first step,

Calculated Features - 0:78:70 - Image Segment Features (cytoplasm)

File

Table "default" - Rows: 13354 | Spec - Columns: 99 | Properties | Flow Variables

Row ID	Bitmask	Label	Source Labeling	Num Pix	Circularity	Perimeter	Convexity	Extend	Diameter	ASM | ...	Contra...	Correla...	Varianc...	IFDM | ...	SumAv...	SumVar...	SumEnt.
Row0#3		3	Labeling[name=01-A-01.ome.tif;source=;dimensions=640,640 (X,...	438	0.385	119.497	0.928	0.581	37.802	0.004	15.79	0.891	72.623	0.291	46.371	274.703	4.06
Row0#4		4	Labeling[name=01-A-01.ome.tif;source=;dimensions=640,640 (X,...	389	0.298	128.083	0.913	0.534	36.878	0.004	45.573	0.742	88.252	0.119	44.409	307.437	4.029
Row0#7		7	Labeling[name=01-A-01.ome.tif;source=;dimensions=640,640 (X,...	316	0.486	90.426	0.898	0.62	35.057	0.016	25.35	0.675	38.992	0.241	33.043	130.62	2.984
Row0#11		11	Labeling[name=01-A-01.ome.tif;source=;dimensions=640,640 (X,...	497	0.266	153.196	0.924	0.458	39.051	0.004	37.308	0.686	59.418	0.234	47.394	200.366	3.947
Row0#16		16	Labeling[name=01-A-01.ome.tif;source=;dimensions=640,640 (X,...	1,050	0.287	214.468	0.96	0.643	53.31	0.004	12.087	0.934	91.498	0.303	35.059	353.904	4.128
Row0#17		17	Labeling[name=01-A-01.ome.tif;source=;dimensions=640,640 (X,...	517	0.255	159.581	0.869	0.362	44.147	0.007	23.658	0.687	37.744	0.276	33.636	127.319	3.661
Row0#18		18	Labeling[name=01-A-01.ome.tif;source=;dimensions=640,640 (X,...	942	0.174	260.693	0.913	0.423	61.774	0.007	28.316	0.676	43.709	0.252	45.365	146.522	3.767
Row0#19		19	Labeling[name=01-A-01.ome.tif;source=;dimensions=640,640 (X,...	338	0.317	115.841	0.934	0.525	32.249	0.007	25.711	0.751	51.544	0.268	38.519	180.466	3.773
Row0#21		21	Labeling[name=01-A-01.ome.tif;source=;dimensions=640,640 (X,...	302	0.265	119.74	0.805	0.378	51.088	0.007	31.902	0.728	58.685	0.263	36.984	202.839	3.863
Row0#23		23	Labeling[name=01-A-01.ome.tif;source=;dimensions=640,640 (X,...	429	0.349	124.368	0.903	0.409	38.275	0.006	13.291	0.868	50.454	0.337	37.633	188.526	3.886
Row0#24		24	Labeling[name=01-A-01.ome.tif;source=;dimensions=640,640 (X,...	1,207	0.147	320.907	0.912	0.333	76.896	0.007	9.584	0.907	51.755	0.427	30.407	197.437	3.856
Row0#25		25	Labeling[name=01-A-01.ome.tif;source=;dimensions=640,640 (X,...	802	0.253	199.439	0.892	0.344	61.911	0.006	18.364	0.818	50.368	0.322	38.787	183.106	3.892
Row0#26		26	Labeling[name=01-A-01.ome.tif;source=;dimensions=640,640 (X,...	1,324	0.197	290.871	0.942	0.463	61.774	0.004	15.274	0.902	77.86	0.363	38.9	296.167	4.153
Row0#32		32	Labeling[name=01-A-01.ome.tif;source=;dimensions=640,640 (X,...	811	0.353	169.823	0.944	0.534	47.043	0.004	18.28	0.828	53.013	0.277	39.557	193.77	3.966
Row0#33		33	Labeling[name=01-A-01.ome.tif;source=;dimensions=640,640 (X,...	869	0.246	210.894	0.93	0.443	57.219	0.005	18.354	0.729	33.817	0.277	37.327	116.913	3.744
Row0#34		34	Labeling[name=01-A-01.ome.tif;source=;dimensions=640,640 (X,...	989	0.253	221.794	0.928	0.423	53.759	0.026	8.594	0.873	33.882	0.569	14.902	126.934	3.168
Row0#35		35	Labeling[name=01-A-01.ome.tif;source=;dimensions=640,640 (X,...	498	0.198	177.924	0.941	0.429	41.593	0.006	29.766	0.648	42.257	0.257	48.864	139.262	3.742
Row0#36		36	Labeling[name=01-A-01.ome.tif;source=;dimensions=640,640 (X,...	861	0.283	195.581	0.947	0.568	42.579	0.005	19.499	0.904	101.663	0.278	33.977	387.154	4.072
Row0#38		38	Labeling[name=01-A-01.ome.tif;source=;dimensions=640,640 (X,...	278	0.199	132.368	0.835	0.242	46.4	0.005	56.168	0.505	56.737	0.221	38.889	170.781	3.809
Row0#40		40	Labeling[name=01-A-01.ome.tif;source=;dimensions=640,640 (X,...	71	0.696	35.799	0.986	0.74	13.153	0.024	35.053	0.643	49.151	0.323	51.684	161.549	3.19
Row0#41		41	Labeling[name=01-A-01.ome.tif;source=;dimensions=640,640 (X,...	298	0.196	138.296	0.928	0.441	32.202	0.006	42.58	0.571	49.66	0.178	43.074	156.06	3.759
Row0#42		42	Labeling[name=01-A-01.ome.tif;source=;dimensions=640,640 (X,...	637	0.289	166.51	0.913	0.419	50.329	0.003	25.467	0.825	72.74	0.268	39.796	265.493	4.103
Row0#46		46	Labeling[name=01-A-01.ome.tif;source=;dimensions=640,640 (X,...	192	0.286	91.841	0.877	0.457	25.06	0.011	52.583	0.388	42.989	0.221	44.887	119.371	3.512
Row0#47		47	Labeling[name=01-A-01.ome.tif;source=;dimensions=640,640 (X,...	913	0.248	214.894	0.915	0.466	52.202	0.007	15.764	0.848	51.806	0.361	39.704	191.46	3.919
Row0#48		48	Labeling[name=01-A-01.ome.tif;source=;dimensions=640,640 (X,...	419	0.202	161.38	0.933	0.306	46.615	0.006	19.985	0.861	72.108	0.271	52.075	268.446	3.885
Row0#49		49	Labeling[name=01-A-01.ome.tif;source=;dimensions=640,640 (X,...	489	0.281	147.782	0.924	0.495	40.311	0.004	43.998	0.642	61.411	0.184	40.021	201.645	3.991
Row0#50		50	Labeling[name=01-A-01.ome.tif;source=;dimensions=640,640 (X,...	941	0.283	204.409	0.958	0.55	60.407	0.005	22.616	0.811	59.726	0.304	41.631	216.286	4.03
Row0#51		51	Labeling[name=01-A-01.ome.tif;source=;dimensions=640,640 (X,...	22	0.576	21.899	0.957	0.733	8.062	0.094	193.5	0.405	162.576	0.42	50.167	456.806	1.979
Row0#52		52	Labeling[name=01-A-01.ome.tif;source=;dimensions=640,640 (X,...	444	0.281	140.853	0.893	0.317	43.6	0.006	23.337	0.788	55.074	0.274	43.458	196.959	3.87
Row0#56		56	Labeling[name=01-A-01.ome.tif;source=;dimensions=640,640 (X,...	723	0.269	183.924	0.954	0.488	49.578	0.005	11.998	0.92	75.325	0.346	44.819	289.301	4.126
Row0#62		62	Labeling[name=01-A-01.ome.tif;source=;dimensions=640,640 (X,...	575	0.323	149.64	0.958	0.604	38.833	0.004	28.132	0.731	52.193	0.25	34.78	180.639	3.912

Fig. 7.14 Output table of *Image Segment Feature* node. Each row corresponds to the numerical measurements of a cells nucleus and cytoplasm

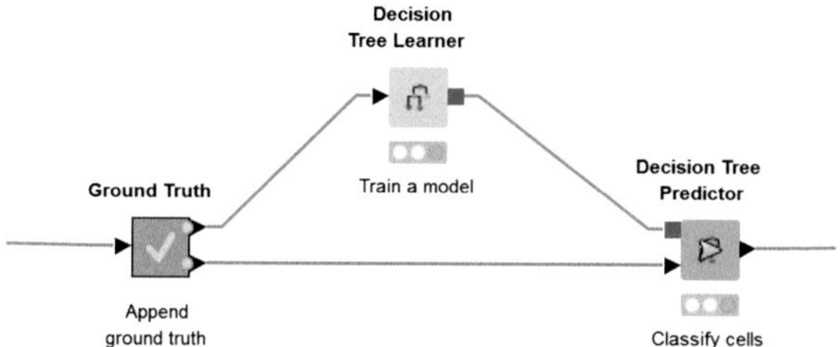

Fig. 7.15 Detailed look into the *Model Learning* meta node

the ground-truth data, which is part of the publicly available high-content screening image data set, is read into KNIME as a text file and joined with the already loaded image data. The ground-truth data contains indications of the classes for several cells, which then serve as the training data for a supervised learning algorithm.[11]

However, if this ground-truth data is not available, users can manually create this information by using the *Interactive Labeling Editor* node. Given the ground-truth and the numerical description of the nuclei and cytoplasm, the *Decision Tree Learner* node can be used to train a decision tree model, which in turn can be applied to cells as yet unseen using the *Decision Tree Predictor* node (see Fig. 7.15). The output of the *Decision Tree Predictor* node comprises an additional column with the classification result.

[11]For details see *Phenotype Classification* workflow at http://knime.imagej.net/aaec.

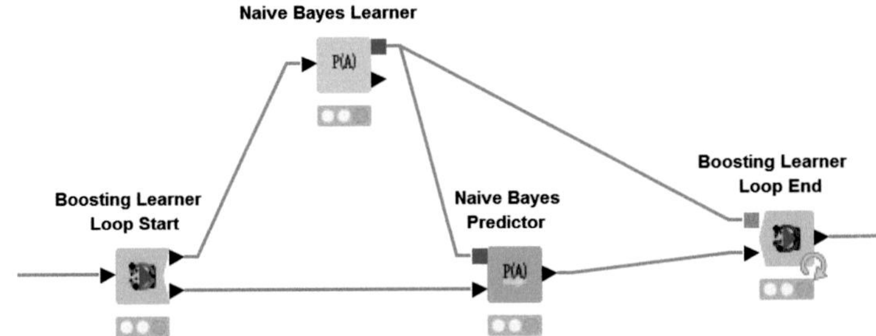

Fig. 7.16 Boosting of a Naive-Bayes learner

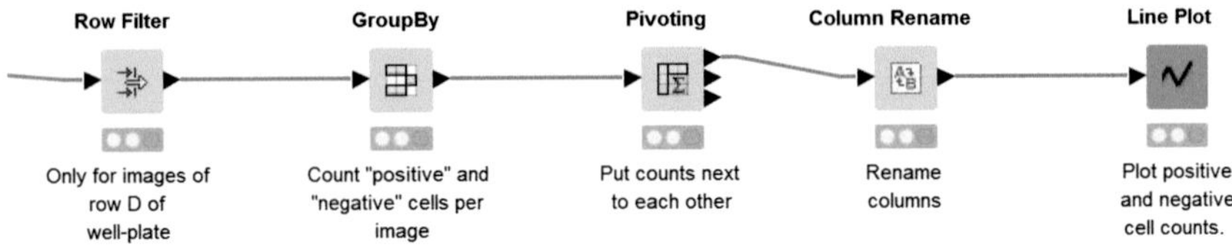

Fig. 7.17 Detailed look into the *Evaluation and Validation* meta node

The Decision Tree model with default configuration settings is used in this example. However, other machine learning techniques can easily be applied instead, for example Support Vector Machines (Scholkopf and Smola 2002), Random Forests (Breiman 2001) or any other algorithm, which are either included in KNIME or available as a KNIME extension, such as Weka, R or Python.

Figure 7.16, for instance, depicts the well-known *Boosting* algorithm, which is offered with the *KNIME Ensemble Learning* plugin and also comprises nodes for *Bagging* or *Stacking*.

7.3.6 Evaluation and Validation

Users often want to manually explore and validate the information extracted from the raw images, as the features or the results of a classification task. KNIME itself offers a wide range of functionalities to visualize numerical data. Scatter plots, line plots, bar plots or histograms are just some examples of those that are offered. Even more plots are available with the R and Python extensions in KNIME. Furthermore, KNIME provides nodes for statistical significance testing, for example *T-Tests* or *ANOVA Testing*.

The *Evaluation and Validation* meta node comprises one example of how to visualize the results from the classification conducted in the *Model Learning* (see Fig. 7.17). The resulting line-plot (see Fig. 7.18) contains the counts of positive and negative cells of images which are part of row *D* in the well-plate. First, the *Row Filter* node removes all images that are not part of row *D*. The subsequent *Group*

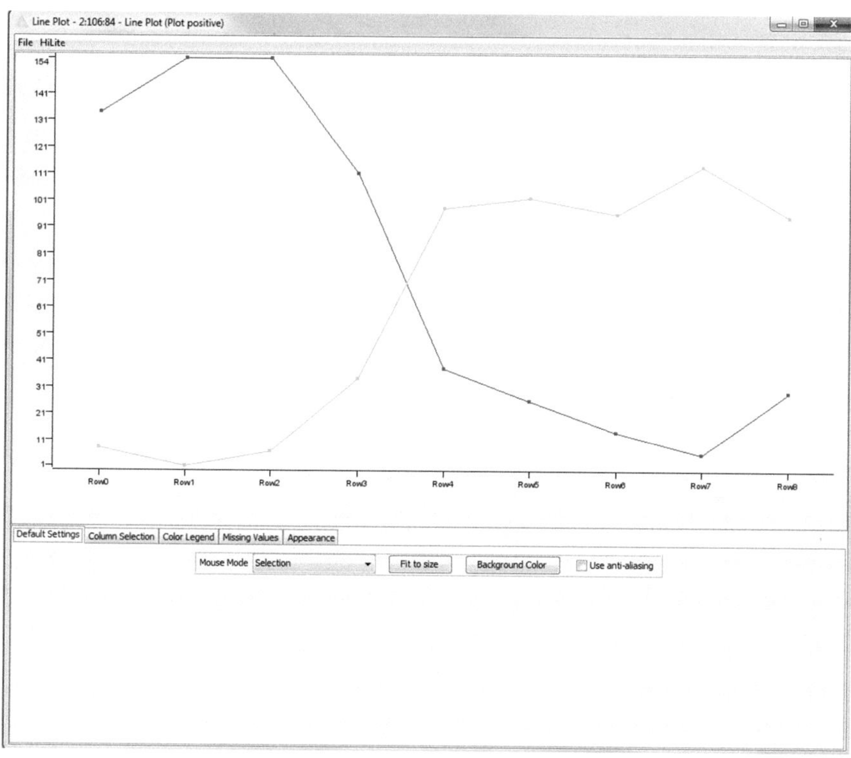

Fig. 7.18 Line-Plot visualizing the classification results of the *Model Learning* meta node

By node counts the number of positive and negative cells for each image in row *D*, while the *Pivoting* node arranges the KNIME table, such that the cell counts appear next to each other. It can be observed that the number of positive cells increases over the columns of row *D* of the well-plate, which meets the expectations.[12]

7.4 Conclusions

In this tutorial the basic concepts of KNIME Image Processing are introduced and the advantages of combining different software packages in a single understandable, multi-domain workflow through KNIME are demonstrated by means of an example workflow for phenotype classification. The applied techniques in this use-case are simple and exemplary. However, the already published workflows in Saha et al. (2013), Lodermeyer et al. (2013), Gunkel et al. (2014), Strauch et al. (2014), and

[12]see http://www.broadinstitute.org/bbbc/BBBC013/ for details on the plate design.

Aligeti et al. (2014) solve simple problems like counting cells or measuring the intensity of segmented cells, as well as more complex tasks involving machine learning and data mining techniques. For instance, in Gunkel et al. (2014) the entire image acquisition process was controlled by a KNIME workflow. The images are analysed on-the-fly and feedback is provided instantly to the microscope. Avoiding the acquisition and storage of uninteresting images, screening costs can be reduced by 90 %.

KNIME has also been successfully applied in other fields of research in life sciences, for instance to pharmacophore identification in classic HTS data (High Throughput Screening) or outlier detection in medical claims. Use cases also exist from entirely unrelated sectors. KNIME workflows for segmentation of customers, churn analysis, market basket analysis, sentiment analysis on social media data (using the KNIME *Text Processing* and *Network Analysis* extensions) as well as credit scoring based on historical data are just a few examples.[13]

The wide range of application areas is a direct result of KNIME's openness. The need for open platforms in classic data analytics is even more pressing now, when data analysts have easy access to an ever-growing number of internal and external data sources. To tackle this challenge they need quick and easy access to best-of-breed tools to intuitively explore new analysis ideas unburdened by the artificial barriers of closed environments. As a result, open platforms are much more powerful than any monolithic application can ever be. Due to the simplicity of mixing and matching inhouse, legacy and external technology within the same intuitive environment, analysts can choose which data and tools they want to use instead of being restricted to the tools available in a proprietary toolbox.

References

Aligeti M, Behrens RT, Pocock GM, Schindelin J, Dietz C, Eliceiri KW, Swanson CM, Malim MH, Ahlquist P, Sherer NM (2014) Cooperativity among rev-associated nuclear export signals regulates HIV-1 gene expression and is a determinant of virus species tropism. J Virol 88(24):14,207–14,221. doi:10.1128/JVI.01897-14. http://jvi.asm.org/content/88/24/14207.full.pdf+html

Allan C, Burel JM, Moore J, Blackburn C, Linkert M, Loynton S, MacDonald D, Moore WJ, Neves C, Patterson A et al (2012) Omero: flexible, model-driven data management for experimental biology. Nat Methods 9(3):245–253

Berthold MR, Cebron N, Dill F, Gabriel TR, Kötter T, Meinl T, Ohl P, Sieb C, Thiel K, Wiswedel B (2008) KNIME: the Konstanz information miner. Springer, Berlin

Breiman L (2001) Random forests. Mach Learn 45(1):5–32

Eliceiri KW, Berthold MR, Goldberg IG, Ibáñez L, Manjunath B, Martone ME, Murphy RF, Peng H, Plant AL, Roysam B et al (2012) Biological imaging software tools. Nat Methods 9(7):697–710

[13]See https://www.knime.org/applications.

Gunkel M, Flottmann B, Heilemann M, Reymann J, Erfle H (2014) Integrated and correlative high-throughput and super-resolution microscopy. Histochem Cell Biol 141(6):597–603. doi:10.1007/s00418-014-1209-y. http://dx.doi.org/10.1007/s00418-014-1209-y

Haralick RM, Shanmugam K, Dinstein IH (1973) Textural features for image classification. IEEE Trans Syst Man Cybern (6):610–621

Kamentsky L, Jones TR, Fraser A, Bray MA, Logan DJ, Madden KL, Ljosa V, Rueden C, Eliceiri KW, Carpenter AE (2011) Improved structure, function and compatibility for cellprofiler: modular high-throughput image analysis software. Bioinformatics 27(8):1179–1180

Khotanzad A, Hong YH (1990) Invariant image recognition by zernike moments. IEEE Trans Pattern Anal Mach Intell 12(5):489–497

Linkert M, Rueden CT, Allan C, Burel JM, Moore W, Patterson A, Loranger B, Moore J, Neves C, MacDonald D et al (2010) Metadata matters: access to image data in the real world. J Cell Biol 189(5):777–782

Ljosa V, Sokolnicki KL, Carpenter AE (2012) Annotated high-throughput microscopy image sets for validation. Nat Methods 9(7):637

Lodermeyer V et al (2013) 90k, an interferon-stimulated gene product, reduces the infectivity of HIV-1. Retrovirology 10(1):11

Otsu N (1975) A threshold selection method from gray-level histograms. Automatica 11(285–296):23–27

Pietzsch T, Preibisch S, Tomančák P, Saalfeld S (2012) Imglib2 - generic image processing in java. Bioinformatics 28(22):3009–3011

Royer LA et al (2015) ClearVolume: open-source live 3D visualization for light-sheet microscopy. Nat Methods 12(6):480–481

Saha AK, Kappes F, Mundade A, Deutzmann A, Rosmarin DM, Legendre M, Chatain N, Al-Obaidi Z, Adams BS, Ploegh HL, Ferrando-May E, Mor-Vaknin N, Markovitz DM (2013) Intercellular trafficking of the nuclear oncoprotein dek. Proc Natl Acad Sci 110(17):6847–6852. doi:10.1073/pnas.1220751110

Schindelin J, Arganda-Carreras I, Frise E, Kaynig V, Longair M, Pietzsch T, Preibisch S, Rueden C, Saalfeld S, Schmid B et al (2012) Fiji: an open-source platform for biological-image analysis. Nat Methods 9(7):676–682

Schölkopf B, Alexander JS (2002) Learning with kernels: support vector machines, regularization, optimization, and beyond. MIT, Cambridge

Strauch M, Luedke A, Muench D, Laudes T, Galizia CG, Martinelli E, Lavra L, Paolesse R, Ulivieri A, Catini A, Capuano R, Di Natale C (2014) More than apples and oranges: detecting cancer with a fruit fly's antenna

Tamura H, Mori S, Yamawaki T (1978) Textural features corresponding to visual perception. IEEE Trans Syst Man Cybern 8(6):460–473

Chapter 8
Segmenting and Tracking Multiple Dividing Targets Using *ilastik*

Carsten Haubold, Martin Schiegg, Anna Kreshuk, Stuart Berg, Ullrich Koethe, and Fred A. Hamprecht

Abstract Tracking crowded cells or other targets in biology is often a challenging task due to poor signal-to-noise ratio, mutual occlusion, large displacements, little discernibility, and the ability of cells to divide. We here present an open source implementation of *conservation tracking* (Schiegg et al., IEEE international conference on computer vision (ICCV). IEEE, New York, pp 2928–2935, 2013) in the *ilastik* software framework. This robust tracking-by-assignment algorithm explicitly makes allowance for false positive detections, undersegmentation, and cell division. We give an overview over the underlying algorithm and parameters, and explain the use for a light sheet microscopy sequence of a Drosophila embryo. Equipped with this knowledge, users will be able to track targets of interest in their own data.

8.1 Introduction

Tracking multiple indistinguishable targets, where each of the tracked objects could potentially divide, is an important task in many biological scenarios.

Manually tracking nuclei in microscopy data is tedious. Human annotators need to take temporal context into account to distinguish cells in regions of poor image quality, and thus manual curation is very time consuming. To allow for high throughput experiments, automated tracking software is in great demand, even though detecting cells especially in high density regions of poor image quality poses an even larger challenge to automated methods than to humans. The reason for this is that humans can intuitively detect and link cells in neighboring time frames by considering, e.g., plausible cell motions, size changes, and the number of cells in

C. Haubold • M. Schiegg • A. Kreshuk • U. Koethe • F.A. Hamprecht (✉)
University of Heidelberg, IWR/HCI, 69115 Heidelberg, Germany
e-mail: fred.hamprecht@iwr.uni-heidelberg.de

S. Berg
Howard Hughes Medical Institute, Ashburn, VA, USA

W.H. De Vos et al. (eds.), *Focus on Bio-Image Informatics*, Advances in Anatomy, Embryology and Cell Biology 219,
DOI 10.1007/978-3-319-28549-8_8

the neighborhood as well as their relative positions. Unfortunately detection and tracking is like a chicken-and-egg problem. For automated methods, it would be extremely helpful to have detected all cells in each frame so that linking between frames becomes a one-to-one matching problem, or one-to-two in the case of divisions. On the other hand, to facilitate detecting and properly segmenting all cells in high density regions, it would be beneficial if the history and fate of the cells—like their positions in the preceding and subsequent frames—could be considered.

The community has developed a range of tracking tools with graphical user interfaces to allow for easy application by experts of other fields who are not trained in image processing. However, due to the large variety of different use cases—varying microscopy techniques, as well as different types of cells being recorded—most of the tools are limited to specific scenarios. Fiji's TrackMate (Schindelin et al. 2012; Nick Perry 2012), for instance, builds on a linear assignment problem (LAP) (Jaqaman et al. 2008) that first links detected objects in consecutive frames and then constructs global tracks from these local links. It can handle merging and splitting targets, but cannot deal with dividing cells. CellProfiler (Carpenter et al. 2006) also uses this LAP tracking algorithm or performs greedy nearest-neighbor tracking. Icy (De Chaumont et al. 2012) has several plugins for tracking, the most recent being a *spot tracking* method building on a multi-hypothesis tracking approach (Chenouard et al. 2013) that can deal with a variety of motion models of non-dividing objects. iTrack4U (Cordelières et al. 2013) is based on a mean-shift algorithm that operates on a fan of triangular regions created around the center of each cell in 2D images, but it cannot handle divisions or merging and splitting of cells.

Maška et al. (2014) review a broad range of further cell tracking approaches, but these do not necessarily provide graphical user interfaces, which often prevents their adoption by biologists.

This tutorial presents the tracking workflow in *ilastik* (Sommer et al. 2011), which tries to make user input as intuitive as possible, while yielding high quality tracking results for a broad variety of use cases by considering a global model when resolving ambiguities (Schiegg et al. 2013) of cells migrating, merging, splitting, and dividing in 2D+t and 3D+t. *ilastik* is an *interactive learning and segmentation toolkit* that uses machine learning techniques to classify pixels and objects by learning from annotations—sparsely placed by the user—to predict the class (e.g., foreground or background) of each unannotated pixel or object. This toolkit contains several workflows, and each workflow is designed to tackle a specific task. We will demonstrate how to obtain a segmentation using *ilastik*'s *Pixel Classification* workflow, and how to track the segmented cells in the *Automated Tracking* workflow. Throughout this tutorial, a crop of a *Drosophila melanogaster* embryo scan[1] (see Fig. 8.1), recorded by the Hufnagel group[2] using selective plane imaging (Krzic et al. 2012), will serve as example.

[1]The cropped dataset can be obtained from the download page http://ilastik.org/download.html, and a larger crop with more information is available at http://hci.iwr.uni-heidelberg.de//Benchmarks/document/Drosophila_cell_tracking/.

[2]European Molecular Biology Laboratory, Heidelberg, Germany.

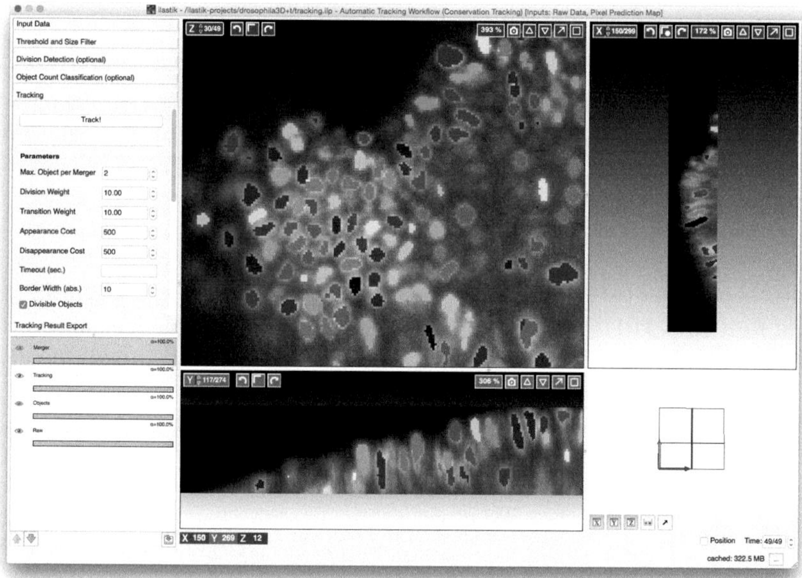

Fig. 8.1 The Drosophila dataset opened in *ilastik*, with color coded segmented and tracked cells. Same color indicates common ancestry since the beginning of the video, but not many cells divided during the short time span. The cropped dataset consists of 50 time steps during the gastrulation stage, with the last frame being displayed

TIP: Dedicated *background* sections will give more details on the used algorithms, and *tips* present experienced users' advice on how to best achieve certain goals in *ilastik*, e.g., through useful keyboard shortcuts.

FAQ: This tutorial refers to *ilastik* version 1.1.7.

FAQ: The recommended system specifications for using *ilastik* are a recent multi-core machine with at least 8 GB of RAM. When working with larger datasets, especially for 3D volumes, more RAM is beneficial as it allows ilastik to keep more images and results cached.

8.1.1 Tracking Basics

Many automated tracking tools use a family of algorithms called *tracking-by-assignment*. These algorithms split the work into two steps: segmentation and tracking (Andriluka et al. 2008; Benfold and Reid 2011; Zhang et al. 2008; Pirsiavash et al. 2011). The segmentation step finds the outline of the targeted objects at each time step (or frame) in the raw data, where each connected component is called a detection, and the tracking step then links detections over time into tracks. Linking cell detections can be done pairwise between consecutive frames (e.g., Kuhn 1955), but can also be approached globally (e.g., Jaqaman et al. 2008; Bise et al. 2011; Schiegg et al. 2013; Padfield et al. 2011; Magnusson and Jaldén 2012). The tracking method in *ilastik* (Schiegg et al. 2013) finds the most probable linking of all detections through a global optimization.

One difficulty of the segmentation step is that the data does not always contain nicely separated objects and is subject to noise or other artifacts of the acquisition pipeline. A good segmentation reduces ambiguities in the tracking step and thus makes linking less error prone. Common segmentation errors are under- and oversegmentation, where the segmentation is too coarse or fine, respectively, as Fig. 8.2 depicts. The tracking algorithm in *ilastik* (Sect. 8.3) can deal with undersegmentation to some extent, but expects that there are no oversegmentations. Thus when creating a segmentation it is important not to split objects of interest into several segments. In the case of an undersegmentation, a detection that contains N objects is called an N-*merger* in the remainder of this tutorial.

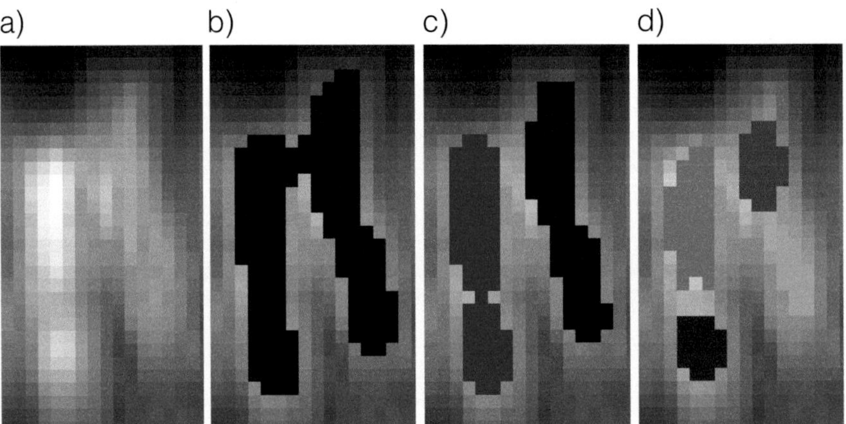

Fig. 8.2 From *left to right*: **(a)** Raw data containing two cells. **(b)** An undersegmentation of the cells means there are too few segments, here only one detection. The optimization might still decide that two cells merged into this single detection, which is designated as a *2-merger*. **(c)** The correct segmentation results in two detections. **(d)** The event in which there are more segments than cells is referred to as oversegmentation. Here both cells are split into a false total of four segments. Best viewed in *color*

Figure 8.3 summarizes the three stages of the pipeline which will be covered in this tutorial. To obtain a segmentation with ilastik, the user trains a classifier by marking a small number of foreground and background pixels. The classifier learns to distinguish the foreground and background class by considering features

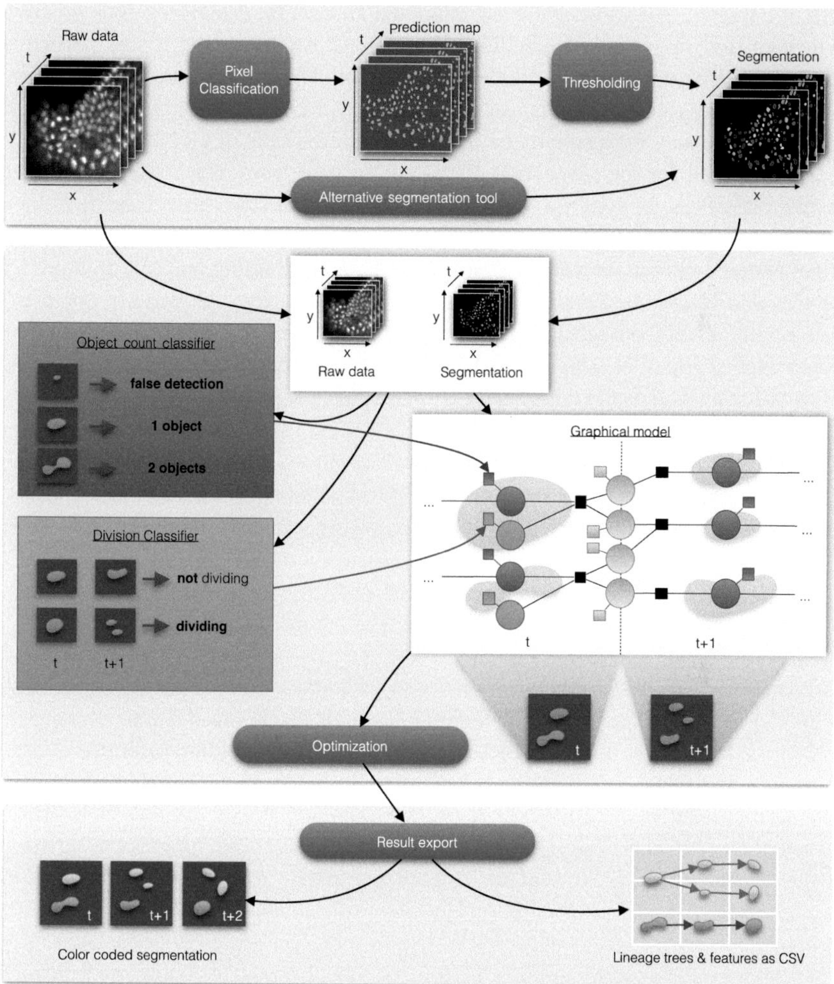

Fig. 8.3 Pipeline overview in three stages: segmentation, tracking, and result export. A segmentation of the raw data can be obtained using pixel classification and thresholding, as detailed in Sect. 8.2, or by any other segmentation tool. The tracking algorithm in *ilastik* (Sect. 8.3) builds a global graphical model of all detections in all time frames. Evidence from the input data is inserted into the model through an *Object Count Classifier* and a *Division Classifier*. See the *Conservation Tracking Background* box for more information on how the model is constructed and how the most probable configuration is found. Finally, results can be exported as images or CSV tables, as shown in Sect. 8.4

which are automatically computed on a per-pixel basis, and then predicts the class of all unannotated pixels. A segmentation can then be obtained by thresholding the predicted probability for the foreground class.

The tracking stage takes raw data and segmentation as input. It builds up a graph of all detections and their possible links between consecutive time frames. Each node in this graph can take one of N possible states, representing the number of cells it contains. All these states are assigned an energy, describing how costly it is to take this very state. A candidate hypothesis has high cost if this particular interpretation of the data is unlikely to be correct. The optimization will then find the overall configuration that has the lowest energy, where the locally most plausible interpretations may occasionally be overridden for the sake of global consistency.

The energies for the respective nodes are derived from probabilities predicted by two classifiers. In contrast to the pixel classification step, these classifiers base their decisions on detection-level features. Possible classes are different numbers of contained cells for the *Object Count Classifier*, or dividing and not dividing for the *Division Classifier*. Finally, the result export stage offers two ways of retrieving the tracking result from ilastik for further analysis, namely a relabeled segmentation where each segment is assigned its track ID, and a spreadsheet containing linking information as well as the computed features of each segment.

Conservation Tracking Background: One of the important features that distinguishes a cell tracker from other multi-object trackers is the ability to detect cell divisions. Candidates for a division are all detections that have at least two possible descendants in the next frame. Figures 8.3(middle) and 8.10 show graphical models where each detection is represented by a (purple) random variable. The nodes are ordered in columns by time step. Possible transitions of an object between two frames are shown as gray nodes linked to the two respective detections through black squared conservation factors. Each detection with at least two outgoing transitions could possibly divide, and is accompanied by a blue division node. If the number of cells entering a detection along its incoming transition nodes exceeds one, the detection represents more than one cell: it is a *merger*. Mergers occur when the segmentation fails to separate cells in clumps. Because we allow mergers, each node in the graph (except for division nodes) can contain more than one cell.

The most probable solution of this graphical model will always follow *conservation laws*, meaning that at each detection the number of cells leaving along outgoing transitions is equal to the number of incoming cells. If the detection is not a merger, then we allow it to divide, meaning the number of outgoing cells can also be two. These laws are represented by the black squares as factors. The conservation laws are the reason why the algorithm

(continued)

used in *ilastik* is called *Conservation Tracking* (Schiegg et al. 2013). The temporal context is important when the segmentation of a single time frame is ambiguous. By building a graphical model for the complete dataset over all time steps, this context can be incorporated into the optimization procedure to improve the tracking results. The graphical model defines a probability distribution, and the optimization finds the *Maximum-a-posteriori* (MAP) solution, which is the most probable state of the overall system, given all current observations.

The structure of the remainder of this tutorial is as follows: Sect. 8.2 first presents how to load data (Sect. 8.2.1), and how to navigate the workflow sidebar (Sect. 8.2.2) and the data views. Then the segmentation by *pixel classification* and *thresholding* is introduced; this step can be omitted if a segmentation is already available. Section 8.3 covers the *automated tracking* workflow, explains how to train classifiers to detect which cells are dividing and how to give hints as to which detections are mergers, followed by a brief overview of the algorithm that is used for tracking. Lastly the result export capabilities are detailed in Sect. 8.4 and we conclude in Sect. 8.5.

8.2 Pixel Classification and Object Segmentation

ilastik provides the *Pixel Classification* workflow to interactively create a segmentation of multidimensional image data. When starting the application, you will be faced with a workflow selection dialog as in Fig. 8.4. Select `Pixel Classification` for now. You will be prompted for a file to save the project to immediately. Save the project as, e.g., `pixel-classification.ilp`. As the name implies, this workflow allows you to classify each pixel, e.g., into foreground and background, but also into more classes. To do this, you first have to load the data, create the desired classes, and give some examples for each class by painting brushstrokes in the data. A color coding will then be displayed over the raw data to indicate which pixel probably belongs to what class. You can refine this prediction by drawing additional annotation strokes.

TIP: We will use `monospaced fonts` to indicate the button/text field/... that should be clicked/edited.

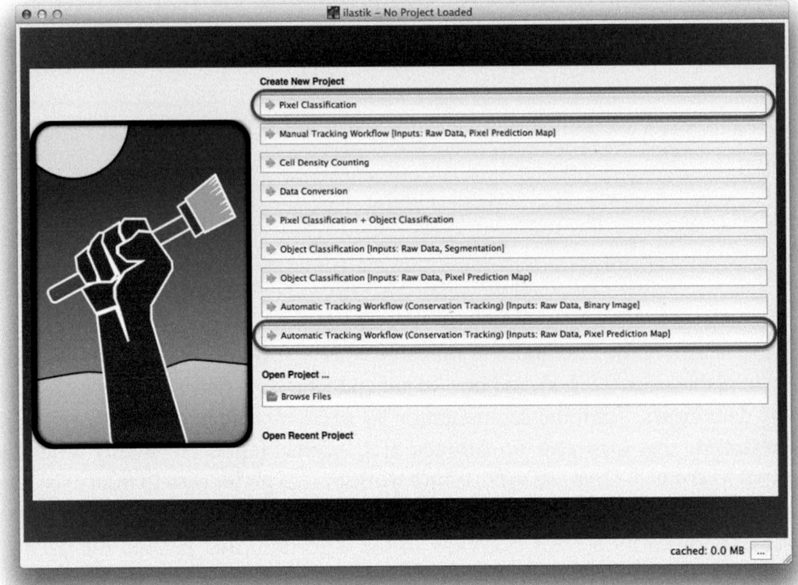

Fig. 8.4 Startup screen of *ilastik*, showing all available workflows. Highlighted in *red* are the workflows covered by this tutorial. In Sect. 8.2 we will use the *Pixel Classification* workflow, and in Sect. 8.3 *Automated Tracking*

8.2.1 Loading Data

The first step when creating a new project is loading data. The *Input Data* applet should already be preselected in the sidebar on the left-hand side. To its right is a table containing *Raw Data* and *Summary* tabs as shown in Fig. 8.5. Select `Raw Data` and click the `Add New` button. *ilastik* can handle most standard image formats like *tiff*, *bmp*, *gif*, *png*, and *jpeg*, as well as *HDF5* files, but not BioFormats (Linkert et al. 2010) yet. If the data of interest consists of only one frame or is stored as a HDF5 volume in a single file, *"Add separate image(s)..."* should be selected, but in the case of a stack with one file per time step or z-slice the option *"Add a single 3D/4D Volume from Sequence..."* should be used. The Drosophila example dataset from Fig. 8.1 is provided as one HDF5 file, so it is opened via `Add separate image(s)...`. As soon as the data is loaded, the first row in the dataset table will be filled in with information about the dataset's axes and shape. If the axes and their dimensions in the *Shape* column do not match, `double click` on the row, or `right click` to select `Settings`. In the dialog that pops up, the `Axes` text box allows to arbitrarily swap the axes around by entering a series of characters indicating spatial dimensions **x**, **y**, and **z**, as well as the time **t** axes and the color channels **c**. The number of characters has to match the number of dimensions in the *Shape* field. Make sure that **t** is present in this list, as tracking requires a

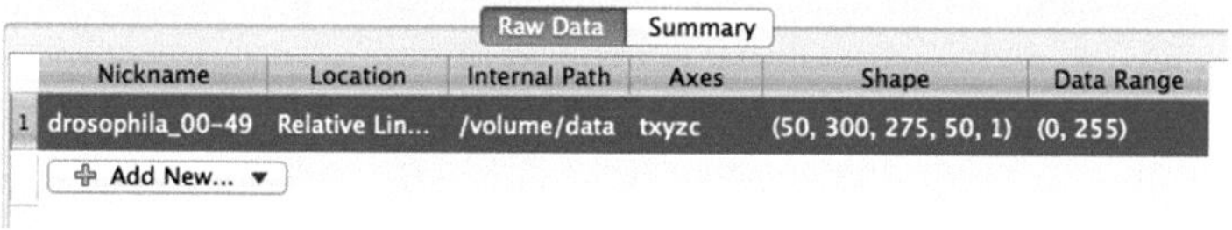

Raw Data | Summary

	Nickname	Location	Internal Path	Axes	Shape	Data Range
1	drosophila_00-49	Relative Lin...	/volume/data	txyzc	(50, 300, 275, 50, 1)	(0, 255)

Add New...

Fig. 8.5 A dataset has been loaded. The axes description specifies how the shape is interpreted. Here the dataset has 50 frames of 3D volumes with a shape of $300 \times 275 \times 50$ pixels, and only one color channel

time axis. If the data was loaded correctly, the image content will be shown in the main window of *ilastik*. For time lapse videos, a small spin box titled *Time* will be displayed in the lower right which indicates the current (zero based) frame index and the highest available frame number.

TIP: On Windows, the data needs to reside on the same drive as the project file.

TIP: Loading a volume from a HDF5 file is faster than loading a tiff stack because HDF5 allows blockwise access. If *ilastik* runs very slowly, try converting the data into HDF5. The easiest way to do this is to select `Copied to project file` for the `Storage` option in the dataset properties window, which you already used to specify the axes.

8.2.2 Navigating ilastik

Depending on whether a 2D+t or 3D+t dataset was loaded, *ilastik* will display one time frame as 2D image, or a split view with *xy*, *yz*, and *xz* slicing planes and a 3D window indicating the relative position of those slicing planes as in Fig. 8.1. In a 3D+t dataset *ilastik* also shows a position marker that indicates which slice is shown in the other planes (note that they have differently colored GUI elements, this color corresponds to the color of the position marker). The position marker can be disabled by a checkbox in the lower right. Double clicking inside one of the slicing views centers the other views on that position in the data. The best user experience is achieved when using a mouse with scroll wheel. To zoom in and out of any of the views use `ctrl + mouse wheel scrolling` (or `cmd + mouse wheel scrolling` on a Mac). Panning can be achieved by holding the `mouse wheel pressed + move`, or dragging the mouse while pressing the `shift`-key. To navigate through time, press `shift + mouse wheel scroll` or use the time spin box in the lower right. Note how the position gets updated in the upper left of each view, and the time frame in the lower right.

Each *ilastik workflow* consists of a series of steps, and each step places its important controls inside an *applet*. A list of *applets* occupies the upper half of the left sidebar in the ilastik window. The aforementioned *Input Data* applet is typically one of the first applets in each workflow, where consecutive steps are ordered from top to bottom. Clicking on the title of the next applet will let it unfold and show its contents, and also update the image views and update the list of visible *layers* in the lower half of the sidebar on the left.

Layers contain data for each pixel to be displayed, e.g., raw data, a segmentation mask, etc., which might be familiar from advanced photo editing software. The visibility of layers can be toggled by clicking the `eye`, and its opacity can be changed by dragging the intensity slider horizontally. `Right clicking` a layer in this list shows a context menu that allows to export this layer, or, for the raw data, adjust its normalization.

8.2.3 *Feature Selection*

The next *applet* in the *Pixel Classification* workflow is `Feature Selection`. As mentioned before, *ilastik* will learn from annotations how to distinguish the different pixel classes (e.g., foreground and background) in the data. This decision is based on a set of *features*, where each of these features is a filter applied to a smoothed version of the raw data. The level of smoothing is also called the scale σ of the feature, and the larger this scale, the more information of the neighborhood around a pixel is taken into account. Click on `Select Features` to bring up the dialog shown in Fig. 8.6, and select a subset of features. Once a set is selected, each chosen feature can be visually inspected in the ilastik data viewer by clicking on the respective feature in the lower half of the left sidebar.

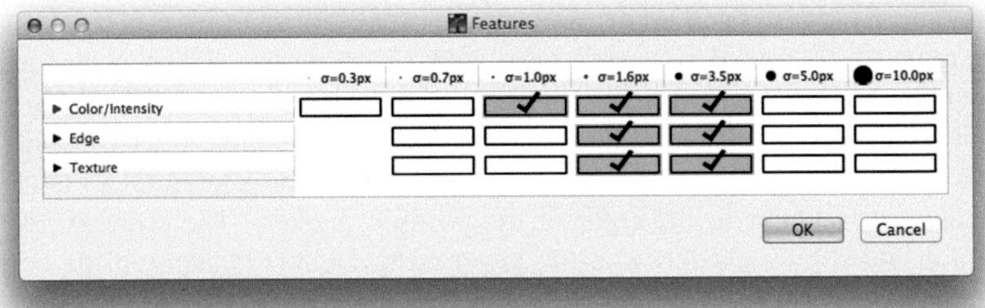

Fig. 8.6 Feature selection dialog. Color, edge, and texture information can be selected on multiple scales σ, indicated by the size of the black disk in the column headers. The size of neighborhood taken into account when computing each feature is also given in pixels, allowing to relate the scale to the image dimensions

TIP: Click on a feature in the list on the lower left to display it for the currently visible region of the data. Good features emphasize the structures of interest.

TIP: The time needed to compute a feature increases with its scale and complexity. In the selection matrix, this means the cheapest feature is in the upper left, while scale and complexity grow towards the lower right. Expensive features are not necessarily more expressive.

TIP: In general one could always select all features and let the algorithm find out which ones it needs to distinguish between the classes, but more features also means more computational effort and thus takes longer. Try to remove the ones that look unnecessary but make sure that the prediction remains good. This could also be achieved by iteratively adding or removing a single feature from the list and assessing the quality of the predictions in the *Live update* mode which will be explained in Sect. 8.2.4.

Background: A classifier learns how to discriminate between different classes from a set of examples. For *Pixel Classification* the classes a pixel can belong to are often foreground and background, but could also represent, e.g., different tissue types. For the discrimination, the classifier considers the selected features of each pixel. While the classifier is trained, it learns how to combine these features such that it predicts the desired class for the training examples as good as possible. The more informative the set of features is, the better the classifier can predict the correct classes.

The classifier we use is a Random Forest (Breiman 2001), an ensemble of decision trees. Each bifurcation of a decision tree represents a binary decision based on a feature, for example: is the smoothed intensity value of this pixel greater than some threshold? Each leaf of the tree is assigned the class of all pixels that it contains during training. For a new pixel p, the decision tree assigns p the class of the leaf it falls into. As a Random Forest consists of an ensemble of trees with randomly selected features at the junctions, it allows averaging this decision over all trees to yield a probability estimate per class.

8.2.4 Training

The selected features specify what information *ilastik* is allowed to take into consideration when classifying a pixel. It now needs to know which classes there are and how they look like.

- Go to the `Training` *applet*, and click on `Add Label` twice to create a foreground and a background class (the terms label and class are used interchangeably here).
- Double click on the classes' names to rename, and double click on the colored square in front of the name to assign a different color. Rename the first (red) label to *"background,"* and the second (green) label to *"foreground."*
- Click on one of the classes in the applet, and make sure that the brush is selected as indicated in Fig. 8.7.
- Example pixels for the classes are now specified by painting them with the brush in the raw data. This process is called *"labeling."* For the Drosophila dataset, place a few labels, for instance, as seen in Fig. 8.7.
- Use the *"Eraser"* tool to remove wrong labels, and adjust the size of brush and eraser to your needs.

After annotating some pixels, *ilastik*'s `Live Update` mode makes it learn from the placed labels, and predict on the unlabeled visible data. The prediction will

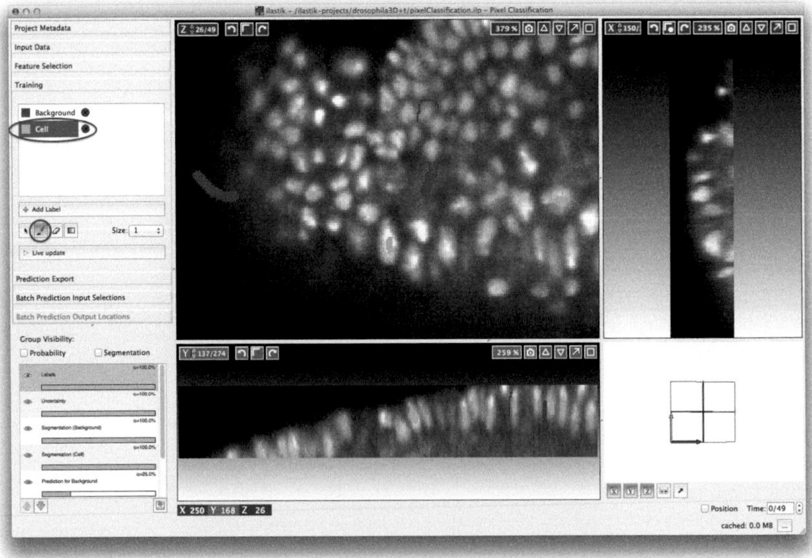

Fig. 8.7 Some labels have been drawn in the raw data by selecting a label in the left sidebar and using the brush tool

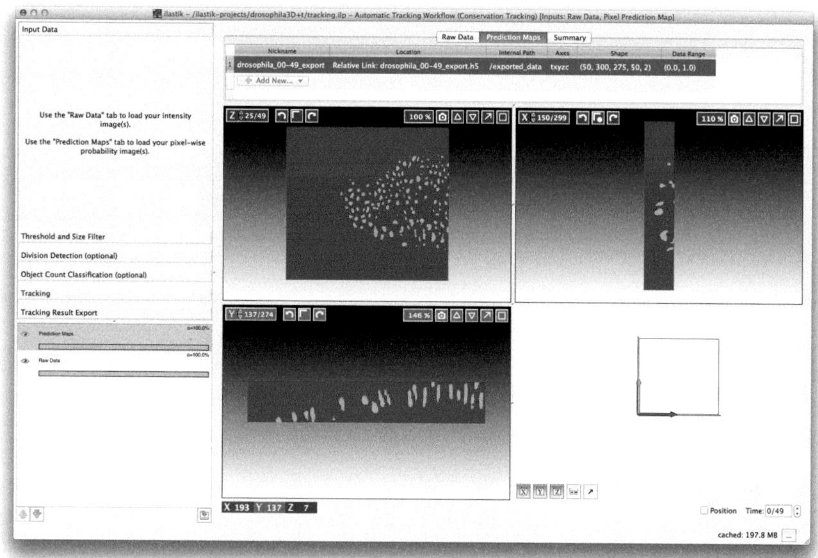

Fig. 8.8 Loading the prediction maps, which were exported from the *Pixel Classification* workflow, into the tracking project

be displayed in the same colors as the labels, but with a lower opacity. The plain predictions (full opacity) should look like in Fig. 8.8.

TIP: The predictions are generated using a machine learning algorithm. Such algorithms try to make decisions based on the examples they have been trained on. In machine learning there is always a tradeoff between a perfect fit to the training data and good predictions on unseen data. The latter, also called generalization, usually works better if the algorithm has not been trained too specifically, which is known as overfitting. Transferring this to the labeling step means: make sure not to give too many examples which all look the same. Experienced users would start with a few general labels such as a thin stroke in the background, and by marking a few pixels of one or two cells. When enabling *live update*, the prediction should already look sensible, but probably the boundary between objects is not properly preserved. To fix this, choose a very thin background brush, paint a line between two closely neighboring targets, and inspect the results. Leaving *live update* enabled, the user can now navigate through the volume and add labels where the prediction is wrong.

TIP: Saving the project frequently helps to prevent data loss. By using `Save Copy As` you can also preserve snapshots of the current state.

TIP: If there are different kinds of objects in the data but only one kind shall be tracked, you can also try to train *'target,' 'other object,'* and *'background'* classes instead of just *'foreground'* and *'background.'*

TIP: Pressing the `i` key toggles between raw input data and the previous view (e.g., predictions).

By default, *live update* displays the probabilities for a pixel belonging to one of our classes as color mixture. To extract a segmentation from the probabilities one has to make a hard decision. The segmentation coming from choosing the class with the highest probability for each pixel individually can be visualized either by clicking on the closed `eye` of the *"Segmentation for foreground"* and *"Segmentation for background"* layers (depends on the name of the labels) or by using the group visibility checkbox for `Segmentation`. As mentioned before, the tracking algorithm can deal with undersegmentations (when there are more than one cell in one segment), whereas oversegmentation (a cell is falsely split into multiple segments) might lead to erroneous tracking, or only one of the segments is used as cell while the others are deemed false detections. To get a good segmentation for tracking, continue this process until most targets are nicely separated.

TIP: Start by labeling the first frame, but not to perfection; and then go to later time steps where the cell density can be much higher. Also make sure that the segmentation still separates cells.

8.2.5 *Export Probabilities*

ilastik offers two options to save the segmentation for further processing. On the one hand, the segmentation can be exported as binary mask, by `right clicking` on the respective segmentation layer and choosing `Export`. On the other hand, one can export the probabilities, which facilitates choosing a different decision strategy for the segmentation later on. Export the probabilities by going to the

Prediction Export applet, and make sure Probabilities is selected in the drop down menu. Click on Choose Export Image Settings to open up a dialog where the desired output format can be chosen, the exported region can be restricted to a region of interest, the axes can be switched, etc. As the probabilities are floating point values, one for each class for each pixel, the export format of choice is a HDF5 file, which is selected by default, and will be called our-dataset-name_Probabilities.h5. Close the dialog box by clicking OK, and then press the Export all button to actually start the export process. Depending on the dataset size and the speed of the workstation, this might take a couple of minutes, because now all selected features have to be computed for all pixels of the dataset. On a recent laptop this takes around 8–10 min for the Drosophila example dataset. You will need these exported probabilities in the next section.

Background: HDF5 is a container format for numerical data in tables and matrices with arbitrary dimensionality and arbitrary data type.

TIP: To segment multiple datasets using the classifier trained above, these datasets could be added here and exported as well.

FAQ: On Windows, in rare cases an error message pops up complaining that the output file is already used by some other process. Make sure to choose a filename which does not exist yet to prevent this. And make sure that input data, output file, and project file reside on the same Windows drive.

8.2.6 Thresholding

Because the first step in the tracking workflow differs depending on whether the segmentation is given as binary mask or probabilities, we will cover this first step now before providing details about the general tracking workflow in the next section. To create a tracking project, save the pixel classification project, close the project, and create a new one. This time, select Automatic Tracking Workflow (Conservation Tracking) [Inputs: Raw Data, Pixel Prediction Map] and save the project as tracking.ilp. In the Input Data applet, load the raw data as in Sect. 8.2.1. This time, a Prediction Maps tab appears between the *Raw Data* and the *Summary* tab.

Select it, choose `Add`, `Add separate image(s)...`, and pick the exported probabilities that were saved as HDF5 file. The *ilastik* window should then look like in Fig. 8.8.

Now go to the `Threshold and Size Filter` applet, where you can specify how to retrieve a segmentation from the probabilities for all classes. The prediction map stores the probability for each class in a different color channel. To identify which of these channels contains the foreground class, toggle the visibility of the different channel layers in the list of layers in the lower left. The second class in the pixel classification walkthrough was foreground, which was green. It should be stored in channel 1 of the prediction map. Enter the correct channel value into the `Input Channel` field in the thresholding applet. Each channel of the prediction maps stores a probability for each single pixel/voxel to be of the class corresponding to the channel, and all probabilities corresponding to one pixel have to sum to one. These probabilities can be smoothed by a Gaussian filter, specified by the `Sigma` values (allowing for anisotropic filtering).

To extract a segmentation, you can threshold the smoothed probabilities at the value θ which is entered in the `One Threshold` option. Every pixel that has a probability for being of the selected class higher than the threshold θ will be used as foreground. The default values for smoothing and thresholding should work well in most of the cases. Click on `Apply` to see the resulting segmented objects, which should look like Fig. 8.9.

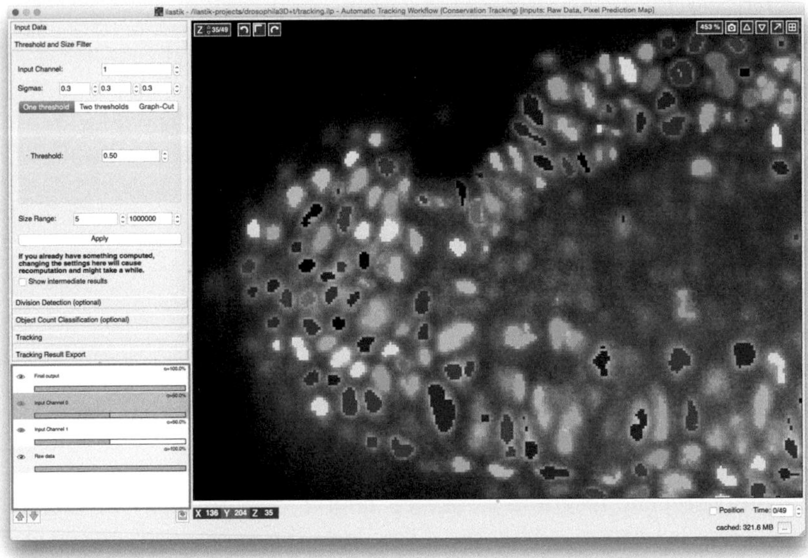

Fig. 8.9 After applying the selected threshold and size filter, the resulting connected components of foreground pixels are assigned random colors

Noise or clutter in the data often produces small detections in the segmentation, which can already be discarded simply by specifying a minimal number of pixels that a detection needs to span to be considered in the following steps. Adjust the `Size Range` minimum and maximum, followed by clicking `Apply`, until all detections that can be discarded judging by size have disappeared.

TIP: Navigate to densely populated regions in the data and ensure that the chosen threshold yields a sensible segmentation there. Increase the threshold if many objects are merged, or decrease the threshold if there are oversegmentations.

TIP: Hover the mouse over the different options to bring up a tool tip giving more details.

8.3 Tracking

Installation: When you download *ilastik* from http://ilastik.org, it will not show the Automatic Tracking Workflow immediately. This is because tracking relies on the external library IBM ILOG CPLEX Optimization Studio,[a] for which you need your own licence. For academic purposes, this licence can be obtained free of charge on IBM's website[b]. This library needs to be installed, and on Linux and Mac OS it needs to be copied into the *ilastik* program folder. Instructions for doing so can be found in the documentation.[c]

[a]http://www-03.ibm.com/software/products/en/ibmilogcpleoptistud

[b]https://www-304.ibm.com/ibm/university/academic/pub/page/membership

[c]http://ilastik.org/documentation/basics/installation.html

The *Automated Tracking* workflow in *ilastik* allows you to automatically track dividing cells or other objects. To access the tracking workflow, a tracking project must be opened in ilastik. This is already the case if you followed the *Pixel Classification* and *thresholding* walkthrough in Sect. 8.2. If you want to use an externally created segmentation of the data, start up *ilastik*, and select `Automatic Tracking Workflow (Conservation Tracking) [Inputs:Raw Data, Binary Image]` from the startup screen. Save the project as `tracking.ilp`. In the `Input Data` applet, load the raw data as described in Sect. 8.2.1. Between the *Raw Data* and the *Summary* tab of the dataset table is a `Segmentation Image` tab, where the binary images should be loaded the same way as the raw data.

This workflow consists of several *applets*,[3] where each applet has the task of preparing some input for the tracking algorithm. The *Object Count Classifier* and *Division Classifier* help the algorithm in finding single cells, mergers, and divisions. Lastly the *Tracking* applet holds a couple of other parameters and is the place to invoke the optimization step.

As shown in Fig. 8.10, which extends on the small illustration in Fig. 8.3, the tracking algorithm internally builds a graph with several (round) nodes for each detection in every timestep t. For each detection in the spatial proximity in the next time frame $t + 1$ it adds a transition node with the appropriate connections. The optimization finds the most probable configuration for the whole graph, taking into account the complete temporal context. It determines how many objects were present in each detection, and via which transitions these objects moved to the succeeding frame's detections. Notice also how the detection in the lower left of Fig. 8.10a, which does not look like a cell at all, is determined to contain no cells in Fig. 8.10c, and thus it is not part of any lineage in Fig. 8.10d. Because the tracking in *ilastik* allows objects to merge into one detection, it is necessary to guide the optimization process by providing information on merged objects and divisions. To this end, an *"Object Count Classifier"* and a *"Division Classifier"* are used. Training these classifiers is achieved by specifying a set of examples for each class. The classifiers can then predict the most probable class for each detection, and present visual feedback that facilitates interactive refinement of the classifier's predictions. Finally, the predicted probabilities for each state of every detection are inserted into the graphical model (as squared factors) and steer the tracking optimization to a sensible result. We will now first explain the two classifiers in detail, and then show how to run the automated tracking. It does not matter in which order you train the *Object Count* and *Division Classifier*. Because the division classifier only applies if your dataset contains dividing cells, we will first look at the object count classification step.

8.3.1 Object Count Classification

As already depicted in Fig. 8.2, the tracking algorithm allows for N-mergers, where N is the number of cells that are merged into the largest detection in the dataset. Go to the `Object Count Classification` applet. This applet tries to classify

[3]For details on applets see Sect. 8.2.2.

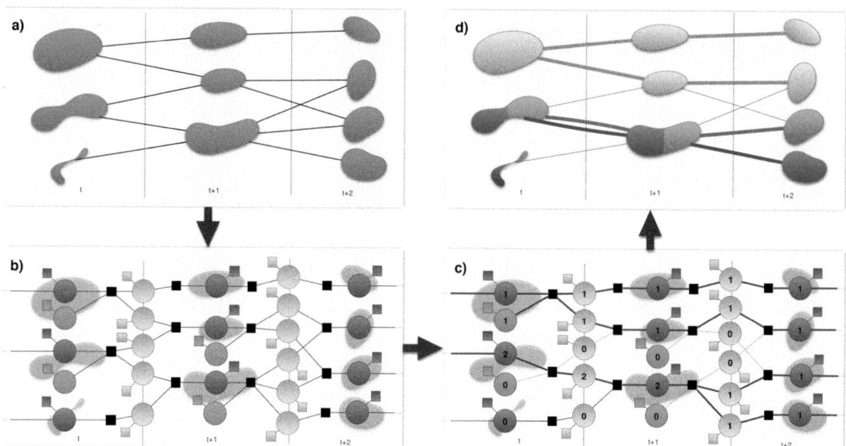

Fig. 8.10 From segmentation to lineage using Conservation Tracking. Counterclockwise: **(a)** First, all detections in the segmentation of frame t get linked to possible successors in $t + 1$ which lie within a user-defined radius. This builds a graph of tracking hypotheses. Note the lower left detection which does not resemble a cell like the others, as well as the big detection in the lower middle. **(b)** A factor graph is constructed by inserting a (*round gray*) transition node for each linking hypothesis in the hypotheses graph. Division random variables are added (*blue circles*) whenever there are at least two possible outgoing transitions from a detection (*purple*). Squares are factors that introduce an energy depending on the state of all connected random variables. The *black squares* model conservation laws, while the colored factors hold the energies (or costs) for different states of the detections, transitions, and divisions. **(c)** The optimization finds the minimum energy configuration of the presented model, which corresponds to the globally most likely solution. Numbers indicate the number of cells in each random variable, while the *red edges* depict the links which are used. **(d)** Extracted lineage trees after tracking and resolving merged detections. Each lineage tree is assigned a unique color. Note that the detection containing the *blue* and *green* cell is split into two segments in frames t and $t + 1$. The oddly shaped detection in the lower left is not included in any lineage because the optimization labeled it as false detection. Best viewed in color

the segmented objects (connected components) to predict the number of cells within each segment. This classifier is again a Random Forest as for *Pixel Classification*, but it bases its decisions on features of a detection, such as size (pixel/voxel count), mean and variance of its intensity, and shape information. The selection of these features can be adjusted by opening the `Select Features` dialog, but the preselected set of features captures information that helps to describe the differences between 1- or 2-mergers for a large variety of cell types. The default set of features (names in the list in *italics*) are the pixel *Count* within an object, the lengths of major axes (*RegionRadii*), as well as the *Mean* and *Variance* of the intensity inside the segment.

TIP: If you can filter your objects based on very special properties like their position (*RegionCenter*), the *Minimum*, *Maximum*, or *Sum* of the intensity inside the segment, you can add these from the list of available features in the `Select Features` dialog. As long as that is not the case, use the ones that are preselected.

- To create classes for all mergers, add new labels using the `Add Label` button, until the list of labels contains the type *"N Objects,"* where N is an estimate on the number of cells combined in the largest merger in the data. Using the provided segmentation of the Drosophila dataset, the largest merger contains two cells, thus the `Add Label` button is pressed three times and $N = 2$.
- For each of these labels, click the label (make sure the `Brush` tool is selected), and then click on detections to mark them as examples for the selected class.
  - If there are small objects left after size filtering, or there are objects which are not cells, these are `False detections`.
  - It is also possible not to give any examples for one of the labels, e.g., do not mark any object as *false detection* if there is no clutter in the segmentation.
  - Find and label around ten detections for each class, until the predictions shown in the `Live update` mode are consistent.
- In the `Live update` mode, the classifier will predict for each detection which label it belongs to; indicate this by coloring the detections in the color that corresponds to the label.
- It is a good idea to check and refine the predictions, so jump to other positions in the data and other time slices (as described in Sect. 8.2.2), and correct the label on detections where the classifier is mistaken.

TIP: Giving more labels in general yields better predictions, so we suggest to specify at least 10 examples per class. Still, as in *Pixel Classification*, too many examples might not generalize well to other data. Do not label thousands of objects.

TIP: Try to balance the number of labels for the different classes. The classifier tries to reduce the number of incorrect predictions with respect to your labels. Consequently, if you label 100 detections as *1 Object*, but only 2 as *2 Objects*, it can achieve only 2 % error by always predicting *1 Object*.

TIP: It is desirable that the segmentation contains many detections of type `1 Object`, and only a few mergers. Thus it will be harder to find examples for the mergers. Try to find a few nevertheless, and use `Live update` to see whether the predictions look sensible.

TIP: Using the provided prediction map of the example dataset, and with a threshold of 0.5, a 2-merger can be found at $115 \times 105 \times 36$ at $t = 30$, as seen in Fig. 8.11.

TIP: You can also `right click` an object and assign any label to it from the context menu.

TIP: Toggle the visibility of the `Objects` layer in the lower left list and use the randomly assigned colors per detection to ensure that multiple cells are merged into one detection (thus have the same color). Alternatively `right clicking` on objects shows the per-frame unique ID for the selected detection.

TIP: Disable `live prediction` to make scrolling through data faster.

8.3.2 Division Classification

If the data contains dividing cells, a division classifier needs to be trained as well. Go to the `Division Classification` applet, which looks very similar, and behaves the same way as the *object count classification* applet. Only the predefined labels *"Dividing"* and *"Not Dividing"* are available for division classification. As mentioned before, a division can only occur where a detection has at least two possible successors in the next time frame. Due to this requirement, the classifier can also take additional features into account. These are, for instance, the size and intensity ratio of the children, as well as the angle that parent and connection lines to children cells span.

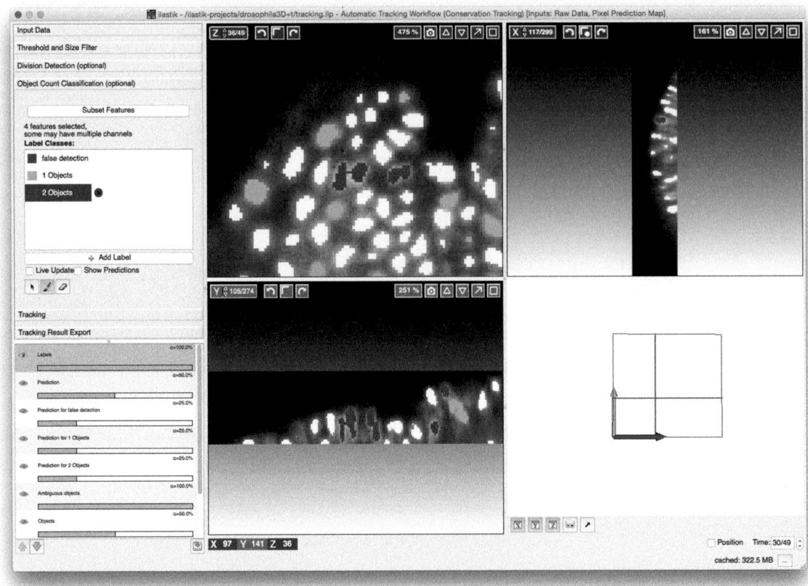

Fig. 8.11 In the *Object count classification* applet, label some detections as *false detection*, *1 object*, or mergers

To place the labels, navigate to a frame where a parent cell is still one detection, but is split up into two children in the next frame, as shown in Fig. 8.12. Use the `Dividing` label for the parent cell. As before, label roughly 10 occurrences for dividing and non-dividing objects, enable *live prediction*, and browse through the data to proof read and correct the predictions.

TIP: 2-mergers which de-merge in the next frame often look similar to divisions to the classifier. By labeling these events properly as *non-dividing* while training both classifiers, the optimization has higher chances to disambiguate them later.

8.3.3 Tracking

To finally run the tracking optimization, go to the `Tracking` applet, as shown in Fig. 8.1. This applet offers to specify a set of parameters to control the tracking algorithm, which is invoked by the `Track!` button. The optimization step finds the most probable configuration of the graphical model consisting of all detections and

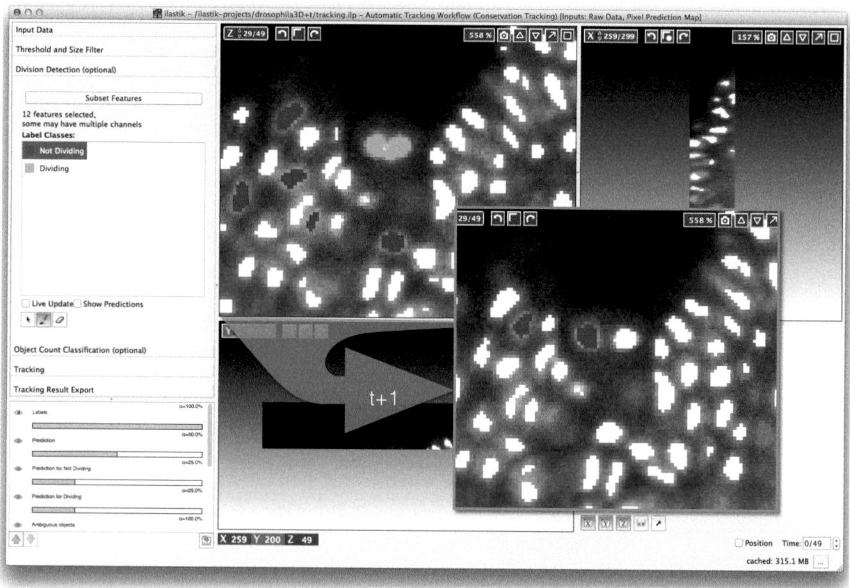

Fig. 8.12 To train the division classifier, a parent cell at time t is labeled as dividing, when in the next time frame $t + 1$ there are two child detections present. It can be useful to label the children as *"Not dividing,"* but it is not mandatory to do so

their possible links as depicted in Fig. 8.10. We derive energies from the predicted classifier probabilities as the negative logarithm. The most probable configuration thus has the minimal energy of the system. The optimization will consider all the energies of detections being mergers or divisions from the classifiers. It also takes into account the distance that an object moved between frames, and penalizes long range transitions. Additionally, objects can appear or disappear, and the optimization incorporates an energy for those events, making them unlikely unless they happen at the border of the dataset. The parameters in the `Tracking` applet, listed in Table 8.1, allow to weigh those different energies against each other. The detection energy (purple square in Fig. 8.10) is always scaled by 10. Thus leaving division and transition weight at the default value 10 will weigh all classifier output equally. The default settings are well balanced and should work for most datasets.

TIP: Hover the mouse over the parameter names to bring up a tool tip that explains the parameter's meaning.

Table 8.1 These parameters can be set in the *Tracking* applet to configure the behavior of the tracking algorithm

Parameter	Meaning
Max objects per merger	Corresponds to the highest number N of objects inside a detection, the biggest N-merger. The complexity of the problem solved by the optimizer grows quadratically with N, which implies much longer runtimes and more RAM usage with higher N. $N > 4$ should be avoided
Division weight	Scales the influence of the energy for division (blue square in Fig. 8.10) of each detection. The higher this parameter, the more costly it is to disagree with the prediction of the *division classifier*
Transition weight	Scales the transition energy (gray square in Fig. 8.10) with respect to other energies. Transitions are penalized exponentially with the distance the cell had to move between frames, and this weight scales the penalty linearly
Appearance cost	Sets the energy of a cell appearance. For instance, this should be very high if no new cells can enter the field of view in a microwell
Disappearance cost	Energy of cell disappearance. Cell deaths are captured as disappearance as well, so if your cells are allowed to die, choose a lower value
Timeout	Restricts the runtime of the optimization algorithm. If this is set too low, the algorithm might not find a feasible solution at all. No value or 0 indicates no timeout
Border width	If a cell appears or disappears within this pixel-distance to the border, the penalty chosen above will linearly diminish towards the outside, such that an appearance at the border of the image is still plausible. Choosing 0 disables this behavior
Divisible objects	Untick this, if there are no divisible objects in the data and the division classifier is not trained. As long as it is enabled, training the division classifier is required
Filter	Restricting the field of view to less time steps or a smaller volume may lead to significant speed-ups of the tracking optimization. Coordinates are in pixels
Size range	Restricts tracking to objects with a pixel/voxel count in the specified range. Especially useful if the segmentation was created externally and contains small debris

Remember that *energy* and *cost* are used interchangeably

TIP: The quality of tracking results is most influenced by the parameters *Division Weight*, *Transition Weight*, *Appearance Cost*, and *Disappearance Cost*. For instance, if you notice that new tracks are started where a cell should have migrated, you could increase the appearance cost and reduce the transition weight to make longer migrations less costly. On the other hand, if cells should appear after the first frame of the sequence, and you have

(continued)

segmented them, but they are not tracked, then you would want to reduce the appearance cost.

Leave all settings as they are, given that *Max Objects per Merger* is set to two (because the highest class was *2 Objects* in *Object Count Classification*). Go to the top of the *Tracking* applet and click on `Track!`. The progress bar at the bottom of *ilastik* will keep bouncing for a while, depending on the size of the dataset and the kind of machine *ilastik* is run on. For the Drosophila example dataset this takes around 1–2 min on a recent laptop.

TIP: Restrict the field of view to only a couple of time frames for tracking, to get quick feedback for assessing whether the chosen parameters yield sensible tracking results, by scrolling down in the `Tracking` applet to the `Filter` section and changing the `From` and `To` values of `Time`. Click on `Track!` above to run the tracking, and inspect the selected time range for tracking errors.

As soon as the progress bar disappears, the optimization is done, and objects within the selected field of view should inherit the color from their predecessor in the previous frame when scrolling through time. Pay attention to the children of a division, which get assigned the same color as the parent cell. Mergers will be assigned as many colors as objects are contained by re-segmenting the detection. A *"Mergers"* layer can be made visible to highlight those original detections that contained more than one object.

FAQ: The tracking parameters that get stored when you save your tracking project are always the ones that you used for the last run of tracking. This is to ensure that the saved parameters are consistent with the stored tracking result.

8.4 Exporting Results

There are two parts of tracking results that get exported, which complement each other. The first part is a set of two spreadsheets, one that contains a list of links of objects between frames, as well as some features that were computed for each object per frame, and a separate table that contains information about each division.

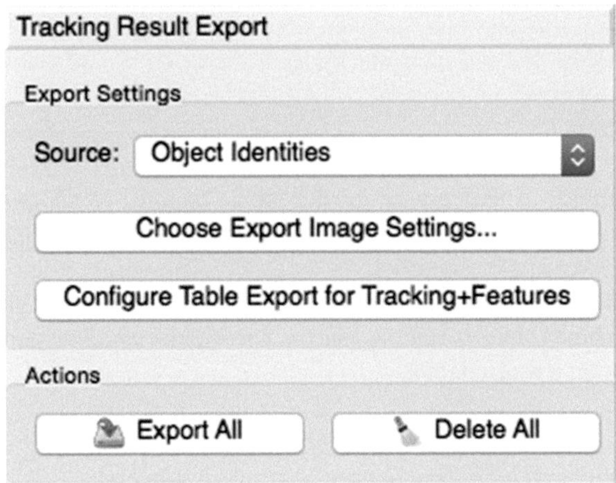

Fig. 8.13 The `Tracking Result Export` applet offers to export the relabeled *Object Identities*, *Tracking Result*, and *Mergers* from the `Export Source` drop down menu. The volume to be exported, as well as the export format can be specified in the dialog (Fig. 8.14) that pops up when selecting `Choose Export Image Settings...`. A table containing all linkings, divisions, and features is exported by default, but can be configured by clicking `Configure Table Export for Tracking+Features`

The other part is the segmentation, where each object has a unique ID per frame, or the ID of the lineage tree it belongs to. We will now look at both parts separately, but cross reference where they are meant to be used together. All export options can be found in the `Tracking Result Export` applet (see Fig. 8.13). The two buttons for configuring the export types will bring up extra dialogs for the settings. To dispatch the actual export, click on "'Export All."'

8.4.1 *Spreadsheet Export*

To configure the spreadsheet export, click the `Configure Table Export for Tracking+Features` button at the bottom of the `Tracking Result Export` applet. In the drop down menu for the export format, select `CSV` for *comma separated value*, and specify a path and file name. CSV files can be read by all spreadsheet software as well as most analysis tools. `Features` allows you to choose which additional features to export that have been computed for each detection. Click on `OK` to save the settings. To actually export the tables and the segmentation, you will have to click on "'Export All."' But before doing so, however, also read about the relabeled segmentation export in Sect. 8.4.2. The export will create one or two files, one called `{dataset_dir}/{nickname}-exported_data_table.csv` and, in case *divisible objects* were enabled during tracking, a file called `{dataset_dir}/{nickname}-exported_data_divisions.csv`.

TIP: The suffixes `_table` and `_divisions` are added automatically, and the placeholders {`dataset_dir`} and {`nickname`} will be filled in depending on the dataset you loaded. The {`nickname`} is the beginning of your loaded dataset filename. If you leave the values at their defaults, the exported files will be created in the same folder as your dataset.

This `yourSelectedName_table.csv` file is one large table that holds linking information between consecutive frames for the whole dataset. A tracked object is assigned a unique *object identifier* (oid) in each frame, which refers to the segmentation (or label image) gray value. However, it also has a *track identifier* (track_id), where a track is a part of a lineage tree between two events like appearance, disappearance, and division. The information on tracks being linked together by divisions can be found in the divisions CSV file. Finally each detection has a *lineage identifier* (lineage_id) which is the same for all objects and tracks that are descendants of the root cell of the lineage tree. Table 8.2 lists the important columns of the tracking CSV file, and Table 8.3 specifies which columns are given per division.

TIP: *ilastik* provides more information on the object features in the *Select Features* dialog of *Object Count Classification.*

Table 8.2 Explanation of the columns found in the tracking export CSV table. All coordinates are given in pixels

Column	Content
object_id	A globally unique running identifier
Timestep	The frame number
labelimage_oid	The object identifier, unique for each object in each frame
track_id1	Unique track identifier (started by appearance or cell division; terminated by disappearance or cell division)
Count	The number of voxels assigned to this object (its size)
Coord<Minimum>_N	The lower left corner of the 3D bounding box around the object ($N = 1, 2, 3$ for dimensions x, y, z)
RegionCenter_N	The center of the segmented object ($N = 1, 2, 3$ for dimensions x, y, z)
Coord<Maximum>_N	The upper right corner of the 3D bounding box around the object ($N = 1, 2, 3$ for dimensions x, y, z)
Additional selected features	For example, mean intensity, variance of intensity, etc.

Table 8.3 Columns in the division export table and their meanings

Column	Content
Timestep	The frame number just before mitosis
parent_oid	The object identifier of the parent cell in the given timestep
track_id	The track identifier of the parent cell in the given timestep
child1_oid	The object identifier of one child cell in the given timestep+1
child1_track1_id	The track identifier of this child cell in the given timestep+1
child2_oid	As above
child2_track1_id	As above

8.4.2 *Relabeled Segmentation Export*

The visual representation of the tracking results which is displayed after running the optimization in the `Tracking` applet can be exported as grayscale segmentation, where black is background, and each object is assigned a value that either corresponds to its lineage or its per-frame object ID. For this, the `Export Source` drop down menu in the `Tracking Result Export` applet offers three choices: *Object Identities* colors each object with its unique ID in that frame (`labelimage_oid` in the table above), *Tracking Result* exports the segmentation where each object is assigned a gray value corresponding to its lineage ID, and *Merger Result* will export only the detections where the optimization decided that it contains more than one object. For visual inspection, *Tracking Result* is usually the most helpful. However, for further analysis in conjunction with the CSV tables from above, the *Object Identities* export provides more valuable information.

Select the desired format as `Export Source` in the applet, then click the `Choose settings` button. The dialog in Fig. 8.14 opens, allowing to select a subregion (note the parentheses: *"[start"* means inclusive, *"stop)"* means exclusive) of the data for exporting. The lower part of the dialog is devoted to the export file format, which should be some kind of *sequence* for tracking results. For the 3D example dataset, this could, e.g., be `hdf5` or a `multipage tiff sequence`. Select your choice and close the dialog box with `OK`. Then click on `Export All` in the applet, or `Export` in the dataset table at the top of the main window to start the actual export process.

8.5 Conclusions and Outlook

This tutorial presented how *ilastik* can be used to generate a segmentation through the intuitive *Pixel Classification* workflow, and how to apply the *Automated Tracking* workflow. By sparsely annotating the data, the user can train several classifiers which allow to segment the data and support the tracking optimization step by disambiguating falsely detected objects and divisions. To facilitate further

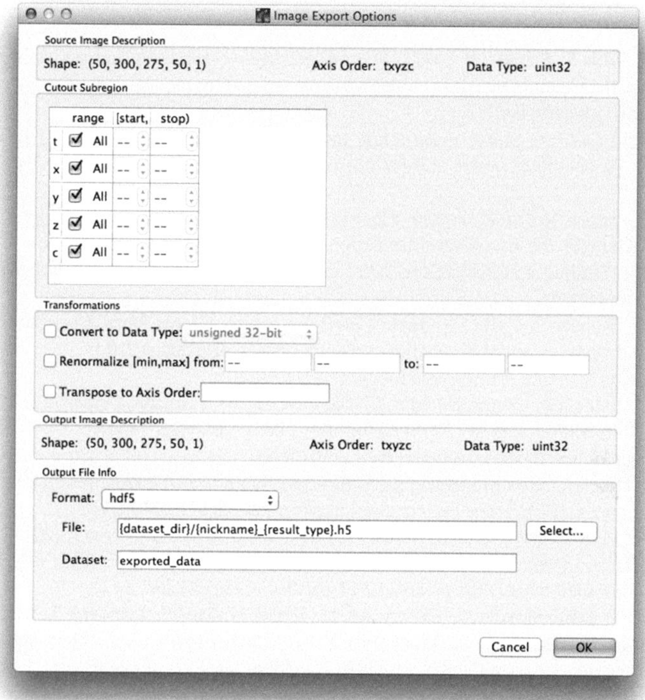

Fig. 8.14 The export settings dialog allows to select a subregion and specify the output file format

analysis of the resulting lineage, we have presented details on *ilastik*'s CSV export file format. ilastik emphasizes intuitive and interactive training to reduce the need of parameter tweaking, while giving state-of-the-art results.

The open source software framework *ilastik* is under active development, and some changes that we are working on are improving the overall performance, learning the remaining tracking parameters from user annotations, as well as a bridge to KNIME (Berthold et al. 2007) to make analysis easier.

Further information about workflows and features, as well as source code and binaries for all major operating systems, can be found on http://ilastik.org/.

Acknowledgements We thank all *ilastik* developers for providing this open source software: Bernhard Kausler, Thorben Kroeger, Christoph Sommer, Christoph Straehle, Markus Doering, Kemal Eren, Burcin Erocal, Luca Fiaschi, Ben Heuer, Philipp Hanslovsky, Kai Karius, Jens Kleesiek, Markus Nullmeier, Oliver Petra, Buote Xu, and Chong Zhang.

Partial financial support by the HGS MathComp Graduate School, the SFB 1129 for integrative analysis of pathogen replication and spread, the RTG 1653 for probabilistic graphical models, and the CellNetworks Excellence Cluster/EcTop is gratefully acknowledged.

References

Andriluka M, Roth S, Schiele B (2008) People-tracking-by-detection and people-detection-by-tracking. In: IEEE conference on computer vision and pattern recognition (CVPR). http://www.mis.tu-darmstadt.de/node/382

Benfold B, Reid I (2011) Stable multi-target tracking in real-time surveillance video. In: IEEE conference on computer vision and pattern recognition (CVPR). IEEE, New York, pp 3457–3464

Berthold MR, Cebron N, Dill F, Gabriel TR, Kötter T, Meinl T, Ohl P, Sieb C, Thiel K, Wiswedel B (2007) KNIME: the Konstanz information miner. In: Data analysis, machine learning and applications. Studies in classification, data analysis, and knowledge organization (GfKL 2007). Springer, Heidelberg

Bise R, Yin Z, Kanade T (2011) Reliable Cell Tracking by Global Data Association. In: IEEE international symposium on biomedical imaging (ISBI), pp 1004–1010

Breiman L (2001) Random forests. Mach Learn 45(1):5–32

Carpenter AE, Jones TR, Lamprecht MR, Clarke C, Kang IH, Friman O, Guertin DA, Chang JH, Lindquist RA, Moffat J, et al (2006) Cellprofiler: image analysis software for identifying and quantifying cell phenotypes. Genome Biol 7(10):R100

Chenouard N, Bloch I, Olivo-Marin JC (2013) Multiple hypothesis tracking for cluttered biological image sequences. IEEE Trans Pattern Anal Mach Intell 35(11):2736–3750

Cordelières FP, Petit V, Kumasaka M, Debeir O, Letort V, Gallagher SJ, Larue L (2013) Automated cell tracking and analysis in phase-contrast videos (iTrack4u): development of java software based on combined mean-shift processes. PLoS One 8(11):e81266

De Chaumont F, Dallongeville S, Chenouard N, Hervé N, Pop S, Provoost T, Meas-Yedid V, Pankajakshan P, Lecomte T, Le Montagner Y, et al (2012) Icy: an open bioimage informatics platform for extended reproducible research. Nat Methods 9(7):690–696

Jaqaman K, Loerke D, Mettlen M, Kuwata H, Grinstein S, Schmid SL, Danuser G (2008) Robust single-particle tracking in live-cell time-lapse sequences. Nat Methods 5(8):695–702

Krzic U, Gunther S, Saunders TE, Streichan SJ, Hufnagel L (2012) Multiview light-sheet microscope for rapid in toto imaging. Nat Methods 9(7):730–733

Kuhn HW (1955) The Hungarian method for the assignment problem. Nav Res Logist Q 2(1–2):83–97

Linkert M, Rueden CT, Allan C, Burel JM, Moore W, Patterson A, Loranger B, Moore J, Neves C, MacDonald D, et al (2010) Metadata matters: access to image data in the real world. J Cell Biol 189(5):777–782

Magnusson KE, Jaldén J (2012) A batch algorithm using iterative application of the Viterbi algorithm to track cells and construct cell lineages. In: International symposium on biomedical imaging (ISBI). IEEE, New York, pp 382–385

Maška M, Ulman V, Svoboda D, Matula P, Matula P, Ederra C, Urbiola A, España T, Venkatesan S, Balak DM, et al (2014) A benchmark for comparison of cell tracking algorithms. Bioinformatics 30(11):1609–1617

Nick Perry JS Jean-Yves Tinevez (2012) Fiji trackmate. http://fiji.sc/TrackMate

Padfield D, Rittscher J, Roysam B (2011) Coupled minimum-cost flow cell tracking for high-throughput quantitative analysis. Med Image Anal 15(4):650–668

Pirsiavash H, Ramanan D, Fowlkes CC (2011) Globally-optimal greedy algorithms for tracking a variable number of objects. In: IEEE conference on computer vision and pattern recognition (CVPR). IEEE, New York, pp 1201–1208

Schiegg M, Hanslovsky P, Kausler BX, Hufnagel L, Hamprecht FA (2013) Conservation tracking. In: IEEE international conference on computer vision (ICCV). IEEE, New York, pp 2928–2935

Schindelin J, Arganda-Carreras I, Frise E, Kaynig V, Longair M, Pietzsch T, Preibisch S, Rueden C, Saalfeld S, Schmid B, et al (2012) Fiji: an open-source platform for biological-image analysis. Nat Methods 9(7):676–682

Sommer C, Straehle C, Kothe U, Hamprecht FA (2011) Ilastik: Interactive learning and segmentation toolkit. In: IEEE international symposium on biomedical imaging: from nano to macro (ISBI). IEEE, New York, pp 230–233

Zhang L, Li Y, Nevatia R (2008) Global data association for multi-object tracking using network flows. In: IEEE conference on computer vision and pattern recognition, (CVPR). IEEE, New York, pp 1–8

Chapter 9
Challenges and Benchmarks in Bioimage Analysis

Michal Kozubek

Abstract Similar to the medical imaging community, the bioimaging community has recently realized the need to benchmark various image analysis methods to compare their performance and assess their suitability for specific applications. Challenges sponsored by prestigious conferences have proven to be an effective means of encouraging benchmarking and new algorithm development for a particular type of image data. Bioimage analysis challenges have recently complemented medical image analysis challenges, especially in the case of the International Symposium on Biomedical Imaging (ISBI). This review summarizes recent progress in this respect and describes the general process of designing a bioimage analysis benchmark or challenge, including the proper selection of datasets and evaluation metrics. It also presents examples of specific target applications and biological research tasks that have benefited from these challenges with respect to the performance of automatic image analysis methods that are crucial for the given task. Finally, available benchmarks and challenges in terms of common features, possible classification and implications drawn from the results are analysed.

9.1 Introduction

Computer analysis of microscopy images has been an indispensable part of biological research for decades. While in the last century semi-automatic methods were sufficient for many applications, this is no longer true in the twenty-first century due to the tremendous increase in the amount of acquired and processed data. Such a large amount of data can only be processed with fully automatic methods tuned for a particular application. Biologists therefore have to rely on the correctness of results obtained by computer analysis. This in turn requires paying proper attention to the quality control of the developed software for automatic image analysis.

M. Kozubek (✉)
Faculty of Informatics, Centre for Biomedical Image Analysis, Masaryk University, Botanická 68a, Brno 60200, Czech Republic
e-mail: kozubek@fi.muni.cz

W.H. De Vos et al. (eds.), *Focus on Bio-Image Informatics*,
Advances in Anatomy, Embryology and Cell Biology 219,
DOI 10.1007/978-3-319-28549-8_9

The common approach to checking the performance of image analysis software is to use benchmark image datasets of known properties and evaluate various analysis methods on the same data. The first widely used benchmark image in image processing history was most likely the famous but controversial Lenna image from 1972 (Rosenberg 1996). The necessity for proper evaluation of image analysis software, comparison of various methods and building on past work became apparent in the 1980s as the number of unlinked papers and theses on the same types of problems started to increase (Price 1986). As soon as the web became available in the 1990s, the need for sharing resources online, including test images, was most felt by the computer vision community (Computer Vision Homepage 1994). Shortly afterwards, the medical imaging community also started to evaluate their image processing methods either on publicly available simulated datasets (Kwan et al. 1999) or on publicly available real datasets accompanied by proper description (Shiraishi et al. 2000). There was even an effort to standardize benchmark-based validation in medical image processing (Jannin et al. 2006). Benchmark images have also been used for quality control during image acquisition, e.g. the so-called Siemens star for optical resolution assessment (ISO/IEC 15775:1999).

Unfortunately, in the bioimaging community, there has been a lack of publicly available reference images for some time, both simulated and real. Some authors validated their methods on their own simulated test data in the 1990s; for example, Lockett et al. (1998) generated a set of artificial spatial objects in the shape of curved spheres, ellipsoids, discs, bananas, satellite discs and dumbbells. However, the first publicly available simulator was released in 2005 by Lehmussola et al. (2005). In 2008, a set of benchmark datasets generated by this tool became part of the first online collection of simulated and real datasets for validation of bioimage analysis methods (BBBC 2008).

Creating the benchmark and making it publicly available online is often not sufficient to attract attention to it. To increase awareness of benchmarks and their usefulness, it is advantageous to make use of the natural human desire to compete and win. This involves competitions (ideally hosted by a well-known conference) that consist in finding the best method for a particular application and testing the methods using a released benchmark. Such competitions (also called contests or challenges) have often been organized by the computer vision community and have been adapted and organized by the medical imaging community since 2007 and by the bioimaging community since 2010 (Fig. 9.1). It is obvious that the number of medical image analysis and bioimage analysis challenges exhibits a growth trend. Most of these (and hopefully also future) challenges can be found at the Grand Challenges web page (van Ginneken and Kerkstra 2015) and the Open Bio Image Alliance web page (2015).

The following sections concentrate on bioimage analysis benchmarks and challenges. First, the design of a benchmark or a challenge is described, including the proper selection of datasets and evaluation metrics (please note that, in this paper,

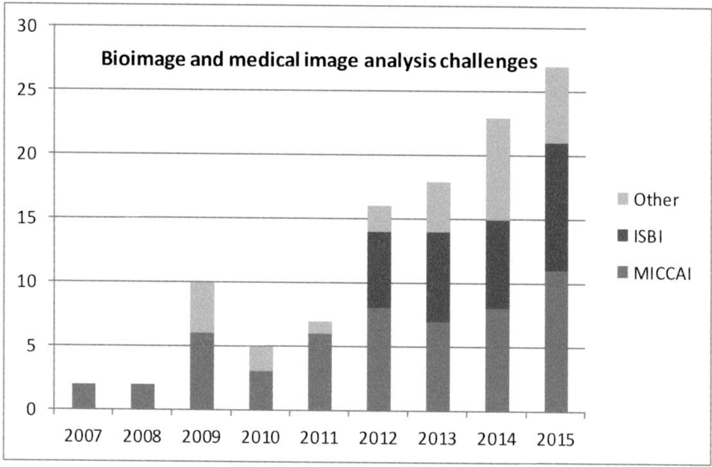

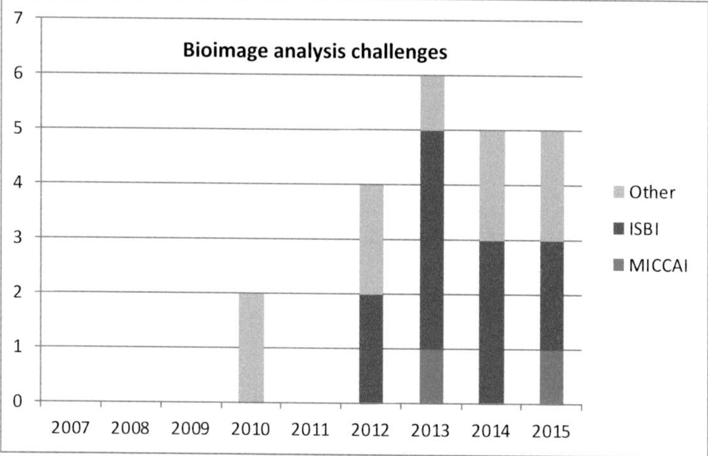

Fig. 9.1 Number of image analysis challenges organized each year so far (*ISBI* International Symposium on Biomedical Imaging, *MICCAI* Medical Image Computing and Computer Assisted Intervention)

the term metric is used in the more general sense as is typical in the software engineering community, not in the strict mathematical sense). Next, the released benchmarks and the challenges organized thus far are reviewed, including their contribution to the relevant biological problem. Finally, the topic is summarized, and the future of the challenges is discussed.

9.2 Benchmark Design

To create a new benchmark, one should think of a target application, select representative datasets (real, synthetic or both) and arrange reliable annotation in the case of real data to create a so-called ground truth *(GT)*, i.e. a correct solution that is used as a reference when testing different algorithms with the benchmark dataset.

9.2.1 Target Application

The mission of the benchmark is to help biological research by cultivating the development of image analysis methods for a particular application. Thus, the selected target application should be in need of such help; in other words, the biological research should still be active in the given area, and the image analysis problem should not be already solved in a satisfactory way. Ideally, the problem should be of interest to a large number of scientists, not just a small group.

Especially suitable target applications are those for which multiple algorithms or software packages have been developed but not compared (or not properly compared using the same benchmark data). The benchmark is then welcomed both by algorithm developers because it reveals the strengths and weaknesses of their methods, which is helpful for fine-tuning, and by users, who obtain some idea of what they can expect from each algorithm and can therefore avoid random or advertisement-based choice.

9.2.2 Dataset Selection

The datasets should be chosen so that they are sufficiently *representative* of the given biological problem. They should cover the variability of the imaged objects (size, shape, texture, density, speed, etc.) as well as the various events, processes and artefacts even if they occur rarely (mitotic or apoptotic events, fluorescence bleaching, dust, uneven illumination, various types of noise, etc.). Moreover, the occurrence of various types of objects or events in the benchmark dataset should be balanced (in a way that corresponds to natural proportions); otherwise, the developed algorithms will adjust to the most frequent object or event types. This is true for both real and synthetic datasets.

If benchmark datasets for the given application already exist, it may still be worth releasing additional datasets if the existing ones are not comprehensive enough (do not fully cover the abovementioned variability or are just not large enough). It is also worthwhile to complement existing datasets with new cell types or new observation modes.

9.2.3 *Real Versus Synthetic Data*

Real data has the advantage of having the best available representation of imaged objects. On the other hand, because real data is blurred and noisy, the correct answer to the biological question is often ambiguous (various experts give different answers). In addition, real data might be available only in limited quantities.

Conversely, *digital synthetic datasets* (also called *simulated data* or *digital phantom images*) can be easily generated at low cost in large quantities at different settings of noise level, cell density, cell speed, etc. The ground truth, i.e. the correct answer to the biological question that is expected from the image analysis algorithm, is known precisely (Fig. 9.2). However, a great deal of attention should be paid to the

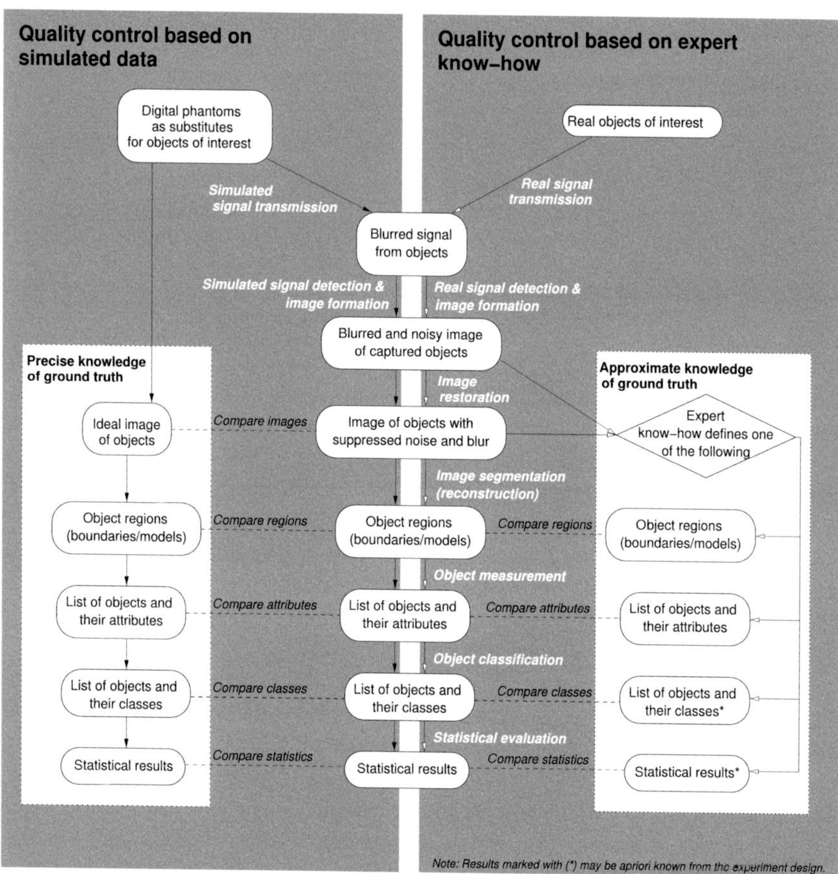

Fig. 9.2 Comparison of benchmarking using real and simulated data: From the image processing point of view, the ground truth derived from expert knowledge is imprecise and laborious to obtain, whereas the phantom-based ground truth is exact and easy to generate in large quantities; on the other hand, from the biological point of view, an expert-based approximate ground truth might be more relevant because it corresponds to real objects, not to synthetic digital phantoms (Reproduced from Svoboda et al. 2009)

similarity between simulated and real data; they should be similar not only visually but also from the point of view of various mathematical characteristics (Svoboda et al. 2009). Because the properties of real and simulated data are complementary, one should ideally use both for benchmarking. In some applications, however, this is not possible, as the ground truth for real data cannot be determined even by an expert (e.g. in the case of deconvolution) and synthetic data remain the only option for quantitative evaluation of algorithm performance.

The third alternative is to use images of so-called physical phantoms, i.e. real nonbiological objects that resemble real biological objects and have similar properties. This option has been exploited in medical imaging, where objects of interest have larger dimensions and are easier to manufacture. In bioimaging, available artificial physical objects are too simple to truly represent reality; therefore, they are employed only for the calibration of microscopy systems; for example, small subresolution beads can be used to measure and correct for chromatic aberration (Kozubek and Matula 2000), and larger spherical beads of size on the order of 10 μm can be used to measure and correct for axial stretching (Ferko et al. 2006) as well as other aberrations and distortions in 3D microscopy (McNally et al. 1997). Such measurements can be useful when simulating microscopy systems while generating images of digital phantoms.

9.2.4 Annotation of Real Data

Because the ground truth for real data is not directly available but is necessary for benchmarking purposes, it must be created manually or semi-automatically by one or more experts in the field. This process is called *annotation*. Due to the subjectivity of expert answers, it is beneficial and more reliable to ask several experts and combine their answers. For binary or multiple choice decisions (especially classification), a reasonable number of experts is three because a *majority voting* scheme can then be applied (the correct answer is the one selected by at least two of them). For answers in the form of a number (especially position determination or boundary delineation), two or three expert answers can be averaged while checking that their variability does not exceed a certain threshold.

Because the human annotation process is tedious and time consuming, it is often simplified as much as possible. Annotators are offered suitable annotation tools (software), for example, automatic boundary tracing from a selected pixel followed by boundary-line dragging with a mouse. Still, some datasets (especially 3D datasets) are beyond the human capacity to annotate precisely; hence, only partial annotation is performed by humans (e.g. only selected z-slices in 3D), and the rest is completed by computer or not annotated. In general, a suitable compromise must be found between annotation quality and quantity.

9.2.5 Releasing Benchmark Data

Benchmark datasets should be made available online with a reliable (ideally nonstop) web server in a compressed and zipped format so that they may be easily downloaded. They should be accompanied by the ground truth data and a description of the acquisition settings and how the real data was annotated. In addition, there should be a clear statement about the data source and its copyright policy. Finally, a suggestion for how to cite the benchmark dataset should be added. To attract more users, it is advantageous to publish the benchmark datasets as part of one of the large collections (see Sect. 9.4) or to organize a challenge based on them.

After the benchmark data is released, attention should be paid to the feedback from those who download and use it. Download statistics should at least be gathered, and a list of citing papers should be created. Publishing and updating the list of citing papers online along with the benchmark data helps attract further interest.

9.3 Challenge Design

Using benchmark datasets for competition purposes is the best way to foster algorithm development for the corresponding biological problem. While preparing a challenge, one should determine how to divide datasets into training and competition datasets, define metrics to compare algorithm results with the ground truth and merge multiple metrics into a single ranking as well as verify that the submitted results are consistent and actually produced by the supplied software.

9.3.1 Training Versus Competition Datasets

Each challenge has a training phase and a competition phase. During the *training phase*, the participants fine-tune their methods to work well on the particular data type using the supplied ground truth. During the *competition phase*, they apply developed software to the competition data (also called the test data), whose ground truth is kept secret. Thus, it is necessary to split the benchmark datasets into training and competition (test) datasets and keep the ground truth for the latter secret. In statistics, such an approach is called the *holdout method* (test data are held out while training), and it is suggested to use about two-thirds of the data for training and one-third for testing. In practice, however, data is not divided in this proportion because organizers prefer to divide it approximately into halves.

The split should be done in a balanced way; i.e. both training and competition datasets should be representative and have similar properties so that the performance of the algorithms is similar. If, for example, the success rate of the participating methods is substantially worse for competition datasets compared to training datasets, the split was not performed properly, and the competition datasets were much harder (or had different properties) than the training datasets, which is very discouraging for the participants. Even if the number of datasets is limited (e.g. the number of patients is limited), it is necessary to provide various datasets during the training phase. If possible, it is preferable to split each of the datasets into two parts (one for training and one for competition) than to split whole datasets into two parts.

From a statistical point of view, it would have been much better to use *cross-validation* instead of the single holdout approach. This involves repeating the holdout process several times, with different data subsets held out for testing. This yields a much better indication of the tested algorithm performance and generalization properties, especially if the amount of data is limited. Unfortunately, this is feasible only for machine learning approaches (typically used for classification tasks) and not for human algorithm design, as it would be too time consuming and people (unlike computers) would not forget acquired knowledge when the training subset is changed.

However, cross-validation might be used in so-called parametric studies when the algorithm and the number of parameters are fixed and only the values of parameters are automatically tuned to best fit the training data. Cross-validation could help determine the influence of parameter settings on the generalization properties of the algorithm, especially which parameters or parameter ranges cause the so-called overfitting, i.e. strong adherence to training data including noise at the expense of being universal and ready for other independent data of the same type.

9.3.2 Evaluation Metrics

Defining proper metrics for the evaluation of algorithm performance is probably the most crucial task when organizing a challenge. Inappropriate metrics may influence algorithm ranking, and the best solution may remain unrecognized.

For some image analysis tasks, there are standard metrics available that are frequently used in various challenges. For example, for pixel or object classification tasks, one can easily define the *confusion matrix* and *accuracy*, in binary cases at least (even if true negatives are not defined), true positives, false positives, false negatives, *precision*, *recall* and finally the *F-score*. For segmentation results (binary mask comparisons), standard metrics are the *Jaccard similarity index* and the *Dice coefficient*. For shape similarity, the standard metric is the *Hausdorff distance*. For position (localization) error, the standard metric is *root-mean-square distance (RMSD)*. For comparing distributions, the standard metric used in challenges is *correlation*, although it might be advantageous to also employ *distribution divergence* (such as Kullback-Leibler, Jensen-Shannon or Bregman) measures in some cases. Finally, algorithm speed and memory consumption can easily be measured and compared if the same hardware is used for each method (typically a server furnished by the organizers with hardware parameters that are common for a given application).

Unfortunately, some types of results do not have a standard metric. For example, it is not easy to compare reconstructed fibre networks, tracking or deconvolution results. For these purposes, challenge organizers usually invent their own metric or a set of metrics.

Especially helpful in these complicated applications are the *graph theory* and related algorithms because fibre network trees or tracking trees can be handled as graphs in the mathematical sense. For example, mapping of a reference set of points to the set of points generated by the tested algorithm (such as in particle tracking or localization microscopy applications) is easily performed using the Kuhn-Munkres algorithm (the so-called Hungarian method known since 1955) applied to bipartite graphs. After such mapping, the abovementioned standard metrics can be computed.

Another useful approach is *Fourier analysis* and computation of various characteristics from the Fourier domain rather than the space-time domain. This means that frequency spectra of both reference images and images generated by the tested algorithm are computed and compared, most frequently using some type of correlation. For example, Fourier shell correlation and Fourier ring correlation metrics have been introduced in deconvolution and localization microscopy applications, respectively (see Sect. 9.5).

9.3.3 Merging Multiple Metrics

Finally, if multiple metrics are computed, their values should be published separately (to assess strong and weak sides of each algorithm) followed by combining them to yield a single value (final score) for each method. Therefore, it is beneficial to normalize each metric and then perform their sum or weighted sum. Alternatively, some metrics can be treated as secondary (additional) for the case of equal values of main metrics. For example, time consumption is typically not the main priority in bioimage analysis, so this metric can either have a low weight or can be applied to two equally performing methods.

Special attention should be paid to combining mutually dependent metrics, typically inversely dependent metrics, that is, two metrics such that if one of them is improved, the other one gets worse. The easiest example is the known precision and recall dependence. Similarly, localization accuracy (typically measured using RMSE) and detection rate (typically measured using the F-score or Jaccard similarity index) are mutually dependent. This dependence is most visible if a scatter plot is generated with these two measures on axes over a sufficient number of algorithms or algorithm settings (see, e.g. Sage et al. 2015). Using the scatter plot, one can compare the behaviour (in terms of both dependent metrics) of a given method with the best-performing methods or theoretical optimum using simple distance measurements.

In practice, merging mutually dependent metrics is often done by simple averaging or weighted averaging if one of them is more important for the application. Sometimes, they are replaced with other metrics that combine the effects of both of them (e.g. the F-score instead of the recall and precision pair). Alternatively, one can fix the value of one of them and measure the other. For example, memory consumption typically grows with decreasing time consumption. It makes sense to fix memory consumption to the common operating memory value of a standard PC and measure time with this restriction. Conversely, for real-time applications (that must be run online during image acquisition), one may want to fix time consumption and measure memory consumption.

In any case, all metrics and their merging process should be precisely defined when announcing the challenge so that participants know how their methods will be evaluated. In addition, it is advantageous to provide source code, plug-in or stand-alone software that performs the computation of the metrics and the final score for a given algorithm output. This ensures that all participants use the same evaluation method as the organizers and saves time.

9.3.4 Format and Verification of Submissions

Further, it is necessary to define the *unified format* for submitting the algorithm results on competition data. Each submission is then checked (automatically) to determine whether it adheres to the prescribed format and whether the submitted data is *consistent*, i.e. whether the results make sense at all (e.g. if they contain basic errors like missing values or out-of-range values). Typically, results are submitted as a simple spreadsheet table with comma-separated values (CSV format) or text files using precisely defined keywords and formatting. Alternatively, a more general extensible markup language (XML) is used.

To prevent cheating in the form of manually edited results, it is advantageous to require submission of analysis software (in addition to results) to verify that the submitted results can be produced by the submitted software. For this purpose, accepted platforms (operating systems plus their minimal versions for stand-alone applications or accepted programming languages) as well as the maximal memory requirements of the submitted software should be specified. Among the most popular accepted programming languages are Java (especially ImageJ/Fiji plugins), MATLAB and Python. These programming languages have the advantage of easy portability due to the independence of the operating system but may be less efficient than a stand-alone application optimized for a specific operating system.

9.3.5 Creating Rankings

After the submitted results are verified, organizers can calculate the quality of the results based on the metric computation tools (the same as those that have been distributed to participants). This yields one number (score) per method per dataset. One can then combine (e.g. average) scores of the same method on different competition datasets and create one ranking for the whole challenge. Often, however, the challenge contains several types of datasets (e.g. several types of cells) of different properties, and there is no algorithm that performs best on all types of datasets. Therefore, it is better to calculate rankings separately for each type of data.

If a single global ranking is required, *individual rankings* can also be combined into a single *global ranking*, e.g. by counting the average rank for each algorithm or counting the occurrences of a particular algorithm among the top three best-performing methods across various data types. The latter approach has the advantage that it naturally copes with the problem of missing scores, i.e. missing results of a particular algorithm for a particular dataset (caused by the fact that users are allowed to supply results only for some datasets). If the former approach is used (score averaging), it is unfair to compute the average of only the scores for supplied datasets because methods that skip the hard datasets are then favoured. There should be some penalty for unsolved datasets without hampering good results for solved datasets.

9.3.6 Challenge Lifecycle

Similar to benchmarks, challenges need a web interface, even one that is sophisticated to address participant *registrations* and *submissions*. The web page should contain (in addition to a description of the challenge, datasets and metrics) a time schedule with deadlines, especially the dates of registration, the release of test datasets with the ground truth, the release of competition datasets, opening and closing submissions, the evaluation period and the announcement of the results. There are special web platforms that simplify this agenda and address challenges in a unified manner (van Ginneken and Kerkstra 2015).

Typically, challenges are associated with a known conference, so the results are announced at a special *conference workshop* where selected participants (authors of the best methods) are given a chance to present their methods. After the conference, the results are usually *published* in a peer reviewed journal. Sometimes, the challenge is *kept open* for further submissions, and sometimes it is *repeated* with some modifications, e.g. using different or enriched datasets. Both open and repeated modes are welcome because they make it possible to build on previous know-how and continuously improve the methods under standardized conditions. The development is accelerated especially if participants are willing to share their codes, not just method descriptions. In the recent call for ISBI 2016 challenges, there is even an explicit support of repeated challenges that encourage open-source solutions.

9.4 Available Benchmarks

The following list of bioimage analysis benchmarks is primarily based on the Open Bio Image Alliance (2015) and is ordered alphabetically. It is possible that there are further benchmarks on the web, so this list should be considered illustrative rather than exhaustive. All benchmark datasets below are freely available for non-commercial research purposes, subject to agreements with licence conditions.

9.4.1 Broad Bioimage Benchmark Collection (BBBC)

This is probably the oldest collection of bioimage benchmark datasets accompanied by a ground truth for each dataset. It was launched in 2008 (BBBC 2008) and appeared in 2012 in *Nature Methods* (Ljosa et al. 2012). It contains mainly real datasets in fluorescence mode but also brightfield as well as DIC. In addition, several simulated datasets are added by means of a Tampere SIMCEP simulator (Lehmussola et al. 2005) and a Brno CytoPacq simulator (Sect. 9.4.4). Four different types of ground truth are available: counts, foreground/background classification, outlines of objects and biological labels from control samples with a known expected biological result. Real samples include, for instance, human HT29 colon cancer cells, human U2OS cells, *Drosophila* Kc167 cells, *C. elegans* live/dead assays or mouse embryos.

9.4.2 Cell Centred Database (CCDB)

This database contains high-resolution 2D, 3D and 4D real datasets from light and electron microscopy, including correlated imaging. Techniques range from wide-field mosaics taken with multiphoton microscopy to 3D reconstructions of cellular ultrastructure using electron tomography. This database is even older than BBBC; it was launched in 2002 (CCDB 2002) and described in detail in *Neuroinformatics* in 2003 (Martone et al. 2003). However, it does not focus on providing a precise ground truth for the datasets, only unverified annotations (either automated or manual), so its use for benchmarking is rather limited. In addition to original real images, it can store, for example, 3D reconstruction, segmented objects (tree structures, surface objects, volume objects and contour objects) or measurements of quantities such as surface area, volume, length and diameter. Subsequently, sophisticated data management and a search engine were also added as well as knowledge engineering tools, e.g. ontologies for annotation and query of microscopic data (Martone et al. 2008).

9.4.3 CellOrganizer and Murphy Lab Data

The group led by Robert Murphy started publishing data from their papers on their web page as early as 1999 (Murphy Lab Data 1999). The primary focus of this database, however, has been to share data and propagate the reproducible research philosophy, not to provide the ground truth for the validation of algorithms. Nevertheless, subsequent datasets that were added to this database had the ground truth available, e.g. hand-segmented 2D nuclear images from an ISBI 2009 paper (Coelho et al. 2009). Since 2012, the main tool for synthetic data generation, CellOrganizer, has been available for download (CellOrganizer 2012). This project uses machine learning to extract models of cell organization from real training image data and generate realistic synthetic cell images based on these models (Buck et al. 2012).

9.4.4 CytoPacq

To provide the bioimage community with an online tool for generating simulated data along with the ground truth, a software package called CytoPacq was developed, launched online in 2008 (CytoPacq 2008) and described in *Cytometry* in 2009 (Svoboda et al. 2009). It contains three parts: a digital cell phantom generator, a simulated light microscopy module and a simulated electronic image detection module. Users can select from several object types (including tissue), select or download their own PSF, influence levels of various noise types, etc. Several benchmark datasets are available for download. Further developments make it possible to generate time-lapse sequences in 3D, which were used in the Cell Tracking Challenge (see Sect. 9.5.10).

9.4.5 Deconvolution in Microscopy

These are 3D datasets with corresponding 3D point-spread function (PSF) models to facilitate and unify the comparison and validation of deconvolution software packages. The test datasets include synthetic hollow bars, a three-channel *C. elegans* embryo and images of fluorescent beads of known size. The benchmark was described in *GIT Imaging & Microscopy* in 2010 (Griffa et al. 2010), in which its application for benchmarking five different deconvolution software packages (two commercial and three open-source packages) was presented including the advantages and disadvantages of individual packages. The benchmark can also be found on the EPFL web pages (Deconvolution Benchmark Datasets, 2010).

9.4.6 JCB DataViewer

In addition to universal databases like CCDB, which takes submissions from all over the world, there are also journal-specific databases containing only images related to papers published in a particular journal. This can be in the form of traditional supplementary data to individual papers or a more convenient web viewer. An example of such a viewer is the Journal of Cell Biology DataViewer announced in 2008 (JCB DataViewer 2008; Hill 2008). It allows viewing, analysis and sharing of multidimensional image data associated with articles published in this journal. Precise ground truths, however, are not available.

9.4.7 SimuCell

Similar to CytoPacq, SimuCell is another open-source framework for specifying and rendering realistic microscopy images; it was announced in 2012 in *Nature Methods* (Rajaram et al. 2012). According to the documentation (SimuCell 2012), it can generate diverse cell phenotypes, heterogeneous populations, and microenvironmental effects; allow users to specify interdependencies among biomarker-, cell-, and population-level phenotypes (e.g. a marker's cellular distribution can be affected by the cell's microenvironment or the localization pattern of another marker); and provide users with a scripting syntax built on top of MATLAB or through the graphical user interface; in addition, intermediate results can define ground truths (e.g. cell boundaries can be used to validate segmentation algorithms). Two script examples are available for download as well. The disadvantage of this framework is that it works only in 2D.

9.4.8 The Cell: An Image Library (CIL)

Similar to CCDB, this library, launched in 2010 (CIL 2010), is a comprehensive public resource database of images, videos and animations of cells, capturing a wide diversity of organisms, cell types and cellular processes. The primary purpose of this database is to advance research on cellular activity, with the ultimate goal of improving human health, and secondary to serve as a tool for education. As in the case of CCDB, it is limited with respect to benchmarking due to the lack of a precise ground truth. Because of the similarity in scope of the CIL and CCDB databases, the authors decided to merge them into a single CIL-CCDB database that is run on the CIL web pages (Orloff et al. 2012).

9.4.9 UCSB Biosegmentation Benchmark

This is one of the oldest bioimage datasets accompanied by ground truth that is accessible on UCSB web pages (UCSB Biosegmentation Benchmark 2008). It was introduced at the International Conference on Image Processing (ICIP) in 2008 and published in detail 1 year later in *BMC Bioinformatics* (Drelie Gelasca et al. 2009). The benchmark contains three levels of scales: subcellular (microtubules), cellular (breast cancer cells, photoreceptors, *Arabidopsis* and COS1 cells) and tissue (retinal layers). Both 2D and 3D data are included. The primary focus is segmentation at different scales. In addition to the datasets, evaluation metrics are also suggested.

9.5 Organized Challenges

Table 9.1 summarizes the challenges in bioimage analysis that have been organized so far. A more detailed description of each challenge with references to corresponding papers and links to web pages follows.

9.5.1 Digital Reconstruction of Axonal and Dendritic Morphology (DIADEM)

The DIADEM challenge (DIADEM Challenge 2010) was the first bioimage analysis challenge launched in April 2009 and concluded in September 2010 at a scientific conference organized at the HHMI Janelia Farm Research Campus. The focus of the challenge was automated neuronal reconstruction. Competitors had 1 year to develop their algorithms and submit reconstructions of the benchmark datasets. Then, a team of judges chose five finalists out of over 120 registrants to compete in the final round at the conference. Six types of 3D real datasets were provided as benchmark data: Cerebellar Climbing Fibres, Hippocampal CA3 Interneuron, Neocortical Layer 1 Axons, Neuromuscular Projection Fibres, Olfactory Projection Fibres and Visual Cortical Layer 6 Neurons. A special metric called the DIADEM metric was developed to assess the reconstruction performance (Gillette et al. 2011). It mainly assesses whether the nodes of the reconstructed trees are in the right position (within a Euclidian distance threshold from the ground truth), how accurate the topological interconnectivity is and how distant the path is from the ground truth path. All details about the challenge (datasets, metrics, best methods, etc.) were published in a special issue of *Neuroinformatics* (volume 9, issue 2–3, September 2011). The challenge helped to advance the image analysis tools that are crucial in neuro-research.

Table 9.1 Challenges in bioimage analysis ordered chronologically based on the year when the final conference workshop took place (launch of the challenge was typically 1 year earlier)

Challenge name (abbreviation)	Main organizers (institution)	Conference (datasets)	Evaluation criteria (metrics)
Digital Reconstruction of Axonal and Dendritic Morphology (DIADEM)	Ascoli (G. Mason Univ), Svoboda (HHMI), Liu (NIH)	Janelia HHMI 2010 (6 types of real 3D datasets)	Reconstruction (DIADEM metric)
Pattern Recognition in Histopathological Images (PRinHIMA)	Gurcan (Ohio State Univ), Madabhushi (Rutgers), Rajpoot (Univ Warwick)	ICPR 2010 (2 types of 2D real datasets: breast cancer for lymphocyte counting and follicular lymphoma for centroblast counting)	Region-based (Dice, etc.), boundary-based (Hausdorff etc.), detection rate (TP/FP)
Segmentation of Neurites in EM Images (SNEMI)	Arganda-Carreras, Seung (MIT)	ISBI 2012 (1 type of real 3D dataset) ISBI 2013 (1 type of real 3D dataset)	Segmentation (1 – max. F-score of Rand index)
Particle Tracking Challenge (PTC)	Meijering (Erasmus MC), Olivo-Marin (Inst. Pasteur)	ISBI 2012 (4 types of simulated time-lapse datasets – 3 types 2D + t, 1 type 3D + t – each at 3 density and 4 SNR levels)	Tracking accuracy (errors after pairing + Jaccard similarity)
Pattern Recognition in Indirect Immunofluorescence (PRinIIF): HEp-2 Cells Classification	Foggia, Percannella, Vento (Univ Salerno), Soda (Univ Rome), Hobson (SNP Brisbane), Wiliem, Lovell (Univ Queensland)	ICPR 2012 (1 type of 2D real dataset: HEp-2 cells, 6 classes) ICIP 2013 (larger dataset provided with 2 new classes) ICPR 2014 (added whole HEp-2 specimen images, 7 classes)	Classification accuracy (overall accuracy and mean class accuracy)
Mitosis Detection in Breast Cancer (MITOS)	Roux (CNRS), Capron (Pitié-Salpêtrière Hosp.)	ICPR 2012 (1 type of real dataset, 5 patients, 3 scan modes – two 40× 2D, multispectral 3D) ICPR 2014 (1 type of 2D real dataset, 4 scan modes 40×/20×)	Mitosis detection (F-score), nuclear atypia score (count of correct – count of opposite scores)

Assessment of Mitosis Detection Algorithms (AMIDA) in Breast Cancer	Veta, Viergever, Pluim, Stathonikos, van Diest (Univ MC Utrecht)	MICCAI 2013 (1 type of real dataset, 23 patients, 1 scanner 40× 2D)	Mitosis detection (F-score), mitotic density (correlation)
Localization Microscopy Challenge (LMC)	Sage, Kirshner, Manley (EPFL), Pengo (CRG), Stuurman (UCSF), Min (KAIST)	ISBI 2013 (1 type of real 2D + t dataset, 4 examples of tubulin images; 4 types of synthetic 2D + t datasets, 3 artificial: eye, snow, seashell + simulated tubulin at 2 SNR + 2 density levels)	Detection rate (F-score, Jaccard), localization accuracy (RMSD)
3D Deconvolution Microscopy Challenge (DMC)	Lefkimmiatis, Vonesch (EPFL)	ISBI 2013 (4 types of 3D synthetic datasets, points, curves, surfaces, dense volumes) ISBI 2014 (the same 4 types but new datasets)	Deconvolution quality (PSNR, NMISE, SSIM, Fourier shell correlation, rel. energy regain)
Cell Tracking Challenge (CTC)	Ortiz de Solórzano (Univ Navarra), Kozubek (Masaryk Univ Brno), Meijering (Erasmus MC), Muñoz-Barrutia (Carlos III Univ Madrid)	ISBI 2013 (6 types of real datasets, 3 of them 2D + t, 3 of them 3D + t; 2 types of simulated datasets, 1 type 2D + t, 1 type 3D + t, each at 6 density/SNR/speed settings) ISBI 2014 (3 new 2D + t and 1 new 3D + t real dataset types added) ISBI 2015 (1 new 3D + t real dataset type added)	Segmentation (Jaccard), tracking (graph comparison), time consumption
Overlapping Cervical Cytology Image Segmentation Challenge (OCCISC)	Bradley (Univ Queensland), Carneiro (Univ Adelaide), Lu (Univ Hong Kong)	ISBI 2014 (1 type of real 2D dataset, 1 type of synthetic 2D dataset) ISBI 2015 (1 type of real 3D dataset)	Cytoplasm segmentation (Dice coefficient), nuclei detection (precision + recall)

(continued)

Table 9.1 (continued)

Challenge name (abbreviation)	Main organizers (institution)	Conference (datasets)	Evaluation criteria (metrics)
Gland Segmentation Challenge (GLAS) for Histology Imaging	Sirinukunwattana (Univ Warwick), Pluim (Eindhoven Univ Tech), Snead (Univ Hosp Coventry and Warwickshire), Rajpoot (Univ Warwick)	MICCAI 2015 (2 types of real datasets – colorectal and breast adenocarcinomas, 1 scanner 20× 2D)	Detection (F-score), segmentation (Dice coefficient), shape similarity (Hausdorff distance)
Image Stitching Challenge (ISC)	Chalfoun (NIST)	BioImage Informatics 2015, NIST (3 types of real 2D datasets – stem cell colonies at 3 levels of difficulty: phase contrast, Cy5 small grid and Cy5 large grid)	Colony centroid location, colony area
Nucleus Counting Challenge (NCC)	Maric (NIST)	BioImage Informatics 2015, NIST (1 type of real 2D dataset – 2-channel fluorescence of brain tissue, 3 datasets)	Nucleus counts, nucleus segmentation masks

HHMI Howard Hughes Medical Institute, *ICIP* International Conference on Image Processing, *ICPR* International Conference on Pattern Recognition, *ISBI* International Symposium on Biomedical Imaging, *MICCAI* Medical Image Computing and Computer Assisted Intervention, *NIST* National Institute of Standards and Technology

9.5.2 *Pattern Recognition in Histopathological Images (PRinHIMA)*

Counting specific cell types in histopathological images is essential in many diagnostic applications. In this challenge, two applications were selected: counting as well as segmenting lymphocytes in breast cancer images and counting centroblasts in follicular lymphoma images. For the first problem, the following measures were used: region-based (Dice coefficient, overlap, sensitivity, specificity and positive predictive value) and boundary-based (Hausdorff distance and mean absolute distance). For the second problem, a true/false detection rate was observed, where true positive was defined so that the error of centroid location was less than 30 pixels (7.5 μm). The challenge including results for five participating groups was published in the ICPR 2010 conference proceedings (Gurcan et al. 2010).

9.5.3 *Segmentation of Neurites in EM Images (SNEMI)*

Two years after the successful DIADEM challenge, another neuroimaging challenge was organized. On this occasion, electron microscopy (EM) datasets instead of light microscopy datasets were used. Participants had to develop image segmentation algorithms to cope with neuronal structures in the real benchmark data, which were 3D stacks of EM images. Although the data were 3D, the evaluation was performed in 2D in 2012 using 2D topology-based segmentation metrics: minimum object splits and merger warping error, foreground-restricted Rand error defined as one minus the maximal F-score of the foreground-restricted Rand index and pixel error defined as one minus the maximal F-score of pixel similarity. One year later, this challenge was repeated with a fully 3D focus, i.e. 3D segmentation algorithms were expected and 3D metrics were used. The scope of the SNEMI challenges nicely complemented that of the DIADEM challenge for the neuroimaging community. Unfortunately, the results have not been published in a paper; however, they are accessible on the challenge web pages (2DSNEMI 2012, 3DSNEMI 2013).

9.5.4 *Particle Tracking Challenge (PTC)*

Reliable particle-tracking algorithms are indispensable in biological research for the quantitative analysis of intracellular dynamic processes. To compare and improve the pool of available methods for this task, four types of synthetic benchmark datasets have been prepared: vesicles with Brownian (random-walk) motion (2D + time), microtubules (larger elongated particles) with directed motion (2D + time), receptor-switching between Brownian and randomly oriented directed motion (2D + time) and virus-switching between Brownian-directed motion with restricted orientation (3D + time). Each of these four scenarios was simulated at 3 density levels and 4 SNR levels, yielding 48 datasets in total. Based on these benchmark datasets, a challenge was organized with metrics to compare estimated and ground truth tracks. Optimal track pairing was found using Munkres' algorithm, and then five measures were computed to assess tracking accuracy: overall degree of matching not taking into account spurious (nonpaired estimated) tracks, penalization of spurious tracks, Jaccard similarity coefficient for track points as well as for entire tracks and RMSD between matching points. A detailed comparison of the methods submitted by 14 teams was described in *Nature Methods* (Chenouard et al. 2014).

9.5.5 *Pattern Recognition in Indirect Immunofluorescence (PRinIIF): HEp-2 Cells Classification*

The indirect immunofluorescence (IIF) methodology is used to detect autoimmune diseases by searching for antibodies in the patient serum. To help physicians with the complicated task of cell classification in IIF images, automated pattern recognition (machine learning) techniques are applied. A typical task is HEp-2 cell classification addressed by this repeated challenge. Classification performance was measured by either overall accuracy at the cell level or mean class accuracy (the average accuracy for particular classes). Thanks to the challenge, the performance of participating methods has been improving with time. The results of the first run were published in *IEEE Transactions on Medical Imaging* (Foggia et al. 2013). In July 2014, a special issue of *Pattern Recognition* appeared with a detailed description of selected methods and improved results. The dataset can be downloaded from the organizers' web pages (HEp-2 Images Dataset 2012).

9.5.6 *Mitosis Detection in Breast Cancer (MITOS)*

Algorithms for mitosis detection and mitotic count computation were the subject of the MITOS 2012 and 2014 challenges. This task is crucial for breast cancer histological imaging. Two 2D scanners were used to acquire real benchmark datasets; in addition, a multispectral 3D scanner was used in the first edition. Correctly detected mitoses (true positives) were defined as those whose centroids were localized within a range of 8 μm from the centroids of the ground truth. Mitosis detection performance was then evaluated using a classic F-score measure. In addition, the nuclear atypia score (which is classified into three grades) was evaluated in the second edition as the difference between the count of correct and the count of opposite scores (an opposite score means that the score difference is 2, i.e. one of the scores is 1 and the other score is 3). Detailed results of the first edition were published in Roux et al. (2013). The second edition can be found on the Grand Challenges website (MITOS-ATYPIA 2014).

9.5.7 *Assessment of Mitosis Detection Algorithms (AMIDA) for Breast Cancer Imaging*

Another challenge focused on mitosis detection in breast cancer histopathology imaging was AMIDA at MICCAI 2013 (i.e. in between the two MITOS challenges). The purpose of this challenge was to eliminate the disadvantage of a low number of patients in the benchmark database at MITOS 2012. While MITOS 2012 provided images from only 5 patients, AMIDA offered image data from 23 patients but from a single scanner. Instead of one annotator, they used two to mark mitoses and two other annotators to solve cases where they disagreed. The threshold for the displacement of mitoses centres between the ground truth and computer results was slightly decreased from 8 to 7.5 μm. Also in this challenge, the F-score was used as the main metric, and, in addition, the correlation of mitotic density between computed results and ground truth was calculated. The whole challenge including the results was described in detail in *Medical Image Analysis* (Veta et al. 2015) and can also be found on the Internet (AMIDA 2013).

9.5.8 *Localization Microscopy Challenge (LMC)*

A rather special task is the processing of time-lapse sequences from super-resolution localization microscopy. In this case, time-lapse imaging does not mean observing moving objects but rather observing different points of static objects sequentially to obtain a single high-resolution image from the whole sequence. Improving relevant algorithms in terms of final image quality (resolution) helps to visualize biological structures more clearly. The Localization Microscopy Challenge tried to compare methods used for this purpose both for real and simulated data (all of them in 2D + time). Four real sequences of tubulin were accompanied by simulated tubulin images at 2 SNR and 2 density levels and further by 3 types of artificial objects: an eye, snow and seashell. For point signals, it is quite easy to compute recall, precision and the F-score as well as the Jaccard index to assess the performance. In addition, localization accuracy was evaluated using an RMSD measure. Unlike other challenges, simulated artificial objects (tubulin networks) were represented here using a continuous-domain model (Sage et al. 2013). The challenge description can be found on the EPFL web pages (LMC 2013) as well as in a recent issue of *Nature Methods* (Sage et al. 2015), where the results of more than 30 software packages are compared and further metrics introduced: Image quality is measured using SNR, image resolution is measured using Fourier ring correlation, and usability is assessed by human and computational time.

9.5.9 3D Deconvolution Microscopy Challenge (DMC)

Another challenge organized by the EPFL Biomedical Imaging Group was the 3D deconvolution microscopy challenge; there were two editions of this challenge, which took place at ISBI in 2013 and 2014. Similar to LMC, simulated 3D artificial objects were used and represented using a continuous-domain model (Sage et al. 2013). Also in this case, improving relevant algorithms in terms of final image quality (similarity to the original non-blurred ground truth) helps to visualize biological structures more clearly. There were 4 types of 3D synthetic datasets – point sources (imitating single molecules, vesicles or mitochondria), curves (imitating microtubules or actin filaments), surfaces (imitating cellular or nuclear membranes) and dense volumes (imitating condensing chromatin or DNA). A variety of different metrics were defined for this challenge: peak signal-to-noise ratio (PSNR), normalized mean integrated squared error (NMISE), structure similarity index (SSIM), Fourier shell correlation (normalized cross-correlation coefficient between two image stacks over corresponding shells in the Fourier space) and relative energy regain (a Fourier-based quality metric that measures the recovery of information at a range of absolute spatial frequency). The description of both editions of the DMC challenge can be found on the EPFL web pages (DMC 2013), but the results are still missing.

9.5.10 Cell Tracking Challenge (CTC)

Time-lapse live cell imaging is probably the most important tool to observe selected processes running within or among cells. Due to the large amount of data acquired, this approach strongly relies on automatic cell segmentation and tracking. Therefore, benchmarking of relevant algorithms is needed. For this purpose, a collection of real as well as simulated datasets of different types was created in the context of the Cell Tracking Challenge at ISBI 2013 and twice extended in the second and third editions in subsequent years. The collection contains mainly real fluorescence datasets (both 2D + time and 3D + time), from low-density isolated cells or cell nuclei to very complex developmental image series. Further, real phase-contrast and differential interference contrast (DIC) datasets are included. The ground truth for real data was created by three experts followed by a major voting scheme. Finally, simulated datasets (both 2D + time and 3D + time) are included with various settings of cell density, cell speed and SNR. Segmentation quality is assessed using the standard Jaccard similarity index and tracking quality using a specially developed measure based on the comparison of calculated and ground truth graphs. The challenge including the results of the first edition was described in detail in *Bioinformatics* (Maška et al. 2014). Further information can be found on the CTC home page (CTC 2013).

9.5.11 Overlapping Cervical Cytology Image Segmentation Challenge (OCCISC)

One of the major problems in automated analysis of cervical cells is detection and segmentation of overlapping cells in Pap smear images. For this purpose, two benchmark datasets were introduced – one based on 2D extended depth of field (EDF) images (16 real + 945 synthetic) and one based on 3D data (17 real multilayer volumes). The 2D benchmark datasets were used for the first edition of the corresponding challenge, while the 3D datasets were used for the second edition. The performance of cell cytoplasm segmentation is measured using the Dice coefficient (DC) over the 'good' cell segmentations, where a 'good' segmentation is considered to be one with a $DC > 0.7$. Also, the object-based false-negative rate was obtained as the proportion of cells having a $DC \leq 0.7$. In addition, a pixel-based evaluation was computed with the true-positive and false-positive rate using the 'good' cell segmentations. Finally, nuclei detection was assessed using object-based as well as pixel-based precision and recall. Unfortunately, only two teams submitted consistent results for each edition. The results for each team along with the method description were published as individual ISBI 2014 and ISBI 2015 conference papers, whose references as well as other details can be found on the challenge web page (OCCISC 2014).

9.5.12 Gland Segmentation Challenge (GLAS) for Histology Imaging

Malignant tumours arising from glandular epithelium, also known as adenocarcinomas, are the most prevalent form of cancer. The morphology of glands is used to assess the degree of malignancy of several adenocarcinomas, including prostate, breast, lung and colon. Reliable morphological characteristics in turn require accurate segmentation of glands. To facilitate comparison and further development of segmentation algorithms for this purpose, a benchmark dataset was created comprising two types of 2D real datasets – colorectal and breast adenocarcinomas. The metric for detection performance was defined using the classic F-score and required the intersection of the segmented object with at least 50 % of the corresponding ground truth object to define true positives. Segmentation quality was defined using the Dice coefficient and shape similarity using the Hausdorff distance between the shape of the segmented object and that of the ground truth object. More information can be found on the challenge web site (GLAS 2015).

9.5.13 Image Stitching Challenge (ISC)

Stitching is a necessary step for automated microscopy applications that rely on the acquisition of neighbouring image tiles (fields of views) and their composition as a single high-resolution image. The challenge dataset is a grid of image tiles of stem cell colonies acquired with 10 % overlap. There are three levels of difficulty: easy, intermediate and hard. The easy level is acquired in feature-rich phase-contrast imaging modality with grid size 10×10. The intermediate and hard levels are acquired in Cy5 (where signal exists only on top of colonies, and mainly background noise fills the rest of the image) with grid size 10×10 and 24×23, respectively. Two criteria were suggested to determine the accuracy of the stitching algorithms: colony centroid location and colony area obtained from the stitched image and from the measured sample (reference value). Unfortunately, precise measures were not announced nor were training data with ground truth provided. The results have been reported at the Bioimage Informatics 2015 conference but have not been published.

9.5.14 Nucleus Counting Challenge (NCC)

This challenge concerns the evaluation of/state-of-the-art nucleus multichannel segmentation algorithms from fluorescent images of brain tissue. The choice of brain tissue is motivated by the complexity and variability of nuclei morphology and staining patterns and intensity in the brain, which requires multiple stains and multichannel segmentation analyses. The provided data consist of three sets of 2-channel fluorescence images of brain tissue. Two nuclear labels for each image dataset are provided: DAPI, which is a ubiquitous stain of all nuclear DNA, and the NeuN nuclear protein, which is specifically expressed only in neurons. Some of the neuronal nuclei stain well for NeuN but not with DAPI. The tasks are to count the total number of cells and neurons visually and then compute a segmentation mask and the number of segmented nuclei in both DAPI and NeuN channels. Two criteria were suggested to determine the accuracy of the results: nucleus counts and nucleus segmentation masks. Unfortunately, precise measures were not announced, and training data with ground truth were not provided. Results have been reported at the Bioimage Informatics 2015 conference but not published so far.

9.6 Analysis of Benchmarks and Challenges

Although the number of benchmarks and challenges in bioimage analysis is still small and their subject matter is very diverse, they share certain features, and it is possible to classify in an approximate way. Analysis of their results is much harder not only due to diversity and the low number of comparison studies (note that some challenges have not published their results) but also the low number of algorithms in most of the comparison studies and incomplete submissions (algorithm results are supplied only for some benchmark datasets, rarely for all of them). Nevertheless, some basic observations can be drawn from the available results.

9.6.1 Analysis of Features and Possible Classification

Both benchmarks and challenges can be characterized by the following attributes: target application, imaging modality, nature of data and dimensionality of data. Challenges can be characterized further by the following additional attributes: validation approach, metric types, supported platforms and lifecycle type.

Target application is the most important characteristic in the classification process. Thus far, benchmarking efforts and organized challenges have concentrated on the following target applications and biological problems:

- Restoration of high-quality images from raw data (Deconvolution and Localization Microscopy Challenges, Image Stitching Challenge)
- Segmentation, classification and tracking of isolated cells and particles (Particle and Cell Tracking Challenges, PRinIIF)
- Segmentation and reconstruction in neuroimaging (the first two challenges, DIADEM and SNEMI)
- Segmentation and classification in histopathology imaging:
  - Counting lymphocytes in breast cancer imaging and counting centroblasts in follicular lymphoma imaging (PRinHEMI)
  - Mitosis detection in breast cancer imaging (MITOS and AMIDA)
  - Cervical cytology imaging (OCCISC)
  - Gland segmentation in adenocarcinoma imaging (GLAS)
  - Nucleus counting in brain tissue (NCC)

Imaging modality is an attribute in addition to the target application that specifies which instrument and which mode of operation are considered for image acquisition. In bioimaging, various types of microscopes are typically used; hence, the high-level classification is light microscopy (LM), electron microscopy (EM) and scanning-probe microscopy (SPM). The basic low-level classification for the most frequent modes is then brightfield LM (BF), phase-contrast LM (PhC), differential interference contrast LM (DIC), fluorescence LM (FM), transmission EM (TEM), scanning EM (SEM) and atomic force SPM (AFM). The most frequently used modality in bioimage benchmarks and challenges has been FM, but others have started to appear as well (e.g. PhC and DIC in CTC challenge or TEM in SNEMI one).

Nature of data can be classified into real data from a real specimen and simulated data generated by a computer based on digital phantoms and virtual image acquisition. Rarely, calibration data based on the real acquisition of physical phantoms (such as microspheres) are used. While most of the benchmarking has been performed using real data, simulated datasets have been used to some extent in approximately one-third of the challenges (5 out of 14; see Table 9.1), especially in the case of the first two target applications above. The other two target applications have been addressed using real data most likely due to the complexity of image data that are hard to simulate; the only exception was cervical cytology imaging, where simulated data were provided and for which an open-source simulation platform has become available in 2015 (Malm et al. 2015). Hopefully, simulators for more complex objects will appear soon as well. Machine learning methods can help in this respect to 'learn' the cell model from real microscopy data (Buck et al. 2012).

Dimensionality of data determines the difficulty of the problem: more dimensions mean more complex tasks and more complicated solutions. The basic categories are 2D, 3D, 2D + time and 3D + time. The occurrence of these categories in the challenges listed in Table 9.1 is 6, 5, 3 and 2, respectively. Hence, 3D imaging keeps pace with 2D imaging, but time-lapse imaging is not as frequent as fixed cell imaging. Another dimension could be the spectral one (wavelength dimension), but typically, one or a few colour channels are imaged, so this dimension is presently used infrequently. Although full spectral imaging has been commercially available in light microscopy for approximately 15 years, it has been used somewhat rarely. The largest number of channels that have been used was 10 for the multispectral 3D dataset in the MITOS 2012 challenge.

Validation approach can be classified into three basic categories: visual validation, the holdout method and cross-validation. Visual validation means that no training data with ground truth is provided – only test data with no ground truth – and the developer fine-tunes the method based on a visual assessment of the algorithm performance. This is naturally the worst validation approach and is used rarely (only in the last two NIST challenges). The holdout method divides data with ground truth into training and competition data and provides the training data along with the ground truth to the developers. This is the standard approach in all bioimage challenges except for those organized by NIST. The best approach is cross-validation, but this is applicable only to machine learning approaches (especially suitable for classification tasks); such bioimage analysis challenges have not been organized thus far.

Metric types can be approximately categorized into confusion matrix-based (TP, FP, TN, FN, precision, recall, F-score, etc.), similarity coefficients (Jaccard, Dice), distance measurements (Hausdorff, RMSD, etc.), distribution comparisons (correlation, distribution divergence), image-based (PSNR, SSIM, etc.), Fourier spectrum-based (Fourier shell correlation, Fourier ring correlation, relative energy regain, etc.), computational (memory usage, time consumption) and human assessment (usability). The first two types have been dominant in bioimage challenges so far (applied in approximately half of them) because they are suitable for segmentation quality assessment, and segmentation has been the dominant task in bioimage analysis. The other metrics are more specialized or secondary. The last metric (suggested in Carpenter et al. 2012) differs from the others because it is not calculated but assessed by a human and, hence, somewhat subjective. Nevertheless, it evaluates important features of software packages like user-friendliness, developer-friendliness and interoperability. It was used with a low weight in LMC.

Supported platforms appearing in bioimage challenges are stand-alone applications for Microsoft Windows, Linux or Mac and programming languages such as Java, MATLAB and Python. Challenge organizers are fortunately liberal in this respect and leave the choice of the platform to the participants.

Lifecycle type of challenge can be one of the following: one-time event (organized once and closed), repeated event (organized periodically with fixed submission dates) or open call (submissions accepted continuously). To keep the information about available methods for a particular task in one place and up-to-date, it is advantageous to use the last type, i.e. not to close the challenge but rather to keep it open for further submissions while keeping the ground truth for competition datasets secret. Unfortunately, this is not a common case, probably because it is too laborious. Thus far, only the Localization Microscopy Challenge has been turned into a permanent online challenge after ISBI 2013, but no new updates have been added to the website so far. The option of repeating the challenge from time to time (not necessarily each year, e.g. MITOS has been repeated every other year so far) is still a good alternative compared to a one-time event. Approximately half of the challenges have been repeated (see Table 9.1; in addition, it was announced that LMC would be repeated in 2016).

9.6.2 Analysis of Available Results

As stated above, the challenges are very diverse, and there are very few statistics available thus far. Moreover, the results usually differ significantly among different datasets from one challenge (easy datasets versus hard datasets). There is often large variability among the results of individual submissions for the same dataset (the worst method performs significantly worse than the best one). Nevertheless, one can at least observe some ranges of values for the most common metrics, i.e. for the confusion matrix-based and similarity coefficients.

For example, the F-score for the best methods typically reaches a value of 0.6–0.8 (MITOS, AMIDA, LMC) and rarely reaches a value of 0.9 for easy tasks (OCCISC) while maintaining a balance between precision and recall; i.e. precision and recall are usually of similar value for the best methods (AMIDA, OCCISC), or precision is slightly favoured (MITOS). The exception is LMC, where precision is strongly favoured (Sage et al. 2015, in which the average for all methods was reported as 0.956 ± 0.09) at the expense of recall (the average for all methods was reported as 0.487 ± 0.15) likely because the loss of some true signals in the application is not crucial (there are enough signals) and simultaneously shining signals are well separated (not clustered) in the image.

Concerning similarity coefficients, the Jaccard index for the best methods typically ranges from 0.3 for very hard tasks (HD data in LMC) to 0.5–0.6 for intermediate tasks (LS data in LMC, part of CTC) and as high as 0.9 for easy tasks (part of CTC). The Dice coefficient can also reach as high as 0.8–0.9 for easy tasks (PRinHIMA, OCCISC).

From these numbers, it is obvious that further improvement in the performance of the methods is desirable. Often algorithms are not able to properly cope with the training data, so one cannot expect good behaviour or the ability to generalize for unseen data. It should be noted, however, that reaching ideal values of precision, recall, the F-score and the Jaccard and Dice measures in the range of 0.95–1.0 is improbable for most applications because even expert annotators are not able to reach mutual agreement in this range. For example, at the AMIDA challenge, it was reported that two expert annotators agreed in just half of the cases when detecting mitoses (649 instances of agreement out of 1344 average annotated cases; see Veta et al. 2015)! If human experts are not able to agree, computer algorithms will never be able to agree either for such tasks.

The statistics regarding the platforms preferred by participants are interesting (as stated above, they are given freedom of choice). The largest number of participants attended LMC (34 software packages) with the following distribution (Sage et al. 2015): stand-alone (10), ImageJ (9), MATLAB (9), Python (5), and Metamorph (1). Obviously, there is currently no preferred platform.

9.7 Conclusion and Future

It is obvious that the benchmarking efforts of the medical imaging community positively influenced the bioimage analysis community in recent years. During the last decade, the number of shared real and synthetic bioimage datasets has substantially grown; in addition, sharing of ground truth data has improved. Moreover, during the last 4 years, five bioimage analysis challenges have been organized each year on average (Fig. 9.1). With a delay of 1–2 years, journal publications based on these challenges started to appear with detailed comparisons of the performance of individual methods.

Not only is the growing number of available benchmarks with ground truth data, challenges and related comparison papers remarkable, but the improvement in the quality of annotations, simulations and target journals is as well. Much attention is paid to careful annotations, preferably by multiple experts, as well as realistic simulations so that synthetic images agree with real images not only visually but also in terms of mathematical descriptors. The comparison studies have become very detailed and are being published along with a great deal of supplementary material in journals such as *Nature Methods*.

These detailed comparisons of existing methods and benchmarking standards for new methods help biologists in the selection of an appropriate analysis method for a particular task and help computer scientists and engineers in related software development. In addition, journal publication policies have changed for the better in the bioimage community – there has been a trend in terms of publishing and sharing image data used in the paper (e.g. JCB DataViewer 2008). Reviewers can better assess a newly developed method of analysis if it is compared to the best methods for a particular data type using standard metrics (and not just to methods and metrics selected on an ad hoc basis).

In spite of all the positive changes, there are still some problems that impede the development process and complicate the effort that was started in the 1980s (Price 1986) to build on previous work. A major problem is still the lack of documentation and source codes for the submitted algorithms either because some participants refuse to provide it or because the organizers postpone releasing detailed results until the related paper is published, which takes 1–2 years as stated above. This delay is unfortunate and a change in journal publication policy would be welcome in this respect so that posting results and method details on a website does not hamper their being published in a journal.

In addition, it would be welcome if more challenge organizers were to assume ownership of the problem, so to speak, after the challenge, and either organize more challenge editions or turn it into an open call. Unfortunately, approximately half of the challenges have been a one-time event. Support for repeated challenges and open-source solutions was announced in the latest call for challenges for the ISBI 2016 conference, so this will hopefully help improve the situation.

Alternatively, it is possible to engage the relevant community during the preparation of the challenge by organizing public brainstorming workshops, allowing dataset contributions from the public and offering centralized resources for running algorithms on benchmark datasets. Such collaborative community effort could attract more attention and help manage more complicated tasks. An example of this approach is the recently announced BigNeuron project (BigNeuron 2015), which tries to build on the work started by the DIADEM challenge 5 years ago and bring it to new levels (Peng et al. 2015).

Benchmarks and challenges have done a good job introducing standards and quality control to the bioimage analysis community; they have also helped to make algorithm development and use somewhat more uniform. The improving quality of bioimage analysis results helps in turn to describe cell morphology and the processes inside the cell more precisely, reveal many of the still unknown mechanisms of life and disease and – together with inputs from bioinformatics, biomechanics and biochemistry – ultimately build a credible model of cell morphology and behaviour (Ortiz de Solórzano et al. 2015).

Acknowledgements The author would like to thank Martin Maška and David Svoboda for their feedback and useful comments. This work was supported by the Czech Science Foundation (Grant No. 302/12/G157).

References

2DSNEMI Challenge (2012) MIT. http://brainiac2.mit.edu/isbi_challenge/. Accessed 17 May 2015

3DSNEMI Challenge (2013) MIT. http://brainiac2.mit.edu/SNEMI3D/. Accessed 17 May 2015

AMIDA (2013) University Medical Center Utrecht. http://amida13.isi.uu.nl/. Accessed 17 May 2015

BBBC (2008) Broad Institute of Harvard and MIT. http://www.broadinstitute.org/bbbc/. Accessed 17 May 2015

BigNeuron (2015) Allen Institute for Brain Science. http://bigneuron.org. Accessed 17 May 2015

Buck TE, Li J, Rohde GK, Murphy RF (2012) Toward the virtual cell: automated approaches to building models of subcellular organization "learned" from microscopy images. Bioessays 34:791–799

Carpenter A, Kamentsky L, Eliceiri KW (2012) A call for bioimaging software usability. Nat Methods 9(7):666–670

CCDB (2002) University of California, San Diego. http://ccdb.ucsd.edu/. Accessed 17 May 2015

CellOrganizer (2012) Carnegie Mellon University, Pittsburgh. http://cellorganizer.org/. Accessed 17 May 2015

Chenouard N et al (2014) Objective comparison of particle tracking methods. Nat Methods 11(3):281–289

Coelho LP, Shariff A, Murphy RF (2009) Nuclear segmentation in microsope cell images: a hand-segmented dataset and comparison of algorithms. In: Proceedings of the 2009 IEEE International Symposium on Biomedical Imaging (ISBI 2009), IEEE, Boston, 28 June–1 July 2009, pp 518–521

Computer Vision Homepage (1994) School of Computer Science, Carnegie Mellon University. http://www.cs.cmu.edu/~cil/vision.html. Accessed 17 May 2015

CIL (2010) American Society for Cell Biology. http://www.cellimagelibrary.org/. Accessed 17 May 2015

CTC (2013) University of Navarra. http://www.codesolorzano.com/celltrackingchallenge/. Accessed 17 May 2015

CytoPacq (2008) Masaryk University, Brno. http://cbia.fi.muni.cz/simulator/. Accessed 17 May 2015

Deconvolution Benchmark Datasets (2010) EPFL. http://bigwww.epfl.ch/deconvolution/. Accessed 17 May 2015

DIADEM Challenge (2010) Howard Hughes Medical Institute. http://diademchallenge.org/. Accessed 17 May 2015

DMC (2013) EPFL. http://bigwww.epfl.ch/deconvolution/challenge/. Accessed 17 May 2015

Drelie Gelasca E, Obara B, Fedorov D, Kvilekval K, Manjunath BS (2009) A biosegmentation benchmark for evaluation of bioimage analysis methods. BMC Bioinformatics 10:368

Ferko MC, Patterson BW, Butler PJ (2006) High-resolution solid modeling of biological samples imaged with 3D fluorescence microscopy. Microsc Res Tech 69(8):648–655

Foggia P, Percannella G, Soda P, Vento M (2013) Benchmarking HEp-2 cells classification methods. IEEE Trans Med Imaging 32(10):1878–1889

Gillette TA, Brown KM, Ascoli GA (2011) The DIADEM metric: comparing multiple reconstructions of the same neuron. Neuroinformatics 9(2-3):233–245

GLAS (2015) Department of Computer Science, University of Warwick. http://www2.warwick.ac.uk/fac/sci/dcs/research/combi/research/bic/glascontest/. Accessed 17 May 2015

Griffa A, Garin N, Sage D (2010) Comparison of deconvolution software in 3D microscopy: a user point of view, part I and part II. G.I.T. Imaging Microscopy 1:43–45

Gurcan MN, Madabhushi A, Rajpoot N (2010) Pattern recognition in histopathological images: an ICPR 2010 contest. In: Ünay D, Çataltepe Z, Aksoy S (eds) Recognizing patterns in signals, speech, images and videos. Lecture notes in computer science, vol 6388. Springer, Heidelberg, pp 226–234

HEp-2 Images Dataset (2012) Mivia Lab, University of Salerno. http://mivia.unisa.it/datasets/biomedical-image-datasets/hep2-image-dataset/. Accessed 17 May 2015

Hill E (2008) Announcing the JCB DataViewer, a browser-based application for viewing original image files. J Cell Biol 183:969–970

JCB DataViewer (2008) Rockefeller University Press. http://jcb-dataviewer.rupress.org/. Accessed 17 May 2015

Jannin P, Grova C, Maurer C (2006) Model for defining and reporting reference-based validation protocols in medical image processing. Int J CARS 1(2):63–73

Kozubek M, Matula P (2000) An efficient algorithm for measurement and correction of chromatic aberrations in fluorescence microscopy. J Microsc 200(3):206–217

Kwan RK-S, Evans AC, Pike GB (1999) MRI simulation-based evaluation of image-processing and classification methods. IEEE Trans Med Imaging 18(11):1085–1097

Lehmussola A, Selinummi J, Ruusuvuori P, Niemist A, Yli-Harja O (2005) Simulating fluorescent microscope images of cell populations. In: Proceedings of the 27th annual international conference of the IEEE Engineering in Medicine and Biology Society (EMBC'05), IEEE, Shanghai, 17–18 Jan 2006, pp 3153–3156

Ljosa V, Sokolnicki KL, Carpenter AE (2012) Annotated high-throughput microscopy image sets for validation. Nat Methods 9(7):637

LMC (2013) EPFL. http://bigwww.epfl.ch/smlm/challenge2013/. Accessed 17 May 2015

Lockett SJ, Sudar D, Thompson CT, Pinkel D, Gray JW (1998) Efficient, interactive, and three-dimensional segmentation of cell nuclei in thick tissue sections. Cytometry A 31:275–286

Malm P, Brun A, Bengtsson E (2015) Simulation of bright-field microscopy images depicting Pap-Smear specimen. Cytometry A 87A:212–226

Martone ME, Zhang S, Gupta A, Qian X, He H, Price DL, Wong M, Santini S, Ellisman MH (2003) The cell-centered database: a database for multiscale structural and protein localization data from light and electron microscopy. Neuroinformatics 1(4):379–395

Martone ME, Tran J, Wong WW, Sargis J, Fong L, Larson S, Lamont SP, Gupta A, Ellisman MH (2008) The Cell Centered Database project: an update on building community resources for managing and sharing 3D imaging data. J Struct Biol 161(3):220–231

Maška M et al (2014) A benchmark for comparison of cell tracking algorithms. Bioinformatics 30(11):1609–1617

McNally JG, Cogswell CJ, Fekete PW, Conchello JA (1997) Comparison of 3D microscopy methods by imaging a well characterized test object. In: Cogswell CJ, Conchello JA, Wilson T (eds) Three-dimensional microscopy: image acquisition and processing IV, San Jose, 8 Feb 1997. Proc SPIE, vol 2984, pp 52–63

MITOS-ATYPIA (2014) Consortium for Open Medical Image Computing. http://mitos-atypia-14.grand-challenge.org/. Accessed 17 May 2015

Murphy Lab Data (1999) Carnegie Mellon University. http://murphylab.web.cmu.edu/data/. Accessed 17 May 2015

OCCISC (2014) University of Adelaide. http://cs.adelaide.edu.au/~zhi/isbi15_challenge/. Accessed 17 May 2015

Open Bio Image Alliance (2015) http://www.openbioimage.org/. Accessed 17 May 2015

Orloff DN, Iwasa JH, Martone ME, Ellisman MH, Kane CM (2012) The cell: an image library-CCDB: a curated repository of microscopy data. Nucleic Acids Res 41:D1241–D1250

Ortiz-de-Solórzano C, Muñoz-Barrutia A, Meijering E, Kozubek M (2015) Toward a morpho-dynamic model of the cell: signal processing for cell modeling. IEEE Signal Proc Mag 32(1):20–29

Peng H, Hawrylycz M, Roskams J, Hill S, Spruston N, Meijering E, Ascoli GA (2015) BigNeuron: large-scale 3D neuron reconstruction from optical microscopy images. Neuron 87(2):252–256

Price K (1986) Anything you can do, I can do better (no you can't). Comput Vision Graph 36:387–391

Rajaram S, Pavie B, Hac NE, Altschuler SJ, Wu LF (2012) SimuCell: a flexible framework for creating synthetic microscopy images. Nat Methods 9(7):634–635

Rosenberg C (1996) The Lenna Story. http://www.lenna.org. Accessed 17 May 2015

Roux L, Racoceanu D, Loménie N, Kulikova M, Irshad H, Klossa J, Capron F, Genestie C, Le Naour G, Gurcan MN (2013) Mitosis detection in breast cancer histological images: an ICPR 2012 contest. J Pathol Inform 4:8

Sage D, Kirshner H, Pengo T, Stuurman N, Min J, Manley S, Unser M (2015) Quantitative evaluation of software packages for single-molecule localization microscopy. Nat Methods 12(8):717–724

Sage D, Kirshner H, Vonesch C, Lefkimmiatis S, Unser M (2013) Benchmarking image-processing algorithms for biomicroscopy: reference datasets and perspectives. In: Proceedings of the 21st European Signal Processing Conference (EUSIPCO), IEEE, Marrakech, 9–13 Sept 2013, pp 1–4

Shiraishi J, Katsuragawa S, Ikezoe J, Matsumoto T, Kobayashi T, Komatsu K, Matsui M, Fujita H, Kodera Y, Doi K (2000) Development of a digital image database for chest radiographs with and without a lung nodule: receiver operating characteristic analysis of radiologists' detection of pulmonary nodules. AJR Am J Roentgenol 174(1):71–74

SimuCell (2012) Altschuler & Wu laboratories, University of California, San Francisco. http://awlab.ucsf.edu/Web_Site/SimuCell/documentation.html. Accessed 17 May 2015

Svoboda D, Kozubek M, Stejskal S (2009) Generation of digital phantoms of cell nuclei and simulation of image formation in 3D image cytometry. Cytometry A 75A:494–509

UCSB Biosegmentation Benchmark (2008) University of California, Santa Barbara. http://bioimage.ucsb.edu/research/bio-segmentation. Accessed 17 May 2015

van Ginneken B, Kerkstra S (2015) Grand challenges in biomedical image analysis. http://grand-challenge.org/. Accessed 17 May 2015

Veta M et al (2015) Assessment of algorithms for mitosis detection in breast cancer histopathology images. Med Image Anal 20(1):237–248

Chapter 10
Bioimage Informatics for Big Data

Hanchuan Peng, Jie Zhou, Zhi Zhou, Alessandro Bria, Yujie Li, Dean Mark Kleissas, Nathan G. Drenkow, Brian Long, Xiaoxiao Liu, and Hanbo Chen

Abstract Bioimage informatics is a field wherein high-throughput image informatics methods are used to solve challenging scientific problems related to biology and medicine. When the image datasets become larger and more complicated, many conventional image analysis approaches are no longer applicable. Here, we discuss two critical challenges of large-scale bioimage informatics applications, namely, data accessibility and adaptive data analysis. We highlight case studies to show that these challenges can be tackled based on distributed image computing as well as machine learning of image examples in a multidimensional environment.

10.1 The Big Data Challenges

There have been substantial advances of bioimage informatics in the last 15 years (Peng 2008; Swedlow et al. 2009; Myers 2012). Now, with the annual conferences of bioimage informatics (http://bioimageinformatics.org) and related topics, as well as the formally added paper submission categories in several computational

H. Peng (✉) • Z. Zhou • B. Long • X. Liu
Allen Institute for Brain Science, Seattle, WA, USA
e-mail: hanchuanp@alleninstitute.org

J. Zhou
Department of Computer Science, Northern Illinois University, Dekalb, IL, USA

A. Bria
Department of Engineering, University Campus Bio-Medico of Rome, Rome, Italy

Department of Electrical and Information Engineering, University of Cassino and L.M., Cassino, Italy

Y. Li • H. Chen
Allen Institute for Brain Science, Seattle, WA, USA

Department of Computer Science, University of Georgia, Athens, GA, USA

D.M. Kleissas • N.G. Drenkow
Johns Hopkins University Applied Physics Laboratory, Laurel, MD, USA

W.H. De Vos et al. (eds.), *Focus on Bio-Image Informatics*,
Advances in Anatomy, Embryology and Cell Biology 219,
DOI 10.1007/978-3-319-28549-8_10

biology and bioinformatics journals (e.g., BMC Bioinformatics and Bioinformatics (Oxford)), more researchers have been attracted to this growing field.

Occasionally, bioimage informatics has been thought to be related to studies that use image analysis and computer vision methods to solve bioinformatics problems in some biology domains such as cell biology and neuroscience (e.g., Danuser 2011; Jug et al. 2014; Mikut et al. 2013). It is however a view that does not necessarily reflect all the intended applications in this field. In a 2012 editorial of the Bioinformatics journal (Peng et al. 2012), bioimage informatics is defined as a category including "Informatics methods for the acquisition, analysis, mining and visualization of images produced by modern microscopy, with an emphasis on the application of novel computing techniques to solve challenging and significant biological and medical problems at the molecular, sub-cellular, cellular, and super-cellular (organ, organism, and population) levels. This also encourages large-scale image informatics methods/applications/software, various enabling techniques (e.g., cyber infrastructures, quantitative validation experiments, pattern recognition, etc.) for such large-scale studies, and joint analysis of multiple heterogeneous datasets that include images as a component. Bioimage related ontology and database studies, image-oriented large-scale machine learning, data mining, and other analytics techniques are also encouraged." In short, we believe that bioimage informatics emphasizes the high-throughput aspect of the image informatics methods and applications.

It is important to stress that the current pace of image data generation has very much exceeded the processing capability in conventional computer vision and image analysis labs. For the four microscopic imaging modalities most used today, namely, bright-field imaging, confocal or multiphoton laser scanning microscopy, light-sheet microscopy, and electron microscopy, it has been very easy to produce big image data with hundreds of gigavoxels or even many teravoxels, where each voxel could correspond to one or multiple bytes of data. This is not only the situation of concerted large-scale projects such as the MindScope project at the Allen Institute for Brain Science (Anastassiou et al. 2015) or the FlyLight project at the Janelia Research Campus of the Howard Hughes Medical Institute (Jenett et al. 2012) but also a commonly confronted scenario in much smaller projects in individual research labs (Silvestri et al. 2013).

In addition to the scale of the image datasets, the complexity of bioimages also makes it very challenging to rely on conventional computational methods to analyze data efficiently. There are two specific challenges. First, in many (if not most) bioimages, there are at least three to five intrinsic dimensions, namely, the X, Y, Z spatial dimensions, the "color" channel dimension that reflects colocalizing patterns, and the time dimension. Further dimensions might also be encountered to include other experimental parameters or perturbations. It is often very hard to navigate through such high-dimensional datasets, let alone detect or mine the biologically or medically relevant patterns from such data.

Second, bioimages often contain a number of different spatial regions corresponding to cells, intracellular objects, or cell populations. Most applications of bioimage segmentation, registration, and classification (Qu et al. 2015) will involve the determination of certain relationships among these objects. For instance, a goal in neuroscience is to quantify the distribution of synapses that connect neurons. To achieve such a goal, the very complicated 3D morphology of a neuron should be reconstructed (traced), and synapses that may have very different shapes should be segmented. In addition, the spatial relationship of neuron(s) and synapses should be characterized. Achieving these complex computational analyses is often technically challenging (Micheva et al. 2010; Kim et al. 2012; Mancuso et al. 2013; Collman et al. 2015).

Here, we discuss briefly two critical challenges of very large-scale bioimage informatics applications, namely, data accessibility and adaptive data analysis, related to the aforementioned concerns of the scale and complexity of bioimage datasets. Some recent advances in bioimage management as well as machine learning for bioimages that address these two challenges are highlighted.

10.2 Big Bioimage Storage and Accessing

Complementary to recent advances in the fields of bioimage acquisition (Khmelinskii et al. 2012; Tomer et al. 2012), visualization (Peng et al. 2010; De Chaumont et al. 2012; Schneider et al. 2012; Peng et al. 2014a, b), and analysis (Kvilekval et al. 2010; Luisi et al. 2011; Schneider et al. 2012; Peng et al. 2014a, b, storage and management of big images form another exciting line of work to address challenges of large-scale bioimage data. Open Microscopy Environment (OME) (Swedlow et al. 2003), BISQUE (Kvilekval et al. 2010), CCDB (Martone et al. 2002), and the Vaa3D-AtlasManager (Peng et al. 2011) are among the existing systems that pioneered different aspects of big data management. Many of these systems provide web services for remote data management. Sometimes, visualization has also been built into the image-serving websites (Saalfeld et al. 2009) to allow browsing through multidimensional images, their individual 2D sections, the maximum/minimum intensity projections, and/or movies of image data.

For cutting-edge applications, large volumes of bioimages are often in the scale of multiple terabytes (Bria et al. 2015). For instance, whole-brain imaging studies of mammalian brains often produce terabytes of raw image data for one single brain. To access such terabyte-sized datasets, it is necessary to produce effective data structures for storage and access. Normally, for 3D datasets, octree or similar data structures are used to organize the big data into many different hierarchical levels. In the coarse level, the image data are downsampled. Each of such coarse level voxels' locations is associated with the higher-resolution image voxels. Therefore, when a user browses through data at different resolution scales, the data can be read from the storage device directly instead of being calculated in real time. The HDF5 format is often used as a convenient way to store such hierarchical data. Custom octree data structures are also considered in many ongoing projects.

However, with the hierarchical organization of large image data, it is still hard to navigate through very large data in the multidimensional space. One key limitation as noticed in recent studies (Peng et al. 2014a, b) is that it often takes too long for a user to manually identify the correct 3D regions of interest (ROIs) to visualize across different resolution scales. One critical new technique, called 3D Virtual Finger, was proposed to generate a 3D ROI with one computer mouse operation (click, stroke, or zoom operations) when the user is operating the rendered images on an ordinary 2D computer display device (e.g., computer screen). Typically, computing such an ROI only takes a few milliseconds; thus, the speed to navigate large images across different resolution scales is the fastest as it can be and is completely limited by the file IO speed of the storage device. The *Virtual Finger* function has been used to develop one of the fastest-known 3D large data visualizers Vaa3D-TeraFly (Bria et al. 2015) for visualizing terabytes of multidimensional bioimage data (Fig. 10.1).

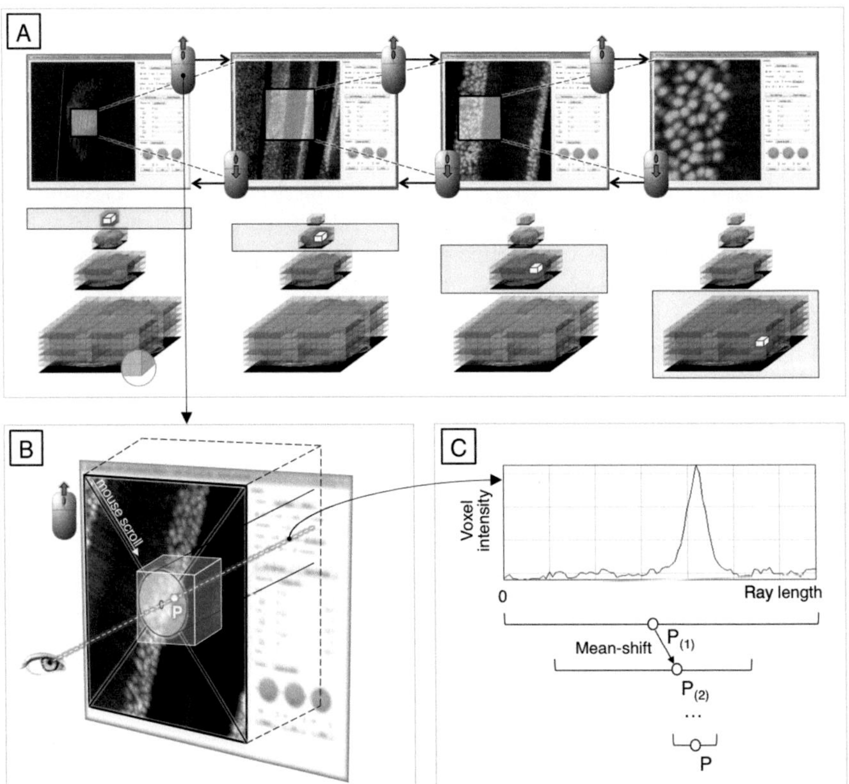

Fig. 10.1 (**a**) Vaa3D-TeraFly 3D exploration through different layers of a ~1 TB image, which is organized using an octree-type multiresolution pyramid, with one computer mouse operation (in the example shown, zoom-in with mouse scroll). (**b**) Virtual Finger is applied to the center of the viewport to generate the 3D ROI for zoom-in. (**c**) The Virtual Finger point-pinpointing algorithm (PPA) is used to find the center P of the 3D ROI (Figure redrawn from (Bria et al. 2015))

In an often-used server-client infrastructure, transferring big bioimage data from one location to another can be tackled by implementing a hierarchical organization of data on the server side and providing the *Virtual Finger* type of random access of multidimensional data on the client side. In this way, it is even possible to consider distributed data storage of very big bioimage repositories on multiple servers or on the cloud and using powerful client-side visualization and analysis software packages, such as Vaa3D (Peng et al. 2010), to effectively fetch data whenever needed. A preliminary prototype was recently implemented during a BigNeuron (Peng et al. 2015a, b) hackathon hosted by the Howard Hughes Medical Institute. Raw images of multiple imaging modalities, including electron microscopy data and fluorescent microscopy data stored at remote servers, can be fetched over the Internet quickly to be visualized and further analyzed in 3D using Vaa3D (Fig. 10.2). These functions provide new ways to share large image datasets for collaborative analysis. In the future, these new tools might be combined with other remote visualization and collaborative software packages, such as Cytomine (Maree et al. 2013) or Arvis Webview (http://webview3d.arivis.com/).

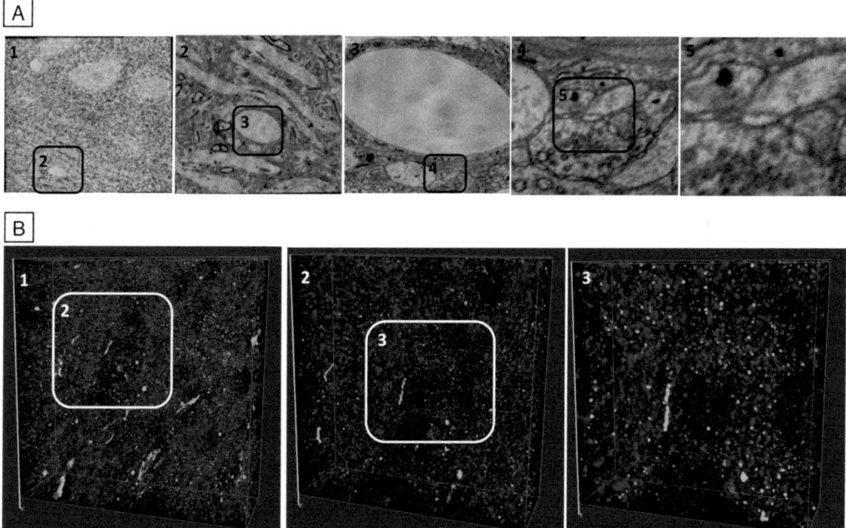

Fig. 10.2 (**a**) Different levels of a 1 TB multiresolution electron microscopy image data (Kasthuri et al. 2015), where the dimensions are 21,504 × 26,624 × 1850 voxels. Several different scales 1 ∼ 5 are shown using Vaa3D's 3D visualization functions. (**b**) Different levels of a 46GB (3409 × 3337 × 70 voxels) multiresolution multichannel (29 channels) fluorescent microscopy image data (Weiler et al. 2014). Three spatial scales 1 ∼ 3 are shown. Both datasets are explored and visualized by Vaa3D in real time across the Internet with raw image data stored on the remote servers of the Open Connectome Project (OCP) (Burns et al. 2013)

10.3 Machine Learning for Multidimensional Bioimages

The substantial complexity of bioimages suggests that many of the developed unsupervised bioimage analysis methods will have limited ability to be generalized for new image data. Even if a specific bioimage data analysis algorithm may remain the same for new datasets, it typically still requires nontrivial efforts to make the algorithm work well for new data. Efforts have been made to use machine learning to help solve these problems. Application examples include screening of gene expression patterns (Long et al. 2009; Zhou and Peng 2011) and many others (e.g., Jones et al. 2009; Kutsuna et al. 2012). Interesting machine learning toolboxes for bioimage informatics have also been built, such as WND-CHARM (Orlov et al. 2008) and ilastik (Sommer et al. 2011). These and other new strategies need to be incorporated into high-throughput analysis pipelines to meet the challenges of increasingly big bioimage datasets.

For example, in neuroscience, high-throughput analyses that categorize, annotate, and quantify high-dimensional multichannel images have been essential for new discoveries. Machine learning-based analysis is attractive for automatic processing of large-volume biological images due to its capability to train models using a small set of examples and then to extend the knowledge to bigger datasets. Meanwhile, given a variety of biological images, it is also important that computational algorithms suit the specific image classification and analysis tasks. Recent studies of pattern recognition and machine learning methods show vast differences on diverse biological image classification problems in terms of effectiveness (Zhou et al. 2013a, b; Li et al. 2015; Sanders et al. 2015). The differences stem from variation in algorithm performance at all stages of the process including feature extractors, selectors, and classifiers, as well as their combinations. In addition, results also indicate that sophisticated algorithms do not always outperform relatively simple alternatives. This implies that blindly designing a very general analysis process with complex and potentially also resource-costly algorithms is not an efficient solution for big bioimages. Instead, adaptive and flexible solutions implemented for particular datasets might be a more pragmatic way to tackle the challenge.

BIOCAT (for BIOimage Classification and Annotation Tool) (Zhou et al. 2013a, b) is one such recent effort toward the goal of adaptive biological image analysis. BIOCAT builds customizable machine learning pipelines for classifying 2D and 3D images as well as ROIs in the images. Being an open, user-friendly, portable, and extensible tool, BIOCAT provides a comparison engine that selects a suitable combination of multiple-dimension feature descriptors, feature selectors, and classifiers. As the result, such combination – termed an *algorithm chain* shown in Fig. 10.3a, b – is expected to provide valuable classification results for the image analysis problem at hand.

To make the machine learning as effective as possible, it is important to be able to easily generate exemplars or training data in multidimensional space. 2D methods had been introduced in ilastik (Sommer et al. 2011) and others. The *Virtual Finger* technique (Peng et al. 2014a, b) has been used for producing such exemplars in

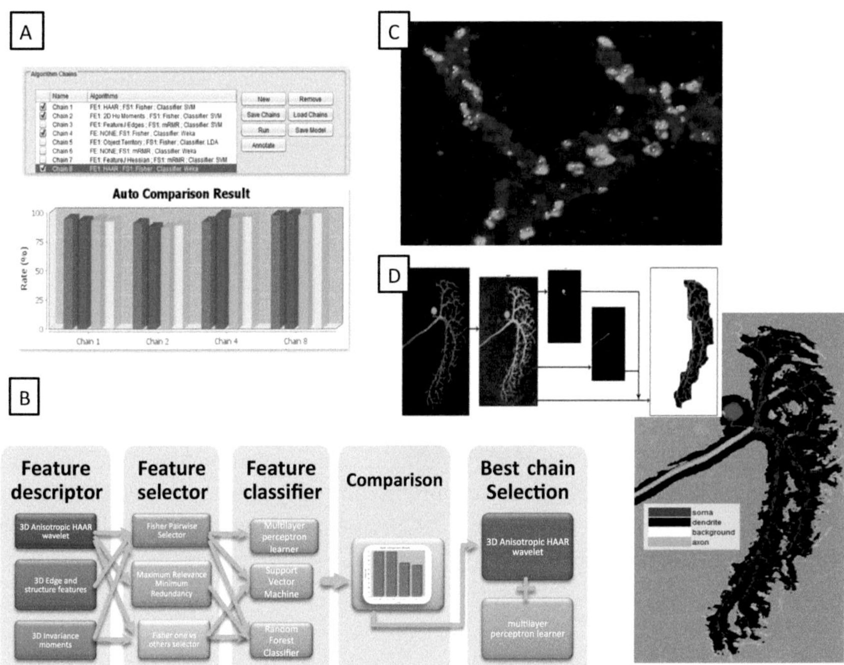

Fig. 10.3 (**a**) BIOCAT screenshots for the learning model selection. (**b**) Illustrative diagram for selecting an algorithm chain for 3D synapse detection. (**c**) Learning-guided 3D synapse quantification. The green channel is the channel of GABAergic synaptic markers. The yellow dots indicate the centers of detected GABAergic synaptic markers located in the axon of lobula plate tangential cells (LPTC) in *Drosophila* brain using 3D confocal images. (**d**) Learning-based dendritic territory identification. The output of the pixel classification from BIOCAT is used to detect different subcellular components of a LPTC neuron (Figure redrawn from (Zhou et al. 2013a, b; Sanders et al. 2015))

multidimensional space at low cost. Powerful machine learning results, even with simple algorithms, have started to demonstrate the success of such exemplar-based adaptive learning for very large image datasets of multiple imaging modalities (Li et al. 2015).

Self-evidence is another technique that complements manually specified exemplars. The key idea is to start from one analysis result of a bioimage to estimate the most confident "regions" of such an analysis. Once the confidence regions are produced, one can design algorithms to pick positive and negative training examples from the image directly, without the need for manually specifying training examples. While it is obvious that manually chosen examples can always be added into the framework as well, having a machine learning algorithm to choose exemplars means that such an algorithm can generalize to very large-scale bioimage datasets easily. These ideas have recently been applied to the tracing of neuron morphologies in 3D. For example, the SmartTracing algorithm (Chen et al. 2015) was recently proposed to start with an almost arbitrarily weak neuron tracing result to generate useful exemplars automatically and eventually improve the tracing results considerably. While this category of self-evidence-based smart algorithms is still in its early days, it might become a mainstream method for very large-scale bioimage informatics applications.

10.4 Conclusions

Continuing innovation in imaging and increasing rates of bioimage data generation have created big data challenges across a wide range of applications. Highly accessible large-scale datasets assisted with easy data visualization are becoming the stepping-stone for bioimage informatics. Developments of sophisticated machine learning tools that can adapt to changing samples and image conditions are essential in quantitatively understanding the rich information offered in those increasing large datasets. Easy exploration of bioimage data before and after analysis will generate insights into the challenges of biology.

References

Anastassiou C et al (2015) Project MindScope: inferring cortical function in the mouse visual system, PNAS (submitted)

Bria A, Iannello G, Peng H (2015) An open-source Vaa3D plugin for real-time 3D visualization of Terabyte-sized volumetric image. International symposium on biomedical imaging: from nano to macro, pp 520–523

Burns R et al (2013) The Open Connectome Project Data Cluster: scalable analysis and vision for high-throughput neuroscience. SSDBM 2013

Chen H et al (2015) SmartTracing: self-learning based neuron reconstruction. Brain Informatics (submitted)

Collman F et al (2015) Mapping synapses by conjugate light-electron array tomography. J Neurosci 35:5792–5807

Danuser G (2011) Computer vision in cell biology. Cell 147:973–978

De Chaumont F et al (2012) Icy: an open bioimage informatics platform for extended reproducible research. Nat Methods 9:690–696

Jenett A et al (2012) A GAL4-driver line resource for Drosophila neurobiology. Cell Rep 2:991–1001

Jones TR et al (2009) Scoring diverse cellular morphologies in image-based screens with iterative feedback and machine learning. Proc Natl Acad Sci U S A 106(6):1826–1831

Jug F et al (2014) Bioimage Informatics in the context of Drosophila research. Methods 68(1):60–73

Kasthuri N et al (2015) Saturated reconstruction of a volume of neocortex. Cell 162:648–661

Khmelinskii A et al (2012) Tandem fluorescent protein timers for in vivo analysis of protein dynamics. Nat Biotechnol 30:708–714

Kim J et al (2012) mGRASP enables mapping mammalian synaptic connectivity with light microscopy. Nat Methods 9:96–102

Kutsuna N et al (2012) Active learning framework with iterative clustering for bioimage classification. Nat Commun 3:1032

Kvilekval K et al (2010) Bisque: a platform for bioimage analysis and management. Bioinformatics 26:544–552

Li X et al (2015) Interactive exemplar-based segmentation toolkit for biomedical image analysis. International symposium on biomedical imaging: from nano to macro, pp 168–171

Long F et al (2009) A 3D digital atlas of C. elegans and its application to single-cell analyses. Nat Methods 6:667–672

Luisi J et al (2011) The FARSIGHT trace editor: an open source tool for 3-D inspection and efficient pattern analysis aided editing of automated neuronal reconstructions. Neuroinformatics 9:305–315

Mancuso JJ et al (2013) Methods of dendritic spine detection: from Golgi to high-resolution optical imaging. Neuroscience 251:129–140

Maree R et al (2013) A rich internet application for remote visualization and collaborative annotation of digital slides in histology and cytology. Diagn Pathol 8(S1):S26

Martone ME et al (2002) A cell-centered database for electron tomographic data. J Struct Biol 138:145–155

Micheva KD et al (2010) Single-synapse analysis of a diverse synapse population: proteomic imaging methods and markers. Neuron 68:639–653

Mikut R et al (2013) Automated processing of Zebrafish imaging data: a survey. Zebrafish 10(3):401–421

Myers G (2012) Why bioimage informatics matters. Nat Methods 9:659–660

Orlov N et al (2008) WND-CHARM: multi-purpose image classification using compound image transforms. Pattern Recognit Lett 29:1684–1693

Peng H (2008) Bioimage informatics: a new area of engineering biology. Bioinformatics 24:1827–1836

Peng H et al (2010) V3D enables real-time 3D visualization and quantitative analysis of large-scale biological image data sets. Nat Biotechnol 28:348–353

Peng H et al (2011) BrainAligner: 3D registration atlases of Drosophila brains. Nat Methods 8:493–498

Peng H et al (2012) Bioimage informatics: a new category in bioinformatics. Bioinformatics 28:1057

Peng H et al (2014a) Extensible visualization and analysis for multidimensional images using Vaa3D. Nat Protoc 9:193–208

Peng H et al (2014b) Virtual finger boosts three-dimensional imaging and microsurgery as well as terabyte volume image visualization and analysis. Nat Commun 5:4342

Peng H et al (2015a) BigNeuron: large-scale 3D neuron reconstruction from optical microscopy images. Neuron. doi:10.1016/j.neuron.2015.1006.1036

Peng H, Meijering E, Ascoli GA (2015b) From DIADEM to BigNeuron. Neuroinformatics 13:259–260

Qu L, Long F, Peng H (2015) 3-D registration of biological images and models: registration of microscopic images and its uses in segmentation and annotation. IEEE Signal Proc Mag 32:70–77

Saalfeld S et al (2009) CATMAID: collaborative annotation toolkit for massive amounts of image data. Bioinformatics 25:1984–1986

Sanders J et al (2015) Learning-guided automatic three dimensional synapse quantification for drosophila neurons. BMC Bioinformatics 16:177

Schneider CA, Rasband WS, Eliceiri KW (2012) NIH Image to ImageJ: 25 years of image analysis. Nat Methods 9:671–675

Silvestri L et al (2013) Micron-scale resolution optical tomography of entire mouse brains with confocal light sheet microscopy. J Vis Exp 80:e50696, doi:50610.53791/50696

Sommer C et al (2011) ilastik: interactive learning and segmentation toolkit. IEEE international symposium on biomedical imaging: from nano to macro, pp. 230–233

Swedlow JR et al (2003) Informatics and quantitative analysis in biological imaging. Science 300:100–102

Swedlow JR et al (2009) Bioimage informatics for experimental biology. Annu Rev Biophys 38:327–346

Tomer R et al (2012) Quantitative high-speed imaging of entire developing embryos with simultaneous multiview light-sheet microscopy. Nat Methods 9:755–763

Weiler N et al (2014) Synaptic molecular imaging in spared and deprived columns of mouse barrel cortex with array tomography. Scientific Data 1, December 23 2014, p 140046

Zhou J, Peng H (2011) Counting cells in 3D confocal images based on discriminative models. Proceedings of the 2nd ACM conference on bioinformatics, computational biology and biomedicine. ACM, pp 399–403

Zhou J et al (2013a) Performance model selection for learning-based biological image analysis on a cluster. Proceedings of the international conference on bioinformatics, computational biology and biomedical informatics. ACM, pp 324–332

Zhou J et al (2013b) BIOCAT: a pattern recognition platform for customizable biological image classification and annotation. BMC Bioinformatics 14:291. doi:210.1186/1471-2105-1114-1291